Supramolecular Architecture

ACS SYMPOSIUM SERIES 499

Supramolecular Architecture

Synthetic Control in Thin Films and Solids

Thomas Bein, EDITOR

Purdue University

Developed from a symposium sponsored
by the Division of Inorganic Chemistry
at the 201st National Meeting
of the American Chemical Society,
Atlanta, Georgia,
April 14–19, 1991

American Chemical Society, Washington, DC 1992

Library of Congress Cataloging-in-Publication Data

Supramolecular architecture: synthetic control in thin films and solids / Thomas Bein, editor

p.   cm.—(ACS symposium series, ISSN 0097–6156; 499)

"Developed from a symposium sponsored by the Division of Inorganic Chemistry at the 201st National Meeting of the American Chemical Society, Atlanta, Georgia, April 14–19, 1991."

Includes bibliographical references and index.

ISBN 0–8412–2460–9

1. Thin films, Multilayered. 2. Layer structure (Solids) 3. Chemistry, Inorganic—Synthesis. 4. Organic compounds—Synthetic.

I. Bein, Thomas, 1954–   . II. American Chemical Society. Division of Inorganic Chemistry. III. American Chemical Society. Meeting (201st: 1991: Atlanta, Ga.) IV. Series.

QC176.9.M84S87   1992
546'.269—dc20

92–14921
CIP

ACS Symposium Series

M. Joan Comstock, *Series Editor*

1992 ACS Books Advisory Board

Foreword

THE ACS SYMPOSIUM SERIES was founded in 1974 to provide a medium for publishing symposia quickly in book form. The format of the Series parallels that of the continuing ADVANCES IN CHEMISTRY SERIES except that, in order to save time, the papers are not typeset, but are reproduced as they are submitted by the authors in camera-ready form. Papers are reviewed under the supervision of the editors with the assistance of the Advisory Board and are selected to maintain the integrity of the symposia. Both reviews and reports of research are acceptable, because symposia may embrace both types of presentation. However, verbatim reproductions of previously published papers are not accepted.

Contents

INDEXES

Preface

DRAMATIC ADVANCES IN MOLECULAR SYNTHETIC CHEMISTRY have led to a high level of control over molecular interactions. However, we are only at the beginning of a more extended design of chemical interactions in two and three dimensions. If we learn how to control the structure, properties, and stability of desired supramolecular assemblies, many areas in materials science and technology, such as microelectronics, optics, sensors, and catalysis, will benefit substantially. Representative areas of research activity include selective monolayer assemblies on electrode surfaces; functionalized pillared layered materials; and assemblies of conductors, semiconductor clusters, or nonlinear optical materials in three-dimensionally ordered hosts such as zeolites.

Twenty-seven contributions from authors who have embarked on research programs in the exciting area of supramolecular architecture are included in this volume. An overview chapter provides a brief guide for the various strategies aimed at structural control of extended assemblies, evolving in their dimensionality from thin films to layered and low-dimensional structures, three-dimensional frameworks, and amorphous networks. I hope that this book serves to disseminate recent research results and to stimulate new and extended activities in this fascinating field of chemistry.

Generous support from the Division of Inorganic Chemistry of the American Chemical Society; Air Products and Chemicals, Inc.; Battelle PNL; The Dow Chemical Company; E. I. du Pont de Nemours and Company; 3M Corporation; and the Petroleum Research Fund has helped to make the symposium and this book possible. Their contributions are greatly appreciated.

THOMAS BEIN
Purdue University
West Lafayette, IN 47907

March 13, 1992

Chapter 1

Supramolecular Architecture
Tailoring Structure and Function of Extended Assemblies

Thomas Bein

Department of Chemistry, Purdue University, West Lafayette, IN 47907

This chapter will provide a brief overview about the various strategies aimed at structural control of extended assemblies. The systems under consideration evolve in their dimensionality from
(i)   Two-Dimensional Assemblies: Thin Films, to
(ii)   Layered and Low-Dimensional Structures, and
(iii)   Three-Dimensional Frameworks and Amorphous Networks.
Representative areas of research activity include selective monolayer assemblies on electrode surfaces, functionalized pillared layered materials, and assemblies of conductors, semiconductor clusters or nonlinear optical materials in three-dimensionally ordered hosts such as zeolites. Reference to articles in this book and to key contributions in the literature is given.

Dramatic advances in the areas of molecular synthetic chemistry such as organic chemistry or organometallics have led to a high level of control over molecular interactions. This becomes particularly clear in the field of molecular recognition where complexation, association and catalysis of organic and organometallic molecules can be fine-tuned almost as precisely as in biological systems, using creative combinations of ionic, hydrogen bonding, and other non-covalent interactions. The work of D. Cram and J.-M. Lehn represents a culmination of these efforts in the past decade. Representative systems include cavitands containing assemblies of aromatic rings,[1] and cryptands and spherands with multiple ligation sites for anions and cations.[2,3] Organic molecular systems have probably been developed to the greatest extent; it is now possible to construct elaborate hierarchical molecules such as starburst dendrimers/arborols,[4] and hyperbranched polymers,[5] assemblies promoting non-covalent molecular association such as "molecular shuttles"[6] or cyclophanes/aromatic complexes,[7] and cyclodextrins[8] that control access and orientation of reactants. Multiple hydrogen bonding for recognition and catalysis in organized systems is being used in Rebek's elegant di- and multiacid structures, and in nucleotide receptors.[9,10,11]
   The dreams of what could be possible very often have been inspired by Nature. This is particularly true for molecular interactions at various levels, for instance enzyme action for energy transfer and catalytic conversions, the senses, and, of course, the brain. However, the natural systems are characterized by enormous *complexity*, that is, by *hierarchies of structures and superstructures built up to functional units* such as

0097–6156/92/0499–0001$06.00/0

cell walls, muscles, nerves, etc. With all the impressive advances at the molecular level, we are only at the very beginning of a more extended control of chemical interactions in two and three dimensions. *Extended molecular interactions are highly important in catalysis, separations, control of energy flow, electrochemistry, optical effects, and ultimately the flow of information at the near-molecular level.* If we learn how to control the structure, properties and stability of desired supramolecular assemblies, many areas in materials science and technology such as microelectronics, optics, sensors and catalysis will benefit substantially. Representative areas of research activity include selective monolayer assemblies on electrode surfaces, functionalized pillared layered materials, and assemblies of conductors, semiconductor clusters or nonlinear optical materials in three-dimensionally ordered hosts such as zeolites.

The aim of this book is to bring together contributions from chemists who have embarked on research programs in this field which is highly challenging in both synthesis and characterization of the resulting structures. This chapter will provide a brief overview about the various strategies aimed at structural control of extended assemblies. Reference to articles in this book is given by indicating one author in parentheses. The systems under consideration evolve in their dimensionality from

(i)   Two-Dimensional Assemblies: Thin Films, to
(ii)   Layered and Low-Dimensional Structures, and
(iii) Three-Dimensional Frameworks and Amorphous Networks.

Many of the structures contain "inorganic" as well as "organic" components; an indication that the desired systems often require a non-traditional approach for their synthesis. Although much of the work may be motivated by very different goals, the common challenge is the *synthetic control of chemical interactions at the "supramolecular" level.* Efforts are underway at different levels of dimensionality, from two-dimensional assemblies on surfaces, to layered materials, and finally complex three-dimensional structures, either crystalline or amorphous. In spite of the diversity, one central theme appears to dominate many strategies to assemble molecular units at higher levels. This theme is the "template" concept. Probably the most famous template of all is DNA itself which contains (almost) all information to duplicate the complex biological structures needed for successful life and reproduction. Templates are also useful and ubiquitous at less ambitious levels, for instance in making cookies, mass-producing cars, microchips, printing books, etc. Because it is very cumbersome to try to assemble significant numbers of moles atom by atom or molecule by molecule (even though recently the "molecule" "IBM" was assembled on a cold Ni(110) substrate by using a scanning tunneling microscope to move individual Xenon atoms into position),[12] some degree of automation in this ambitious endeavor is highly desirable. The template approach is extremely useful for this purpose, as will be demonstrated below and in the contributions throughout this book. The question then is how to motivate the species of interest to form the hierarchical structures desired.

Two-Dimensional Assemblies: Thin Films

Crystallization can be one of these techniques, however, to date it has been exceedingly difficult to predict and control the three-dimensional packing of larger molecules.[13] In contrast, <u>two-dimensional</u> crystallization on surfaces such as organothiols on gold surfaces leaves one degree of freedom for design: the surface. Sulfur has a high affinity for gold and forces the alkane chains of the alkane thiol into a densely packed structure.[14] The spatial requirements of the alkyl chain match well with the optimal packing arrangement of sulfur on the gold substrate. The polymethylene chains can be terminated with many different functionalities to modify the wetting and/or reactivity of the newly formed surface (G. M. Whitesides). Thus, complex assemblies of functional groups in ordered organic films where redox centers are buried at different distances in the organic layer permit to control electron transport as a function of substrate size.[15] These systems are promising model systems for modified electrodes, studies of friction, protein adhesion, and in the more distant future possibly microelectronic or optoelectronic devices.

At a slightly "thicker" scale, it could be shown that classical electronic functions such as diodes and transistors can be realized with thin layers of conducting polymers on microelectrode arrays.[16] Attempts are now being made to reduce the dimensions of these arrays down to the molecular level. The proposed scheme for the design of a molecular diode is based on "orthogonal selfassembly",[17] that is, preferential attachment of certain functional groups on different substrates (e.g., thiols on gold, isonitriles on platinum). If a microelectrode array with closely spaced alternating metal electrodes M_1, M_2, one of them reversibly blocked, is exposed to molecules L_1--L_2, the molecules are expected to attach to one metal and bend over to the second after it has been unblocked.

Ordered thin films of layered metal phosphonates - inorganic analogs to Langmuir-Blodgett films - have been built using surface-attached phosphonates that were reacted with zirconium salts to replicate zirconium phosphate crystallization on a surface (T. E. Mallouk),[18] and to afford multilayers with electron donating (e.g.,tetrathienyldiphosphonic acid) and accepting interlayer functions (H. E. Katz). A related approach leads to mixed valence charge transfer compounds such as Ru(II)-pyrazine-Ru(III) assembled in multilayers (B. I. Swanson). Thin metal complex-containing layers can also be deposited using the Langmuir-Blodgett technique, for example with the system stearic acid/Ru(II)bipyridyl (M. K. De Armond). By modeling Nature's basic concepts for biomineralization, the group of P. C. Rieke has explored a variety of surface nucleation phenomena to grow thin crystalline layers of calcium carbonate, iron oxide, CdS and other materials on substrates such as modified polymers and functionalized, self-assembled monolayers. This approach mimics the functions of nucleation proteins and in some cases even permits to control the orientation of the crystallites deposited on the substrate.

Layered and Low-Dimensional Structures

If extended layered structures are considered, the template approach becomes even more obvious. Here the template is often an (inorganic) host, sometimes layered, that accommodates guest molecules in specific orientations, at certain distances, or stabilized against diffusion in cavities. Modified layered materials are therefore promising candidates for three-dimensional assemblies. The design of functionalized pillared metal compounds with, e. g., organic ion-exchange and ligating functions, is a main area of activity in A. Clearfield's group. In addition to clays, these compounds include group 4 and 14 phosphates, titanates, and antimonates. Enormous flexibility exists in the design of the pillaring groups, including diphosphonates with aromatic "rods" and sulfonate ion exchange capabilities, polyethers for ion conductivity, and polyimine chains as complexing agents within the galleries of the pillared material.[19] Furthermore, the use of <u>different</u> precursors, such as simple phosphonic acid and one substituted with a ligating function offers the construction of mixed galleries which can enhance diffusion or reactivity in the layered material. The group of M. E. Thompson explores the intralayer reactivity of acyl chloride zirconium phosphonate with amines and alcohols to give layered amides and esters. Layered (clay) systems become porous and more stable under demanding conditions if intercalated with inorganic pillars such as metal oxide clusters,[20] and they can incorporate porphyrin catalysts when synthesized from silicate gels containing water-soluble cationic metalloporphyrins (K. A. Carrado). If clays are intercalated with long-chain quaternary alkylammonium ions, attractive catalysts for three-phase reactions such as conversion of alkylbromides are formed. The surface of the layered material plays the role of a mediator at the liquid/liquid interface, bringing incompatible reactants into close proximity (T. J. Pinnavaia).

Inclusion polymerization of conducting polymers such as polyaniline or polypyrrole in layered structures such as FeOCl or V_2O_5 opens the way to new classes of hybrid systems (M. Kanatzidis). If the host lattice involves redox reactions such as in V_2O_5, mobile carriers are created on the host network which contribute to the overall conductivity of the system, to varying degrees such that n-type, p-type, and metallic behavior can be observed. Aniline could also be oxidatively intercalated into the layered proton conductor $HFe(SO_4)_2$ xH_2O (D. J. Jones).

"Soft chemistry" strategies are important for the design of metastable low-dimensional and open framework solids, for example by using redox processes (e.g., deintercalation of Li(I) from $LiVS_2$ to obtain VS_2)[21] or acido-basic reactions, in order to avoid thermodynamic control typical for high temperature processes (J. Rouxel). A classical example is the synthesis of a new form of TiO_2 from the low-dimensional $K_2Ti_4O_9$ with K^+ ions between titanium oxide ribbons.[22] Acid ion exchange and heating replaces K^+ with hydroxyl groups that are eliminated at higher temperature such that the Ti blocks are sealed to a new titanium dioxide. In another example, layered FeOCl can be used as a precursor to lamellar iron phosphonates by anchoring organophosphonates onto the FeO layers (P. Palvadeau). Electrocrystallization is another approach to low-dimensional molecular solids, where one of the ions is produced by redox reactions at the electrode to crystallize with a counterion present in solution (M. D. Ward). One finds that nucleation, morphology and stoichiometry can be controlled through manipulation of the electrochemical growth conditions.

Superlattices, consisting of several lattices stacked in a regular fashion with long-range order in the z-axis, are of great interest for their unusual quantum electronic and optical effects.[23] The design of these systems has typically required vacuum deposition methods, but extended artificial superlattices are now accessible through controlled electrocrystallization (J. Switzer). Alternating layers of different Tl-Pb oxides as thin as 3 nm could be deposited by changing the electrochemical (galvanostatic) deposition conditions in the same solution. The near-perfect match of the lattice parameters of these systems with varied Pb/Tl ratio is the basis for the successful epitaxial growth - another type of structural templating.

Multilayered repeating structures with potential applications in nonlinear optics have been grown on glass surfaces using a sequence of silane coupling agent, chromophore, and structural reinforcement with siloxane/polyvinylalcohol layers.[24] The scanning tunneling microscope allows unprecedented resolution in imaging and direct-writing of nanostructures (B. Parkinson). Surface modification techniques using the STM tip include large voltage pulses to field emit atoms from the tip or disrupt the substrate, deposition of matter from solution or gas phase, and the use of van der Waals forces to move atoms and pattern layered materials. This is exemplified in the interesting "etching" behavior of layered MoS_2 and $SnSe_2$ where single molecular layers can be sequentially removed in areas as small as 20 nm.

Three-Dimensional Frameworks and Amorphous Networks

Based upon their well-defined, crystalline pore architecture with channel sizes at molecular dimensions, zeolites are becoming increasingly attractive as hosts for a variety of species and purposes. This increasing activity is also critically dependent upon recent advances in analytical techniques such as synchrotron X-ray powder diffraction, neutron diffraction, EXAFS, and solid state NMR. Researchers have now also begun to explore the fundamental kinetic aspects of intrazeolite model reactions such as CO and PMe_3 substitution at $Mo(CO)_6$ (G. A. Ozin) that was found to be first order. Resulting reaction parameters such as the entropy of activation permit conclusions about the nature of the transition state stabilized in the unique zeolite cavity. Intrazeolite diffusion and assembly of reactants is also key to understanding of electron transfer processes between photoexcited electron donors $Ru(bpy)_3^{2+}$ and viologen acceptors (T. E. Mallouk). These systems can be described as *molecular diodes* and *photodiodes*. The quenching behavior indicates that electron transfer involves diffusion of the acceptor. More elaborate assemblies show that charge recombination can be slowed considerably by using the structural templating function of the zeolite host.

The interplay between acentric (zeolite) hosts and molecules with large hyperpolarizabilities as well as the stabilization of nanometer size II-VI, III-V semiconductor clusters in zeolites promise to become important strategies for assemblies with non-linear optical properties (G. D. Stucky). The sodalite structure, consisting of a three-dimensional periodic arrangement of truncated octahedra, is a highly crystalline host with great compositional variation for the packaging approach. Clusters encapsulated in these cages include Cd_4S, Zn_4S, Na_4^{3+}, and Zn_3GaAs. Molecule-sized silver-halo clusters stabilized in different zeolites, in particular in sodalites,[25] show promise for optical data storage and sensor applications, while intrazeolite molybdenum oxide clusters[26] derived from $Mo(CO)_6$ are further examples of quantum-size materials stabilized by the zeolite host (G. A. Ozin). In these cases the hosts determine cluster structures and optical properties which are dramatically different from the corresponding bulk properties and even from those of colloidal semiconductors.

The molecule-sized zeolite channels are ideal hosts for stabilizing oligomers of thiophene[27] and single chains of conducting polymers such as polyaniline, polythiophene, polypyrrole and pyrolized polyacrylonitrile that are promising candidates for molecular electronic assemblies (T. Bein). The materials when recovered from the zeolite hosts are in a conducting state, while microwave absorption measurements indicate that in the absence of electrostatic host-guest interactions even the encapsulated species have significant carrier mobility.

Three-dimensionally ordered frameworks with structures sometimes derived from simple prototypes (e.g., PtS, diamond, rutile) have been constructed using an imaginative combination of metal coordination chemistry with certain geometries and rod-like or plate-like building blocks (Robson). This versatile approach leads to a large variety of fascinating structures with channels, cavities, and interpenetrating networks. For example, $Cd(CN)_2$. $1.5 H_2O$) . t-BuOH forms hexagonal channels (filled with the solvents) consisting of interconnected square-planar and tetrahedral centers at ratio 1:2. Some of the new frameworks contain porphyrin stacks, for example in (tetrapyridylporphinato)palladium . $2Cd(NO_3)_2$. hydrate, and variations of this theme may afford interesting catalytic applications.

Another type of three-dimensional assembly using not coordination but hydrogen bonding forces is more related to the molecular recognition studies in solution mentioned above (G. M. Whitesides). The reaction of melamine with isocyanuric acid forms a stable, 1:1 complex that features a two-dimensional honeycomb structure where each hexagonal unit of the lattice consists of a ring of three alternating cyanurate and melamine rings. This structure shows surprising temperature stability (450 °C) and is just one of many related, tape-like arrangements that form from similar complementary molecules via hydrogen bonding.

New approaches to the synthesis of known materials are in high demand for several reasons. Traditional solid state synthesis requires high temperature and frequent grinding to allow sufficient interdiffusion of the reactants, thus the products are usually those with highest thermodynamic stability. In contrast, the deposition of alternating atomically thin layers of reactants allows interdiffusion at much lower temperature, and the nucleation of the resulting amorphous phase becomes the rate-limiting step such that metastable phases can be formed when their nucleation rate is faster than that of thermodynamically stable phases (D. C. Johnson). If the heat for the solid state reaction is chemically stored in the precursors, reaction times can be radically shortened from days to seconds (R. B. Kaner). Thus, layered molybdenum disulfide can be made in a metathesis reaction between $MoCl_5$ and Na_2S, where the reaction is driven by the stability of resulting sodium chloride. Related reactions can be devised for many other compounds such as metal and main group chalcogenides, carbides, oxides, silicides, and solid solutions.

Structural control can also be achieved in porous amorphous inorganic materials such as silicate glasses derived from alkoxysilanes. The structural parameters include pore size hierarchies described by fractal dimension, and pore volumes. Recent advances in sol-gel processing have made it possible to control porosity over wide ranges of narrow pore size distributions by varying the size and mass fractal dimension of the precursor polymer, aging conditions, the relative rates of drying and condensation, and the associated capillary pressure exerted on the gel network during drying.[28] The porous glasses are versatile hosts for a variety of organic dyes and reagents that allow the design of chemical sensors, enzyme-based catalysts, and optical devices (D. Avnir; Y. Haruvy). The latter groups have developed a trapping approach that holds the guest molecules in the gel-glass yet maintains access to a significant fraction of the guest molecules from solution.

In summary, fascinating new avenues towards the control of chemical interactions and physical properties in two- and three-dimensional systems are presently being explored. It is hoped that this book serves to disseminate recent research results and to stimulate new and extended activities in this young field of chemistry.

Literature Cited

1   Cram, D. C. Angew. Chem. Int. Ed. Engl. 1988, 27, 1009.
2   Lehn, J.-M. Angew. Chem. Int. Ed. Engl. 1990, 29, 1304.
3   Cram, D. C. Angew. Chem. Int. Ed. Engl. 1986, 25, 1039.
4   (a) Tomalia, D. A.; Naylor, A. M.; Goddard, W. A. Angew. Chem. Int. Ed. Engl. 1990, 29, 138. (b) Newkome, G. R.; Baker, G. R.; Saunders, M. J.; Russo, P. S.; Gupta, V. K.; Yao, Z.; Miller, J. E.; Bouillion, K. J. C. S. Chem. Commun. 1986, 752.
5   Kim, Y. H.; Webster, O. W. J. Am. Chem. Soc. 1990, 112, 4592.
6   Anelli, P. L.; Spencer, N.; Stoddart, J. F. J. Am. Chem. Soc. 1991, 113, 5131.
7   Diederich, F. Angew. Chem. Int. Ed. Engl. 1988, 27, 362.
8   Breslow, R. Science (Washington, DC) 1982, 218, 532.
9   Rebek, Jr., J. Science (Washington, DC) 1987, 235, 1478.
10 Rebek, Jr., J.; Wolfe, J.; Nemeth, D.; Costero, A. J. Am. Chem. Soc. 1988, 110, 983.
11 Galan, A.; de Mendoza, J.; Toiron, C.; Bruix, M.; Deslongchamps, G.; Rebek, Jr., J. J. Am. Chem. Soc. 1991, 113, 9424.
12 Eigler, D. M.; Schweitzer, E. K. Nature, 1990, 344, 524.
13 Etter, M. C. Acc. Chem. Res. 1990, 23, 120.
14 Laibinis, P. E.; Whitesides, G. M.; Allara, D. L.; Tao, Y.-T.; Parikh, A. N.; Nuzzo, R. G. J. Am. Chem. Soc. 1991, 113, 7152, and references cited therein.

15   Rubinstein, I.; Steinberg, S.; Tor, Y.; Shanzer, A.; Sagiv, J. Nature 1988, 332, 426; 1989, 337, 514.

16   (a) Chao, S.; Wrighton, M. S. J. Am. Chem. Soc. 1987, 109, 2197; (b) ibid. 6627. (c) Turner Jones, E. T.; Chyan, O. M.; Wrighton, M. S. J. Am. Chem. Soc. 1987, 109, 5526.

17   Chemical & Engineering News 1991, May 27, p. 24.

18   Lee, H.; Kepley, L. J.; Hong, H.; Mallouk, T. E. J. Am. Chem. Soc. 1988, 110, 618.

19   Clearfield, A. Chem. Rev. 1988, 88, 125.

20   Pinnavaia, T. J.; Landau, S. D.; Tsou, M.-S.; Johnson, I. D.; Lipsicas, M. J. Am. Chem. Soc. 1985, 107, 7222.

21   Murphy, D. W.; Carides, J. N.; di Salvo, F. J.; Cros, C.; Waszczak, J. V. Mat. Res. Bull. 1977, 12, 827.

22   Marchand, R.; Brohan, L.; Tournoux, M. Mat. Res. Bull. 1980, 15, 1129.

23   (a) Synthetic Modulated Structures, Chang, L. L.; Giessen, B. C., Eds. Academic Press, Orlando, FL, 1985. (b) Interfaces, Quantum Wells, and Superlattices, Leavens, C. R.; Taylor, R., Eds., NATO Series B: Physics, Vol. 179, Plenum: New York, 1988.

24   Li, D.; Ratner, M. A.; Marks, T. J.; Zhang, C.; Yang, J.; Wong, G. K. J. Am. Chem. Soc., 1990, 112, 7389.

25   Ozin, G. A.; Stein, A.; Stucky, G. D. J. Am. Chem. Soc., 1990, 112, 904.

26   Ozin, G. A.; Özkar, S. J. Phys. Chem., 1991, 95, 5276.

27   Caspar, J. V.; Ramamurthy, V.; Corbin, D. R. J. Am. Chem. Soc., 1991, 113, 600.

28   Brinker, C. J.; Scherer, G. W. Sol-Gel Science: The Physics and Chemistry of Sol-Gel Processing, Academic Press, San Diego, CA, 1990.

RECEIVED April 7, 1992

TWO-DIMENSIONAL ASSEMBLIES: THIN FILMS

Chapter 2

Designing Ordered Molecular Arrays in Two and Three Dimensions

John P. Folkers, Jonathan A. Zerkowski, Paul E. Laibinis, Christopher T. Seto, and George M. Whitesides[1]

Department of Chemistry, Harvard University, Cambridge, MA 02138

Two- and three-dimensional assemblies — self-assembled monolayers (SAMs) and hydrogen-bonded co-crystals, respectively — show substantial changes in their supramolecular structures with seemingly minor changes in the structures of their molecular/atomic constituents. The structure of SAMs obtained by adsorption of alkanethiols onto silver and copper are indistinguishable, although the atomic radii of silver and copper are different. The structure of the SAM on silver is different from that on gold, although the atomic radii of these metals are essentially the same. A macroscopic property of the SAMs, wetting, is not affected by these structural differences. Co-crystals formed from derivatives of barbiturates and melamines form hydrogen-bonded tapes in the solid state. These tapes provide a template for studying the packing forces within crystals. The three-dimensional arrangement of the tapes in the crystals changes markedly in response to subtle differences in the steric and electronic structures of the molecular constituents.

The construction of large ensembles of molecules is a current challenge for molecular science. Although the design and synthesis of macromolecular ensembles in solution is well advanced *(1)*, development of corresponding techniques for the organic solid-state are more difficult and less well developed. We have started a program in designing solid-state structures based on inorganic and organic coordination chemistry.

Solution-phase patterns of reactivity in inorganic and organic chemistry are a starting point for the design of solid-state materials. We use both coordination and hydrogen bonds to design solids. In this paper we survey two approaches to the formation of solid-state structures: the application of inorganic coordination chemistry to the formation of self-assembled monolayers, and the design of three-dimensional crystals with controlled structures using networks of hydrogen bonds.

Self-Assembled Monolayers

Background. Self-assembled monolayers (SAMs) form by the spontaneous adsorption of ligands from solution onto the surface of a metal or metal oxide *(2,3)*.

[1]Corresponding author

0097–6156/92/0499–0010$06.00/0

The processes involved in these adsorptions are related to coordination chemistry in solution, but occur in two dimensions. Metal surfaces can be viewed as planes of metal atoms having vacant coordination sites. Appropriate ligands coordinate (or, in the terms of surface science, adsorb) to a metal surface and form an ensemble that we and others refer to as a self-assembled monolayer. Adsorbates containing polymethylene chains are the most commonly studied because they often form oriented, highly ordered SAMs. Some of the presently available systems that yield SAMs include alkanoic acids on oxidized metal surfaces (especially aluminum) (4), alkyl amines on oxidized surfaces of chromium and platinum (5), isonitriles on platinum (6), sulfides (7), disulfides (8–10), and thiols (11,12), on gold, and thiols on silver (13–15). All of these systems have analogs in classical coordination chemistry (16).

SAMs on Gold. SAMs formed by the adsorption of alkanethiols onto gold surfaces are presently the best characterized of these systems. Our studies have employed evaporated gold films as substrates. The resulting surfaces of the gold are polycrystalline, and these crystallites are oriented in a way that presents predominantly the (111) crystal face (9). This face has the lowest surface free energy. Unlike most metals, gold does not form an oxide under ambient conditions. With gold as a substrate, the SAMs form on a polycrystalline metal rather than on an amorphous overlayer of oxide. These gold films do not require special handling or cleaning, and the resultant monolayers are stable to cleaning and manipulation.

The structure of SAMs derived from adsorption of n-alkanethiols $(HS(CH_2)_nCH_3; n = 10-21)$ on gold has been determined using a variety of techniques (11,12,15,17–22). The adsorbed species is believed to be a gold(I) alkanethiolate $(RS^-Au(I))$ rather than an alkanethiol (RSH) (15,22); the mechanism of formation of the thiolate, however, has not been established. Transmission electron microscopy and diffraction (TEM) experiments determined the arrangement of thiolates on a gold (111) surface (17). On this surface, the sulfur atoms occupy three-fold hollow sites and form a hexagonal lattice. The lattice of sulfur atoms is displaced 30° relative to the gold (111) hexagon, and the intramolecular distances are $\sqrt{3}$ times larger than the interatomic distance of gold (17,20). This structure is referred to as $(\sqrt{3}x\sqrt{3})R30°$ (Figure 1a). The terminal methyl groups of the SAM also form a hexagonal lattice with the same intermolecular spacing as the sulfur atoms, as shown by low-energy helium diffraction (18) and scanning tunneling microscopy (21).

Polarized infrared external reflection spectroscopy (PIERS) has been very useful in determining the structural details of these SAMs (11,15,20,21). PIERS results show that SAMs of alkanethiolates on gold are nearly analogs of two-dimensional crystalline alkanes. The alkane chains are predominantly in a trans zig-zag conformation; the few gauche conformations are concentrated near the ends of the chains (15,18,23). The alkane chains are tilted ~26° relative to the surface normal. This tilt, which allows the chains to be close-packed, is a direct result of the lattice spacing of the sulfur atoms.

The high selectivity of the gold surface towards sulfur-containing groups allows a wide variety of functional groups to be incorporated into the assembly (3,12,19,20,24,25). In many cases, the highly-ordered structure of the SAM is preserved and two-dimensional ensembles of organic functional groups are formed. We have employed these functionalized monolayers in the study of wetting (3,12,24,25).

The wetting properties of the SAM are due almost entirely to the tail group (X) of the alkanethiol $(HS(CH_2)_nX)$. For example, $X = CH_3$ yields a SAM that is oleophobic (advancing contact angle of hexadecane, $\theta_a(HD)$, $\approx 50°$) and hydrophobic $(\theta_a(H_2O) \approx 115°)$; $X = CO_2H$ or OH yields SAMs that are hydrophilic $(\theta_a(H_2O) < 15°)$ (12,24,25). The wetting properties of these SAMs indicate that the tail group are

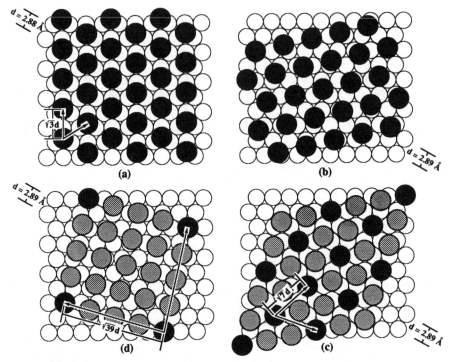

Figure 1. Schematic illustration of several of the lattices formed by the adsorption of sulfur-containing species onto gold and silver. In these figures, the small, open circles represent metal atoms in the (111) plane, black circles define the unit cell of sulfur atoms, and gray circles represent other sulfur atoms within the unit cell. (a) The $(\sqrt{3} \times \sqrt{3})R30°$ lattice formed by the adsorption of long-chain alkanethiols onto gold (17,20); the surface species in this structure is $RS^-Au(I)$ (15,22). The area per sulfur atom is 21.5 Å². (b) Incommensurate lattice formed by the adsorption of octadecanethiol onto silver (31); the surface species is $RS^-Ag(I)$ (14,15). The area per sulfur atom in this structure is 19.1 Å². (c) The $(\sqrt{7} \times \sqrt{7})R10.9°$ lattice formed by the adsorption of hydrogen sulfide onto silver (32) and by the adsorption of dimethyl disulfide onto silver (33); the surface species are S^{2-} and $CH_3S^-Ag(I)$, respectively. The area per sulfur in this structure is 16.9 Å². This area is too small to accommodate an alkane chain with a cross sectional area of 18.4 Å². (d) The $\sqrt{39}\,R16.1° \times \sqrt{39}\,R16.1°$ lattice of a monolayer of silver sulfide that is also formed by the reaction of hydrogen sulfide with silver (111) (32). The area per sulfur in this structure is 20.4 Å².

present at the monolayer/vapor interface. SAMs can be prepared containing two components to provide greater control over the interfacial properties of the SAM (12,24–27). For example, SAMs derived from mixtures of thiols terminated in methyl and hydroxyl groups yield intermediate wetting properties that can be tuned to specific values by controlling the composition on the surface (see below) (25,26).

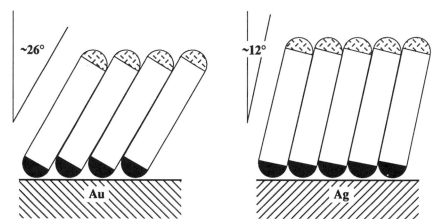

Figure 2. Illustration of the different cant angles for SAMs of alkanethiolates on gold and silver.

In addition to their utility in studies of wetting, SAMs comprising n-alkanethiolates on gold are useful model systems for studying protein adsorption to surfaces (*27*), X-ray-induced damage to organic materials (*28*), and electron transfer from fixed distances (*11,20,29*). In collaboration with Mark Wrighton's group (MIT), we have also developed a molecule-based pH sensor by incorporating an electroactive, pH sensitive group (*para*-quinone) and an electroactive reference compound (ferrocene) into the SAM (*30*).

SAMs on Silver. Silver surfaces are also highly reactive towards the adsorption of thiols (*13–15*). Silver and gold form face-centered cubic lattices with nearly identical interatomic spacings (2.88 Å and 2.89 Å for Au and Ag, respectively). Crystallites on evaporated silver surfaces also orient to present predominantly the (111) face (*15*). One would therefore predict that the structure of the SAMs formed on the two metals would be virtually indistinguishable.

Spectroscopic characterization of SAMs derived from alkanethiols adsorbed on silver using PIERS (*13,15*) and surface-enhanced Raman spectroscopy (SERS) (*14,22*) has been carried out in several laboratories; these results are in excellent agreement. In this section we focus on the results that we have obtained in collaboration with Ralph Nuzzo (AT&T Bell Labs) and David Allara (Penn State) using PIERS (*15*). Porter and co-workers have also done an excellent study on this system using PIERS (*13*).

We found the structure of SAMs derived from the adsorption of alkanethiols on silver to be related to, but different from the structure of these SAMs on gold (*15*). On both metals, the species on the surface is a thiolate, the SAM is oriented, and the alkyl chains are present primarily in a trans zig-zag conformation. On silver, however, the polymethylene chains are oriented closer to the surface normal (12° on silver vs. 26° on gold; Figure 2) and contain a lower population of gauche bonds. These observations suggest that the thiolates are more densely packed on silver than on gold (*i.e.*, the spacing between neighboring sulfur atoms is smaller on silver than it is on gold).

It is clear from the PIERS results that the sulfur atoms do not adopt the same commensurate ($\sqrt{3}\times\sqrt{3}$)R30° structure that is formed on gold (111). A recent study using both grazing in-plane X-ray diffraction and low-energy helium diffraction has shown that the structure of octadecanethiol adsorbed on silver (111) is more tightly packed than that on gold (~19.5 Å^2 vs. 21.5 Å^2) (31). On silver (111), the thiolates are arranged in a hexagonal array with nearest neighbor distances of ~4.7 Å, but the lattice is incommensurate with the underlying silver lattice (Figure 1b) (31).

We had hypothesized that the structure of thiolates on silver might be the same as one of the structures observed for an overlayer of silver sulfide formed by the reaction of H$_2$S with silver (111) (15). Of the possible structures, ($\sqrt{7}$x$\sqrt{7}$)R10.9° (Figure 1c) and $\sqrt{39}$ R16.1°x$\sqrt{39}$ R16.1° (Figure 1d) seemed most probable (32). The former structure is observed for methyl thiolate on silver, but the packing density of this lattice is too high to accommodate a trans-extended alkyl chain (33). The latter structure has a lower packing density, and could accommodate a trans-extended alkyl chain with a low cant angle, relative to the surface normal (15,32).

Hypothesizing that the structure of alkanethiolates on silver is similar to a surface layer of silver sulfide was, however, not unreasonable. The oxide that forms on the surface of silver upon exposure to air disappears upon formation of a SAM (15). This observation suggests that thiolates have replaced the oxide. Clean silver surfaces will also desulfurize aromatic thiols, sulfides, and disulfides (34); these reactions result in the formation of a layer of silver sulfide. We have also observed that our surfaces incorporate S^{2-} species after prolonged exposure to thiol (15); the properties of the monolayers, however, exhibit little change. Thus, in these instances, the SAM may actually rest on a substrate of Ag$_2$S. Figure 1 shows that several structures are formed when hydrogen sulfide reacts with silver (111) to form a monolayer of silver sulfide; it is therefore possible that the structure formed by the adsorption of alkanethiols onto silver may depend on the experimental conditions, especially the length of time for adsorption.

SAMs of alkanethiolates on silver can also accommodate the introduction of many different tail groups (25). As with gold, the contact angles of water on these SAMs span a large range of wettabilities. Although the structures of SAMs on silver and gold differ in density, cant angle, and relation to the underlying substrate, the wetting properties of SAMs with common terminal functional groups on these two metals are almost indistinguishable (Figure 3). Wetting is, therefore, insensitive to the structural differences that exist between SAMs formed on gold and those formed on silver.

The formation of alkanethiolate SAMs on silver is sensitive to the degree of oxidation of the silver prior to exposure to the adsorbate. Stearic acid forms monolayers on silver oxide surfaces (35). While molecules of the formula HS(CH$_2$)$_n$CO$_2$H adsorb to silver preferentially via the sulfur end as long as the surface has not oxidized significantly (25), both termini adsorb on silver that has been exposed to air for relatively brief periods (>5 min). These latter SAMs exhibit higher contact angles of water than those formed on silver with no oxide. To overcome these problems, exposure of the unfunctionalized silver substrates to air should be minimized; once the SAM is formed, the substrate is much less susceptible to oxidation (15).

SAMs on Copper. Like its congeners, copper also adsorbs alkanethiols that form oriented SAMs attached to the surface as thiolates (15). These samples are especially difficult to obtain in high quality, and the samples that we have examined always contained copper(I) oxide. We find this system to be extremely sensitive to the details of preparation, particularly the extent of exposure of the metal film to dioxygen (formation of a thick copper oxide) or to solution (formation of copper sulfide). Optimization of the procedure produced high-quality samples with PIERS spectra indistinguishable from those obtained on SAMs on silver (15). The structure of the SAMs on copper is, therefore, probably the same as that on silver: The axis of the trans-extended hydrocarbon chain is oriented close to the surface normal. Since the SAMs we characterized formed on an oxidized surface, we are hesitant to make claims

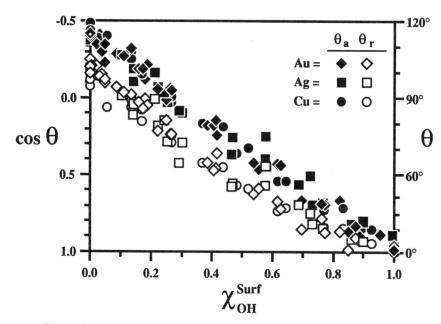

Figure 3. Advancing (filled points) and receding (open points) contact angles of water on mixed monolayers of $HS(CH_2)_{11}CH_3$ and $HS(CH_2)_{11}OH$ on gold (diamonds), silver (squares), and copper (circles). The x-axis is the mole fraction of hydroxyl-containing thiolates in the SAM as determined by X-ray photoelectron spectroscopy. The data are plotted as the cosine of the contact angles since these values are related to the interfacial free energy.

about the positions of the thiolates relative to the copper lattice. Unlike silver and gold, diffraction studies have not yet been carried out on these SAMs. On the basis of other evidence not presented here (*15*), we hypothesize that the arrangement of sulfur atoms on copper is related to copper sulfide.

Even though these SAMs form on copper oxide, they can still accommodate a wide range of polar and non-polar tail groups forming both hydrophobic and hydrophilic SAMs (*25*). As with gold and silver, SAMs derived from mixtures of thiols on copper have wetting properties that can be "tuned" to any value between those of the pure SAMs (Figure 3). The only difference between the wetting properties of mixed SAMs on the three coinage metals is that the hysteresis (the difference between the advancing and receding contact angles) increases as the substrate is changed from gold to silver to copper. This increase is probably due to an increase in the roughness of the substrate caused by oxidation of the substrate before formation of the SAM (*25*).

Summary. Self-assembled monolayers of alkanethiolates on surfaces of gold, silver, and copper have helped to illustrate differences in the chemistry of these surfaces and have clarified the relationship between the structure of a monolayer and its wetting properties (*15,25*). We are presently examining other ligands and substrates to identify surface coordination chemistries that will lead to new self-assembling systems (*36*). One of the goals of this project is to apply *the differences* between the coordination chemistries of different surfaces to the formation of "orthogonal" monolayers (37) — systems that will simultaneously form different SAMs on different metal surfaces from a solution containing a mixture of ligands. These differences in adsorption can then be used to form patterned, two-dimensional organic ensembles.

Hydrogen-Bonded Networks

Background. In a self-assembled monolayer, the surface imposes a direction and orientation to the molecular ensemble, and generates a general structural motif. The three-dimensional ordering required for bulk crystallization of organic molecules is, however, usually too complicated to predict or control. A primary reason for this complexity is the great number of orientations available, in principle, to most small molecules. A number of researchers have searched for simple patterns relating molecular composition and solid structure. Leiserowitz, Etter, and McBride have respectively produced important studies on systematic functional group crystallization patterns (38), hydrogen-bond preferences that can be used predictively (39), and relationships between substituents, packing, and solid-state reactivity (40). Desiraju has comprehensively reviewed work on crystal engineering (41).

Our approach to studying the packing forces that determine three-dimensional order in crystals has been to compare the solid-state structures of a series of molecules constrained by the presence of certain functional groups. The functional groups we use are hydrogen-bond donors and acceptors, because hydrogen bonds are significantly stronger than most other interactions between small, neutral organic molecules (42). By restricting the number of orientational degrees of freedom of the individual molecular components with hydrogen bonds, we hoped to form crystals in which the molecules packed in regular, easily visualized arrays, and in which the substructures were relatively invariant to changes in substituents.

Stimulated by an interest in the sheet network structure proposed for the 1:1 complex between melamine and isocyanuric acid (Figure 4) (43), we chose to study co-crystals of derivatives of melamine (M) and barbituric acid (B). These components often form 1:1 cocrystals (eq 1). These crystals are interesting for two reasons. First,

$$\tag{1}$$

B   M

these compounds are easily synthesized, and allow for the incorporation of a wide range of substituents into four sites (R_{1-4}) in each M·B dimer pair. Second, the hydrogen bonds between these two components significantly restrict the number of orientations the molecules can adopt in the crystals, thus limiting the number of probable substructures.

Crystalline Substructures. Figure 4 indicates that a number of structures are conceivable based on the two alternative arrangements of the three-fold hydrogen bond pattern connecting the melamine and barbiturate (isocyanurate) units. Isomerism around these sets of bonds leads to various possible substructures in the solid state, from the straight tape at the bottom of the figure to the cyclic hexamer at the top. Unless the substituents are tailored to fill interstitial voids efficiently, we suspect that

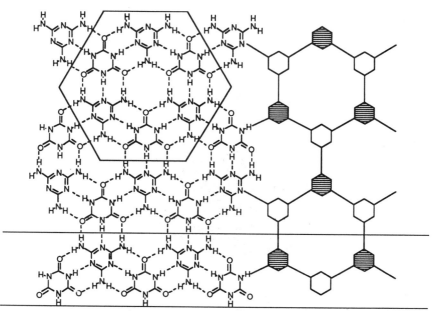

Figure 4. Portion of the proposed infinite hydrogen-bonded sheet of the complex between melamine and isocyanuric acid. Lines indicate substructures that might be obtained crystallographically by substituting parent compounds to prevent infinite hydrogen bonding.

the cyclic hexamer will rarely occur in crystals, due to the awkwardness in close packing such a shape. Several forms of tapes should, however, be accessible. We will use the nomenclature T = 1 to indicate the straight chain, where one M·B dimer unit is propagated infinitely by translation within a tape, T = 2 for a "crinkled" tape (see examples) with two dimers expressing the simplest tape unit, and so forth (Figure 5).

Once the molecules are organized at the level of tapes, they will probably undergo further assembly to some intermediate substructure. Stacking the tapes together in a venetian blind-like arrangement should give favorable close packing. Within such a stack, or sheet, the tapes could adopt a head-to-head or head-to-tail orientation. The latter would give a tape dimer that cancels dipoles; the dipoles in the former kind of stack could also cancel if the stack were adjacent to another stack in a head-to-head fashion, such as by using mirror or inversion symmetry. Analogously to our T designations, we call the head-to-head sheet S = 1, since only one orientation is translationally propagated within a stack, and the sheet consisting of head-to-tail dimers S = 2. Again, further complications are possible.

The final step for crystal construction is stacking together sheets of tapes. (We should note here that we do not expect tapes or sheets to exist as independent entities in solution; they are merely intellectual constructs intended to help visualize crystalline packing.) At this stage, sheets could line up with all their individual tapes parallel, or twisting could occur between sheets. In the absence of a strong force, such as extra hydrogen bonds that anchor a twisted configuration, we suspect that a parallel arrangement caused by the lining up of infinite ridges and valleys of substituents will be most favorable.

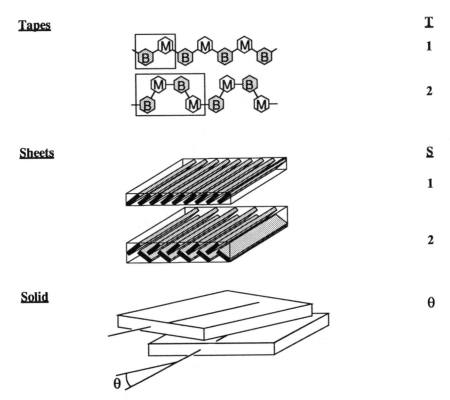

Figure 5. Crystalline substructures obtainable in 1:1 co-crystals of derivatives of barbituric acid (B) and melamine (M). T = the number of B·M dimers that constitute a translational repeat unit along a tape (boxed); S = the number of tapes that constitute a translational repeat unit in a sheet; θ = the angle between tape axes in adjacent sheets. The structures in this figure are representative examples of possible geometries and not an exhaustive list of all possible orientations of M and B.

Our approach can thus be summarized as follows: We design molecules which have firm constraints on their packing freedom, imposed upon them by intermolecular hydrogen bonds. These molecules then should repeatedly provide us with recognizable substructures of tapes and sheets in their crystals. While we expect these patterns to be formed consistently, the range of substituents available is great, so there may be significant variation between examples. We can make small perturbations on the substituent patterns and observe how, due to steric or electronic factors, the tape-tape or sheet-sheet interactions change.

Representative Examples. We have examined a large number of combinations of M and B (42). Most seem to form 1:1 micro-cocrystals, many of which are not large enough for single-crystal diffractometry. Nonetheless, we have been able to obtain

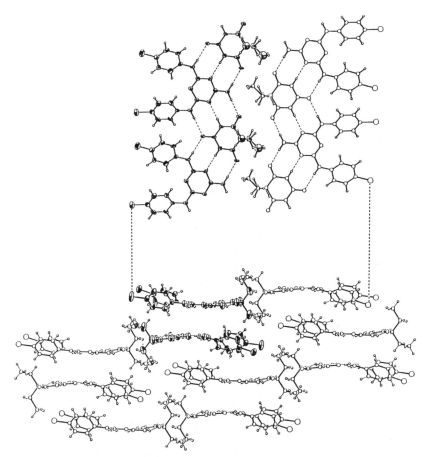

Figure 6. Bottom: End-on view of sheet packing in complex of N,N'-bis(4-chlorophenyl)melamine and barbital. A head-to-tail dimer of tapes is highlighted using thermal ellipsoids. Top: View of two tapes from the top.

structures for 21 crystals; all exist as some variant of the tape motifs summarized in Figure 5. Here we summarize only three examples to demonstrate how small changes in molecular structure can lead to large changes in tape or sheet packing. We emphasize that we have not systematically searched for polymorphisms in all of these crystallizations. We do not therefore know if the differences reflected in these three structures represent large differences in crystal energies, or smaller kinetic differences influenced by the conditions of crystallization.

The complex between N,N'-bis(4-chlorophenyl)melamine ($R_3 = R_4 = p\text{-}C_6H_4Cl$) and diethylbarbituric acid (barbital; $R_1 = R_2 = CH_2CH_3$) packs as shown in Figure 6. At the top, the straight-chain nature of the tapes (T = 1) can be seen, and the bottom end-on view shows tilted sheets. Each stack consists of head-to-tail dimers of tapes (S = 2), with one of these dimers drawn using thermal ellipsoids.

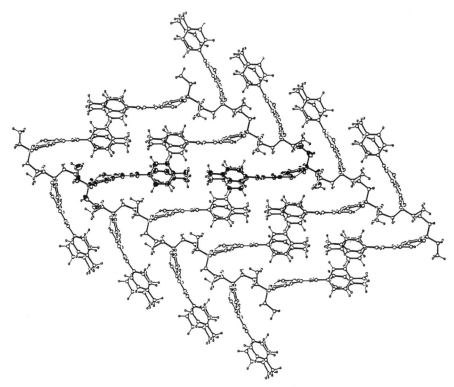

Figure 7. End-on view of sheet packing in complex of N,N'-bis(4-methylphenyl)melamine and barbital showing S = 1 packing. A head-to-head dimer of tapes is highlighted.

Changing the para-substituent from Cl to CH$_3$ (Figure 7) does not change the T = 1 tape format (not illustrated), but does produce a large change in the sheet architecture. Now the cancellation of dipoles is of a different type — head-to-head (S = 1) — and is inter-stack, rather than operating within one stack. One head-to-head dimer is shown in Figure 7 with thermal ellipsoids. Even though the Cl to CH$_3$ mutation should involve only a minimal change in volume, the steric demands might force apart head-to-tail dimers and necessitate this S = 1 packing.

Finally, moving the Cl substituent to the meta-position (Figure 8) causes a kink to appear in the chain. This crinkled form is stabilized by an intra-chain CH----O interaction, and the spacing of the tapes is apparently such as to allow incorporation of a molecule of solvent, here a well-ordered THF. The sheet architecture, while more difficult to see due to the thickness and waviness of the tapes, is still observed to be of the S = 2 dipole-cancelling type. Here, however, the "head-to-tail" dipoles are better termed "end-to-end". Within one crinkled tape (Figure 8, highlighted), the meta-chloro vectors all point in the same direction, to one end of the tape, for example toward the right of the page. In any one sheet, neighboring tapes above and below (shown in light bonds) the reference tape have their vectors pointing the other way, toward the left of the page.

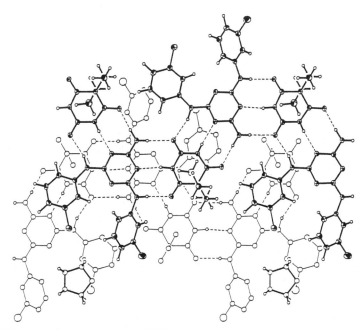

Figure 8. Two tapes of the N,N'-bis(3-chlorophenyl)melamine/barbital complex viewed from the top. The nearer tape is highlighted, and the lower one is drawn in light bonds with alkyl and aryl protons removed for clarity. Two crystallographically equivalent THF solvate molecules are shown; others have been removed.

Summary. From the results presented above (and others not discussed here), we can conclude that our basic hypothesis has been proved. The strategy for controlling the structures of crystals, provided by the network of hydrogen bonds present in the melamine/isocyanuric acid complex, but employing substituted derivatives of the parent heterocycles, has yielded crystalline substructures that can be obtained repeatedly. The motif "Tapes → Sheets → Solids" is general, and accommodates a number of pendant groups. By varying these groups, we can perturb (but not yet predict or control) the crystalline packing. Since the nature of most of the tape remains unaltered, we intend ultimately to correlate differences in packing with differences between these substituents. As we observe relationships between molecular and crystal structures over several self-consistent series (e.g. halides, alkyl chains), and with the aid of force field calculations, we hope to be able to predict crystal structures.

Acknowledgements. This research was supported by the Office of Naval Research, the Defense Advanced Research Projects Agency, and by the National Science Foundation (Grant CHE-88-12709 to G.M.W., Grant CHE 80-00670 to Harvard University for the purchase of a Siemens X-ray diffractometer, and Grant DMR-89-20490 to the Harvard University Materials Research Laboratory). J.P.F. acknowledges the National Institutes of Health for a training grant in biophysics (1989-1990). We are grateful to our many collaborators: Ralph G. Nuzzo (AT&T Bell Laboratories), David L. Allara (Penn State), Mark S. Wrighton and James J. Hickman (MIT), and Derk A. Wierda (Harvard).

Literature Cited

1. For leading references, see: Lehn, J. -M. *Angew. Chem. Int. Ed. Engl.* **1988**, *27*, 89-112. Ringsdorf, H.; Schlarb, B.; Venzmer, J. *Angew. Chem. Int. Ed. Engl.* **1988**, *27*, 113-158. Cram, D. J. *Angew. Chem. Int. Ed. Engl.* **1988**, *27*, 1009-1020. Rebek, J., Jr. *Angew. Chem. Int. Ed. Engl.* **1990**, *29*, 245-255. Sauvage, J. -P. *Acc. Chem. Res.* **1990**, *23*, 319-327.
2. Swalen, J. D.; Allara, D. L.; Andrade, J. D.; Chandross, E. A.; Garoff, S.; Israelachvili, J.; McCarthy, T. J.; Murray, R.; Pease, R. F.; Rabolt, J. F.; Wynne, K. J.; Yu, H. *Langmuir* **1987**, *3*, 932-950.
3. For some recent reviews, see: Bain, C. D.; Whitesides, G. M. *Angew. Chem. Int. Ed. Engl.* **1989**, *101*, 522-528. Whitesides, G. M.; Laibinis, P. E. *Langmuir* **1990**, *6*, 87-96.
4. Bigelow, W. C.; Pickett, D. L.; Zisman, W. A. *J. Colloid Sci.* **1946**, *1*, 513-538. Allara, D. L.; Nuzzo, R. G. *Langmuir* **1985**, *1*, 45-52, 52-66 and references therein.
5. Bartell, L. S.; Ruch, R. J. *J. Phys. Chem.* **1956**, *60*, 1231-1234. Bartell, L. S.; Ruch, R. J. *J. Phys. Chem.* **1959**, *63*, 1045-1049. Bartell, L. S.; Betts, J. F. *J. Phys. Chem.* **1960**, *64*, 1075-1076.
6. Hickman, J. J.; Zou, C.; Ofer, D.; Harvey, P. D.; Wrighton, M. S.; Laibinis, P. E.; Bain, C. D.; Whitesides, G. M. *J. Am. Chem. Soc.* **1989**, *111*, 7271-7272.
7. Troughton, E. B.; Bain, C. D.; Whitesides, G. M.; Nuzzo, R. G.; Allara, D. L.; Porter, M. D. *Langmuir* **1988**, *4*, 365-385.
8. Nuzzo, R. G.; Allara, D. L. *J. Am. Chem. Soc.* **1983**, *105*, 4481-4483. Li, T. T. -T.; Weaver, M. J. *J. Am. Chem. Soc.* **1984**, *106*, 6107-6108. Nuzzo, R. G.; Zegarski, B. R.; Dubois, L. H. *J. Am. Chem. Soc.* **1987**, *109*, 733-740.
9. Nuzzo, R. G.; Fusco, F. A.; Allara, D. L. *J. Am. Chem. Soc.* **1987**, *109*, 2358-2368.
10. Bain, C. D.; Biebuyck, H. A.; Whitesides, G. M. *Langmuir* **1989**, *5*, 723-727.
11. Porter, M. D.; Bright, T. B.; Allara, D. L.; Chidsey, C. E. D. *J. Am. Chem. Soc* **1987**, *109*, 3559-3568.
12. Bain, C. D.; Troughton, E. B.; Tao, Y. -T.; Evall, J.; Whitesides, G. M. *J. Am. Chem. Soc.* **1989**, *111*, 321-335.
13. Walczak, M. M.; Chung, C.; Stole, S. M.; Widrig, C. A.; Porter, M. D *J. Am. Chem. Soc.* **1991**, *113*, 2370-2378.
14. Bryant, M. A.; Pemberton, J. E. *J. Am. Chem. Soc.* **1991**, *113*, 3629-3637.
15. Laibinis, P. E.; Whitesides, G. M.; Allara, D. L.; Tao, Y. -T.; Parikh, A. N.; Nuzzo, R. G. *J. Am. Chem. Soc.* **1991**, *113*, 7152-7167.
16. For general examples, see: Cotton, F. A.; Wilkinson, G. *Advanced Inorganic Chemistry*; 5th Ed.; Wiley-Interscience: New York, New York, 1988. For Au(I)-sulfur interactions in solution, see: Schmidbaur, H. *Angew. Chem. Int. Ed. Engl.* **1976**, *12*, 728-740 and references therein. For a recent structural determination of RS⁻-Ag(I), see: Dance, I. G.; Fisher, K. J.; Banda, H.; Scudder, M. L. *Inorg. Chem.* **1991**, *30*, 183-187.
17. Strong, L.; Whitesides, G. M. *Langmuir* **1988**, *4*, 546-558.
18. Chidsey, C. E. D.; Liu, G. -Y.; Rowntree, P.; Scoles, G. *J. Chem. Phys.* **1989**, *91*, 4421-4423.
19. Nuzzo, R. G.; Dubois, L. H.; Allara, D. L. *J. Am. Chem. Soc.* **1990**, *112*, 558-569.
20. Chidsey, C. E. D.; Loiacono, D. N. *Langmuir* **1990**, *6*, 682-691.
21. Widrig, C. A.; Alves, C. A.; Porter, M. D. *J. Am. Chem. Soc.* **1991**, *113*, 2805-2810.

22.   Bryant, M. A.; Pemberton, J. E. *J. Am. Chem. Soc.* **1991**, *113*, 8284-8293.
23.   Hautman, J.; Klein, M. L. *J. Chem. Phys.* **1989**, *91*, 4994-5001.
24.   Bain, C. D.; Whitesides, G. M. *Langmuir* **1989**, *5*, 1370-1378.
25.   Laibinis, P. E.; Whitesides, G. M. *J. Am. Chem. Soc.* submitted.
26.   Bain, C. D.; Evall, J.; Whitesides, G. M. *J. Am. Chem. Soc.* **1989**, *111*, 7155-7164. Bain, C. D.; Whitesides, G. M. *J. Am. Chem. Soc.* **1989**, *111*, 7164-7175.
27.   Pale-Grosdemange, C.; Simon, E. S.; Prime, K. L.; Whitesides, G. M. *J. Am. Chem. Soc.* **1991**, *113*, 12-20. Prime, K. L.; Whitesides, G. M. *Science (Washington, D.C.)* **1991**, *252*, 1164-1167.
28.   Bain, C. D. Ph.D. Thesis, Harvard University, Sept. 1988. Laibinis, P. E.; Graham, R. L.; Biebuyck, H. A.; Whitesides, G. M. *Science (Washington, D.C.)* accepted.
29.   Chidsey, C. E. D. *Science (Washington, D.C.)* **1991**, *251*, 919-922. Miller, C.; Cuendet, P.; Grätzel, M. *J. Phys. Chem.* **1991**, *95*, 877-886. Miller, C.; Grätzel, M. *J. Phys. Chem.* **1991**, *95*, 5225-5233.
30.   Hickman, J. J.; Ofer, D.; Laibinis, P. E.; Whitesides, G. M.; Wrighton, M. S. *Science (Washington, D.C.)* **1991**, *252*, 688-691.
31.   Fenter, P.; Eisenberger, P.; Li, J.; Camillone, N., III; Bernasek, S.; Scoles, G.; Ramanarayanan, T. A.; Liang, K. S. *Langmuir* **1991**, *7*, 2013-2016.
32.   Schwaha, K.; Spencer, N. D.; Lambert, R. M. *Surf. Sci.* **1979**, *81*, 273-284. Rovida, G.; Pratesi, F. *Surf. Sci.* **1981**, *104*, 609-624.
33.   Harris, A. L.; Rothberg, L.; Dubois, L. H.; Levinos, N. J.; Dhar, L. *Phys. Rev. Lett.* **1990**, *64*, 2086-2089. Harris, A. L.; Rothberg, L.; Dhar, L.; Levinos, N. J.; Dubois, L. H. *J. Chem. Phys.* **1991**, *94*, 2438-2448.
34.   Sandroff, C. J.; Herschbach, D. R. *J. Phys. Chem.* **1982**, *86*, 3277-3279.
35.   Harris, A. L.; Chidsey, C. E. D.; Levinos, N. J.; Loiacono, D. N. *Chem. Phys. Lett.* **1987**, *141*, 350-356.
36.   Folkers, J. P.; Laibinis, P. E.; Whitesides, G. M. unpublished results.
37.   Laibinis, P. E.; Hickman, J. J.; Wrighton, M. S; Whitesides, G. M.; *Science (Washington, D.C.)* **1989**, *245*, 845-847. Hickman, J. J.; Laibinis, P. E.; Auerbach, D. I.; Zou, C.; Gardner, T. J.; Whitesides, G. M.; Wrighton, M.S. *Langmuir* submitted.
38.   Leiserowitz, L.; Hagler, A. T. *Proc. R. Soc. London* **1983**, *A388*, 133-175. Leiserowitz, L. *Acta Crystallogr.* **1976**, *B32*, 775-802.
39.   Etter, M. C. *Acc. Chem. Res.* **1990**, *23*, 120-126.
40.   McBride, J. M.; Segmuller, B. E.; Hollingsworth, M. D.; Mills, D. E.; Weber, B. A. *Science (Washington, D.C.)* **1986**, *234*, 830-835.
41.   Desiraju, G. R. *Crystal Engineering. The Design of Organic Solids;* Elsevier: Amsterdam, 1989.
42.   Zerkowski, J. A.; Seto, C. T.; Wierda, D. A.; Whitesides, G.M. *J. Am. Chem. Soc.* **1990**, *112*, 9025-9026. Zerkowski, J. A.; Seto, C. T.; Whitesides, G. M. unpublished results.
43.   Seto, C. T.; Whitesides, G. M. *J. Am. Chem. Soc.* **1990**, *112*, 6409-6411. Wang, Y.; Wei, B.; Wang, Q. *J. Crystallogr. Spectrosc. Res.* **1990**, *20*, 79-84.

RECEIVED January 16, 1992

Chapter 3

Synthesis and Deposition of Electron Donors, Acceptors, and Insulators as Components of Zirconium Diphosphonate Multilayer Films

H. E. Katz, M. L. Schilling, S. Ungashe, T. M. Putvinski, and
C. E. Chidsey

AT&T Bell Laboratories, 600 Mountain Avenue, Murray Hill, NJ 07974

The deposition of three organodiphosphonates as multilayered zirconium salts is described. The organic cores of the compounds include quaterthienyl, dicyanodipyridylquinodimethane, and biphenyl, which are electron donating, accepting, and insulating, respectively. Structures of the multilayers are proposed based on ellipsometric and absorbance data. Preliminary electrical measurements indicate that the films are nonconductive. The compounds are potential components of heterostructures in which photoactivated charge transfer may be observed.

Current and anticipated uses of organic thin films include those that may be classed as "passive", such as for electrical insulation or protection against corrosion, and others in which the films "actively" contribute to an electrical or optical response. The effectiveness of these films depends upon achieving dense substrate coverage with layers of defined thicknesses of tens to hundreds of angstroms. Many of the latter, "active," applications rely additionally upon the spacing and directional sequence or orientation of multiple chemical components comprising the film.

Several options are available for producing films with the required architectural control. For example, the sequence of components in a multilayer may be established by subliming compounds onto a substrate in a particular order.[1] Unfortunately, control of orientation and thickness with this method is difficult. The Langmuir-Blodgett technique [2] offers firmer control of layer thickness and orientation, as does organotrichlorosilane-based self-assembled multilayer deposition.[3] However, all of these procedures lead to films that are held together primarily by ordinary covalent bonds or van der Waals interactions, and therefore lack the desirable attributes of strength and temperature stability that are associated with other material classes such as inorganic crystals and ceramics.

Mallouk and coworkers recently demonstrated the layer-by-layer deposition of zirconium α,ω-alkylidenediphosphonates on Au and Si substrates, forming multilayers whose schematic structure is illustrated in Figure 1.[4] The resulting films

0097–6156/92/0499–0024$06.00/0
© 1992 American Chemical Society

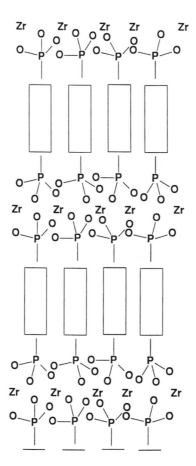

Figure 1. Schematic diagram of Zr organodiphosphonate multilayer structure.

are markedly stable because the interlayers consist of Zr^{+4} ions strongly and multiply coordinated to phosphonate groups, analogously to high-melting, chemically robust zirconium phosphate and phosphonate bulk layered solids. We wished to take advantage of this new multilayer deposition method and the resulting architecture to incorporate conjugated moieties with well-defined conformations and electronic properties. These new molecular subunits may allow us to explore the bulk passivating properties of the multilayers, as well as in observing microscopic electron transfer phenomena in a rigorously defined arrangement.

This paper focuses on three organodiphosphonates that have been synthesized and deposited as zirconium phosphonate multilayers. Compounds 1 and 2 (Figure 2) comprise an electron donor-acceptor pair, and were designed to possess minimal conformational freedom, two antiparallel C-P bonds at each end, and no ionic groups in the middle of the molecules that might interfere with multilayer formation. These structural features are shared by 3, which is envisioned as a spacer or barrier molecule separating electroactive layers, and as a building block for insulating multilayers.

Syntheses and Electrochemical Properties

The reaction sequences leading to the three compounds are shown in Figure 2. While the syntheses of 1 and 3 are straightforward, that of 2 required three steps for which there was little precedent in the literature. The reaction in which 2 is actually formed may be viewed as a dehydrogenative Swern oxidation in which the dimethylsulfoxide (DMSO) oxidant is activated by the trimethylsilyl phosphonate groups also present in the substrate.

Electrochemical experiments were performed on the soluble octamethyltetraamide and tetraethyl ester precursors to 1 and 2, respectively. The quaterthiophene displayed a partially reversible oxidation at +1.1 V vs SCE in 0.05 M Bu_4NPF_6-THF. The quinodimethane exhibited two fully reversible reductions at -0.32 and -0.71 V under the same conditions. Thus, the energy required to transfer an electron from 1 to 2 is expected to be ca. 1.4 eV (11,000 cm^{-1}), considerably less than the energy available from photoexcitation of the compounds. Compound 2 as a multilayer on Au displayed a hint of the predicted reversible electrochemistry, but much less than the theoretical current was observed due to the insulating nature of the system (vide infra).

Deposition and Characterization of Multilayers

The protocol described by Mallouk for long chain aliphatic diphosphonic acids was modified to enable the deposition of more rigid diphosphonic acids with Zr^{+4}. Surfaces with high densities of phosphate/phosphonate functionality were prepared by immersing gold-coated silicon or mica in 1.5 mM ethanolic 8-mercaptooctylphosphonic acid, or by treating silicon oxide surfaces with (3-aminopropyl)trimethoxysilane in octane at reflux, followed by $POCl_3$ and a tertiary amine in CH_3CN. These $PO_3^=$-terminated surfaces were then zirconated with 5 mM aqueous $ZrOCl_2$. Using 1 as the substrate for optimization of conditions, aqueous DMSO was judged to be the most suitable solvent from which to deposit layers of the diacids. It was necessary to maintain a temperature of >80 °C and a pH of 3-4 to ensure complete coverage of the zirconated surfaces by the aromatic compounds in a reasonable time.

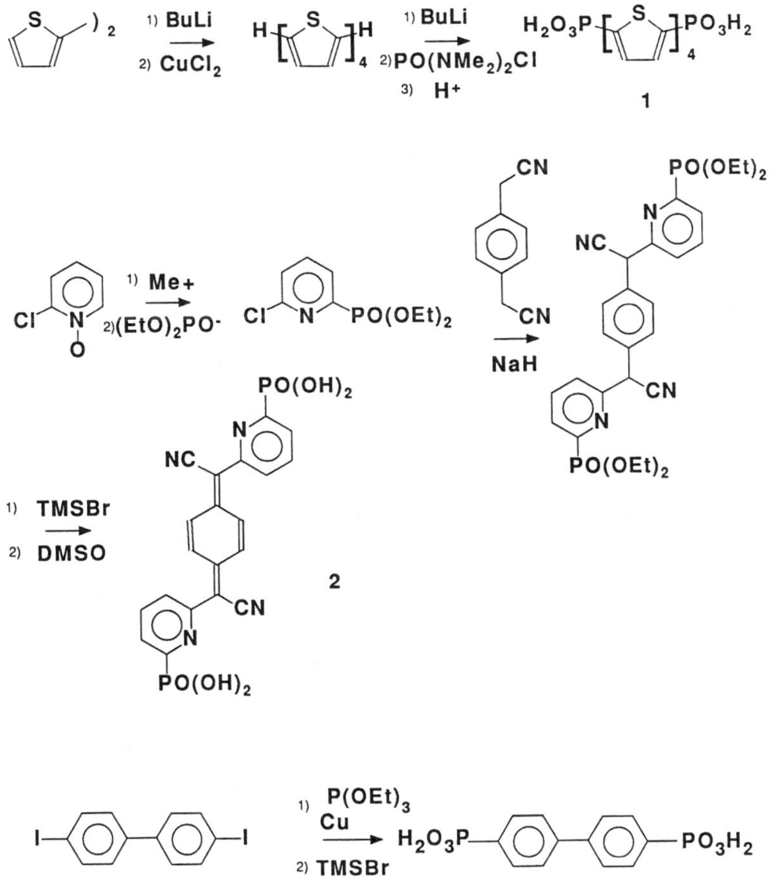

Figure 2. Syntheses of organodiphosphonates.

Layers of **1** and **3**, which are constrained to pack as extended molecular structures, were completely formed in 5-20 minutes, giving the expected ellipsometric thicknesses, 21 and 15 Å, respectively. An idealized structure for the multilayer array of **1** is shown in Figure 3. Slight defects in this structure could arise from C-C bond rotation or tilting of the long axes of the molecules.

On the other hand, **2**, despite possessing only two conformationally significant rotational degrees of freedom, may exist in conformations where the C-P bonds are not parallel to the long axis of the molecule, and may have particularly electron-deficient phosphonate groups. This species required 1 hour for each deposition and gave layers of only half the 20 Å thickness expected if the molecules were fully extended. A possible structure for this array appears in Figure 4; the upper limit of the layer thickness for this particular arrangement is 14 Å. Twisting the pyridyl rings away from normality to the surface would further reduce the thickness. The layer-to-layer reproducibility of the monolayer depositions in the construction of multilayers is illustrated in the plots of Figure 5, in which ellipsometric thickness as a function of layer number defines nearly straight lines.

Multilayers of **1** and **2** were further characterized by UV-vis spectroscopy on glass; Figure 6 shows absorbance vs layer number for these systems. From the slope of the plot for **1**, it was determined that each layer contained 20 Å of material with extinction coefficient $3.4 \times 10^4 \, M^{-1} cm^{-1}$ (derived from solution spectra) and number density corresponding to an appropriate crystallographic[5] concentration of 2.7 M. A similar treatment of the absorbance data from 5 layers of **2** (deposited atop 5 layers of **1**), with extinction coefficient $4.8 \times 10^4 \, M^{-1} cm^{-1}$, gave a concentration of **2** of only 1.2 M, assuming the ellipsometric thickness of 11 Å/layer. Considering that the molecules are of similar molecular weight and allowing for the fact that the density of **2** is less than that of **1** due to the absence of sulfur atoms, the low calculated concentration for layers of **2** is indicative of poorer packing, more severe tilting, and/or a greater number of discontinuities than is reflected in Figure 4. The possibility that the extinction coefficients of the molecules in solution do not exactly correspond to the extinction coefficients in the films adds a further degree of uncertainty to the above analyses.

Some preliminary electrical measurements have been performed on multilayers of **1** and **3**. A coating of as few as 3 layers of **1** on indium-tin oxide-coated glass formed an insulating barrier to a four-point probe. Multilayers of **1** on glass, when exposed to Br_2 vapor, displayed a weak absorbance at 680 nm, roughly in the range expected for quaterthiophene radical cation,[6] but did not show increased sheet conductivity. An Hg drop electrode resting on 10 or 20 layers of **3** on Au formed a capacitor. The dielectric constant of the multilayer was estimated to be 4, in rough agreement with bulk layered Zr phosphonates,[7] and the resistivity was ca. 10^{13} ohm cm. Further experiments are in progress to determine these values more precisely. Understanding the mechanisms of conductivity and dielectric response in these materials will aid in the interpretation of data that may be acquired in the future on photoactivated systems.

Conclusions

Fully conjugated diphosphonic acids containing electron-rich, electron-poor, and electrochemically unreactive chromophores have been synthesized and shown to form

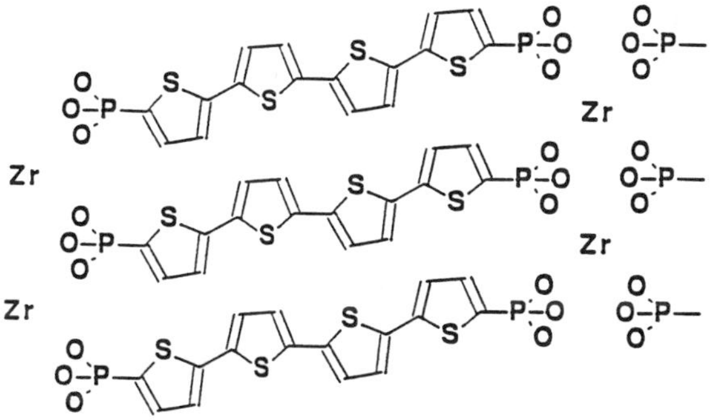

Figure 3. Layer arrangement of 1-Zr.

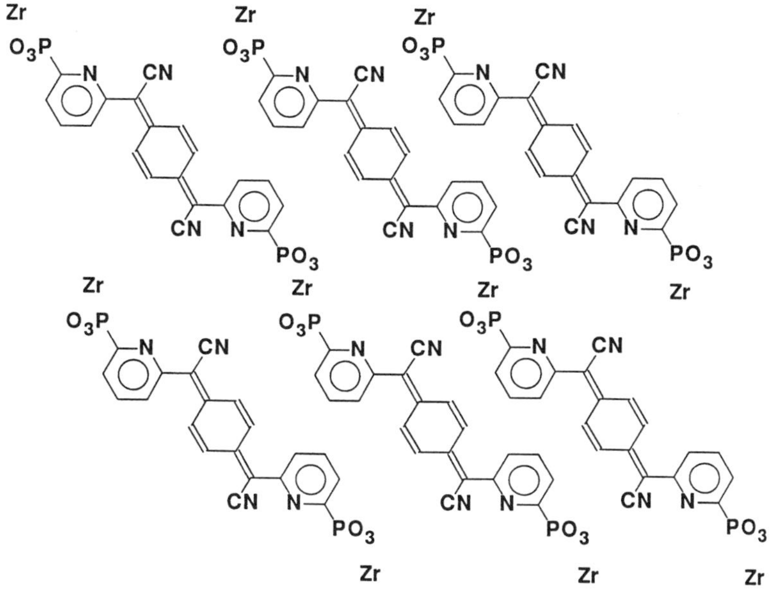

Figure 4. Layer arrangement of 2-Zr

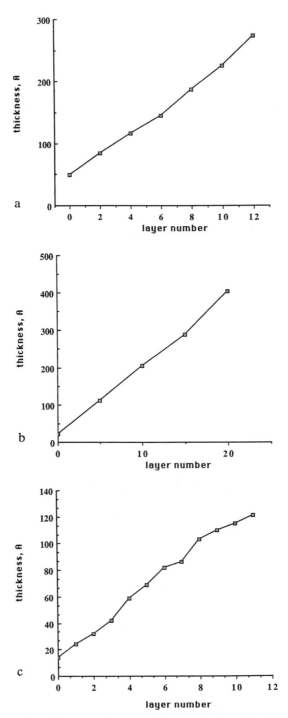

Figure 5. Ellipsometric thickness vs layer number for (a) **1** on Si, (b) **3** on Au, and (c) **2** on Au.

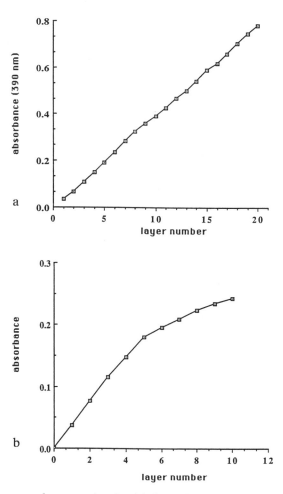

Figure 6. Absorbance vs layer number for (a) **1** on glass, and (b) 5 layers of **2** (layers 6-10) on 5 layers of **1** (layers 1-5) on glass. Absorbances for layers of **1** are recorded at its λ_{max} of 390 nm, and absorbances for layers of **2** are cumulated based on readings at its λ_{max}, 440 nm. Note that the second sample constitutes an acceptor-donor heterostructure.

stable, densely packed Zr-based self-assembled multilayer assemblies. The films electrically and chemically passivate underlying layers. We plan to investigate optically induced charge transfer in heterogeneous multilayers containing various arrangements of compounds such as **1-3**, with particular attention to the environmental stability, kinetics, decay, and mechanism (conduction vs tunneling, distance dependence) of the charge transfer, and the possible utilization of the charge separation and recombination events to produce useful optical and electrical signals.

Literature Cited

1. So, F.F.; Forrest, S.R.; Shi, Y.Q.; Steir, W.H. *Appl. Phys. Lett.* **1990,** *56,* 674-676.
2. Popovitz-Biro, R.; Hill, K.; Landau, E.M. Lahav, M.; Leiserowitz, L.; Sagiv, J.; Hsuing, H.; Meredith, G.R.; Vanherzeele H. *J. Am. Chem. Soc.* **1988,** *110,* 2672-2674.
3. Maoz, R.; Netzer, L.; Gun, J.; Sagiv, J. *J. Chim. Phys.* **1988,** *85,* 1059-1065; Tillman, N.; Ulman, A.; Penner, T.L. *Langmuir* **1989,** *5,* 101-111.
4. Lee, H.; Kepley, L.J.; Hong, H.-G.; Mallouk, T.E. *J. Am. Chem. Soc.* **1988,** *110,* 618-620.
5. Van Bolhuis, F.; Wynberg, H.; Havinga, E.E.; Meijer, E.W.; Staring, E.G.J. *Synth Met.* **1989,** *30,* 381-389; Visser, G.J.; Heeres, G.J.; Wolters, J.; Vos, A. *Acta Cryst.* **1968,** *24B,* 467-473.
6. Fichou, D.; Horowitz, G.; Garnier, F. *Synth. Met.* **1991,** *39,* 125-131; Caspar, J.V.; Ramamurthy, V.; Corbin, D.R. *J. Am. Chem. Soc.* **1991,** *113,* 600-610.
7. Casciola, M.; Costantino, U.; Fazzini, S.; Tosoratti, G. *Solid State Ionics* **1983,** *8,* 27-34.

RECEIVED January 16, 1992

Chapter 4

Synthesis and Properties of Novel Low-Dimensional Ruthenium Materials

Self-Assembled Multilayers and Mixed Inorganic–Organic Polymers

DeQuan Li, Sara C. Huckett, Tracey Frankcom, M. T. Paffett, J. D. Farr, M. E. Hawley, S. Gottesfeld, J. D. Thompson, Carol J. Burns, and Basil I. Swanson

Inorganic and Structural Chemistry Group INC–4, Los Alamos National Laboratory, Los Alamos, NM 87545

Low-dimensional ruthenium materials, built either from molecular self-assemblies, by using multidentate ligands and $Ru(H_2O)_6{}^{2+}$ or $Ru_2(O_2CR)_4$ polymeric building blocks, are reported. The molecular self-assemblies and their derivative Ru films have been characterized by ellipsometry, FTIR-ATR, UV absorption spectroscopy, XPS, SIMS and STM. Polymers built from $Ru_2(O_2CR)_4$ and 2,5-dimethyldicyanoquinonediimine (DMDCNQI) have been synthesized and characterized by elemental analysis and IR and UV-Vis spectroscopies. For DMDCNQI, these results indicate the formation of chains of oxidized $Ru_2(O_2CR)_4$ centers symmetrically bridged by radical DMDCNQI anions. The magnetic properties of these polymers as a function of tuning via the carboxylate R-group are discussed.

The design and synthesis of low-dimensional materials has become an active area of research in recent years. This is due in part to their potential applications in areas such as energy conservation, sensing, superconductivity (1), and electromagnetic shielding materials (2). A basic understanding of these purposely tailored systems with respect to their magnetic, metallic, and optical properties will help design macromolecular architecture constructions based on molecular constituents (2).

Two approaches to building macromolecular constructions are described in this paper. The first is to develop strategies for producing surface-bound linear chains using self-assembly techniques. With this method, 1-D materials can be built with a large degree of control over the specific sequence of components. These techniques have been successfully employed to generate thin film nonlinear optical (NLO) materials (3). In addition, the production of mixed inorganic-organic polymers built from molecular components has been investigated. These approaches offer tremendous flexibility with respect to the organic and inorganic building blocks and, therefore, a chemically controlled local environment resulting in desired physical, magnetic, and optical properties. The intrinsic relationships between molecular structures and materials properties can then be probed by studying these low dimensional molecular assemblies.

0097–6156/92/0499–0033$06.00/0

Ru Molecular Self-Assemblies

The design, construction and molecular architecture of artificial super-molecular self-assembled arrays with planned structures and physical properties has attracted growing interest recently (4-8). Covalently bound self-assembled monolayers (9) or multilayers (10) such as alkylsilanes on glass or alkylthiols on gold offer a potential starting point for fabricating highly ordered multifunctional thin films such as covalently bound multilayers for nonlinear optical materials (11,12). The central advantage of using self-assembly schemes is that the weak physical interaction between interfaces (as in the case of Langmuir-Blodgett films) is replaced with covalent bonds. Therefore, it would be an instructive and unique challenge to sequentially construct an aligned interlocked mixed valence Ru (II/III) supralattice structure on well defined surfaces in a self-assembled manner thereby generating ideal structures for low-dimensional materials.

We report here the construction of covalently bonded self-assembled monolayers of N-[(3-trimethoxysilyl)propyl]ethylenediaminetriacetate (TMPEDTA) and their Ru (II) pyrazine self-assembly derivatives.

Experimental. All procedures were carried out under an Ar atmosphere and solvents were degassed before use. All solutions are aqueous unless otherwise noted. TMPEDTA was obtained from HULS and used without any further purification.

The fused quartz substrates were ultrasonically cleaned in a 10% detergent solution for 10 minutes and then refluxed in a 1% tetrasodium ethylenediamine-tetraacetate solution for 10 minutes followed by another 10 minutes sonication at ambient temperature. Finally, the substrates were thoroughly rinsed with deionized water, acetone and then exposed to Ar plasma for several hours.

Synthesis of $[Ru^{II}(H_2O)_6] \cdot 2tos$ (tos = p-toluenesulphonate). The hexaaquaruthenium(II) complex was synthesized according to a slightly modified literature procedure (13,14).

Functionalization of the Quartz Substrate with a TMPEDTA Monolayer. A *cleaned* SiO_2 substrate was incubated at pH = 2-3 (adjusted with concentrated HCl, a catalyst for Si-OMe bond hydrolysis) with a 5.0×10^{-3} M, N-[(3-trimethoxysilyl)propyl]ethylenediaminetriacetate (TMPEDTA) solution at 70°C for 3 days. After thoroughly rinsing with deionized water, the quartz substrate was immersed in a pH = 5.5 HAc/KAc buffer for 2-3 hours with 10 minutes gentle sonication every 30 min. The substrate was then cleaned by sonication (1 minute) in water and finally rinsed with water and acteone.

Formation of the $[TMPEDTA]Ru^{II}(H_2O)$ and $[TMPEDTA]Ru^{II}Pz$ Molecular Self-Assemblies. The quartz substrate coated with the TMPEDTA monolayer was immersed in a 10 ml degassed solution of $Ru^{II}(H_2O)_6 \cdot 2tos$ (3.6×10^{-2}M). The temperature of the purple solution was kept at less than 30°C for 6 hours with 20 minutes periodic sonication. The resulting substrates with the covalently attached monolayer of $[TMPEDTA]Ru^{II}(H_2O)$ were quickly rinsed with deionized water and then dipped into a 3.6×10^{-2} M pyrazine (Pz) solution. The solution again was sonicated 5 times with 1 hour duration.

Low Temperature (< 50°C) Ru Mirror Formation. The covalently modified TMPEDTA quartz substrate was immersed in a 10 ml degassed solution of $Ru^{II}(H_2O)_6 \cdot 2tos$ (3.6×10^{-2}M). The temperature of the purple solution was slowly increased to 35-40°C and maintained overnight. A shiny, smooth, metallic mirror was formed on the quartz substrate.

Surface Analysis. Auger electron spectroscopy (AES) and X-ray photoelectron spectroscopy (XPS) were performed with commercial cylindrical mirror electron energy analyzers. Electron beam conditions for AES were typically 3KeV and 1 nA total beam current in a spot size $0.1 mm^2$. X-Ray irradiation for XPS was performed using an unmonochromatic Mg source (hν=1253.6 eV). Narrow scans were performed with a band pass of 50eV and satellite and inelastic background were removed from spectra using established procedures. AES depth profiling was performed using 3KeV Ar ions with a nominal sputter removal rate of ~100Å min^{-1} for Ru. Secondary ion mass spectroscopy (SIMS) was performed in both static and dynamic modes using either 5 KeV Ar or Xe mass filtered ion beams of appropriate beam current density. Mass analysis of the sputtered material was performed using a quadrupole instrument with provision for detection of positive, negative, and sputtered neutrals species. Insulating surfaces (e.g., quartz substrates) were examined using appropriate application of a low energy ($E_i \leq 100eV$) electron source for both SIMS and XPS.

Results and Discussion. The synthetic approach is illustrated in Scheme I. A monolayer of TMPEDTA, a multidentate ligand, was introduced onto a *cleaned* fused quartz surface via a siloxane bond linkage. The monolayer formation was confirmed by observation of UV absorptions at 213 and 277 nm. The FTIR-ATR spectrum shows a C=O stretch at 1730 cm^{-1} and two CH$_2$ bands at 2848 and 2924 cm^{-1} (shoulder at 2957 cm^{-1}) corresponding to the υ_s(CH$_2$) and υ_a(CH$_2$) vibrations, respectively. The TMPEDTA layer was functionalized with Ru by reaction with $Ru^{2+}(H_2O)_6 \cdot 2tos$ at room temperature. Throughout this reaction the monolayer surface was maintained at pH≈5.5 by immersing the ligand-functionalized substrate in an HAc/KAc buffer solution; this helps to avoid degredation reactions of the $Ru^{II}(H_2O)_6 \cdot 2tos$ reagent.

A pyrazine (Pz) bridging ligand was then coordinated to the surface Ru metal center by substitution of the last Ru aquo ligand. Surface characterization of the [TMPEDTA]Ru^{II}Pz modified substrate was done using XPS and static SIMS. The 1s transitions of N, C, and O were monitored along with the Ru 3p and 3d transitions. In addition, a rough estimate of the attached layer thickness of 10Å was obtained from attenuation of the Si 2p and 2s transitions. The position of the Ru 3d$^{5/2}$ transition (281.7eV binding energy) of the monolayer attached [TMPEDTA]Ru^{II}Pz substrate was 0.5 eV higher in binding energy than that obtained from the pressed powder of the [TMPEDTA]Ru^{II}Pz salt (when referenced to the C 1s peak) (Figure 1). The difference in the Ru 3d$^{5/2}$ binding energies between the [TMPEDTA]Ru^{II}Pz salt and covalently attached monolayer indicates a slightly altered electronic environment for the latter entity. Note that the intensity of Ru 3d$^{5/2}$ transition is only ~10% that of the C 1s transition located at 284.6 eV binding energy. Positive ion SIMS spectra displayed the presence of RuO, Ru, SiO, Si, K, Na, O, N, CH, and C ions. Negative ion SIMS did not indicate any significant contribution due to surface Cl.

In an attempt to generate a Ru(II/III) mixed-valence self-assembly, the [TMPEDTA]Ru^{II}Pz monolayer was further reacted with a K$_2$[Ru^{III}Cl$_5$(H$_2$O)] solution at ~ 30°C. XPS spectra of the Ru 3d$^{5/2}$ region indicates that the Ru concentration has roughly doubled relative to the [TMPEDTA]Ru^{II}Pz modified substrate. In addition,

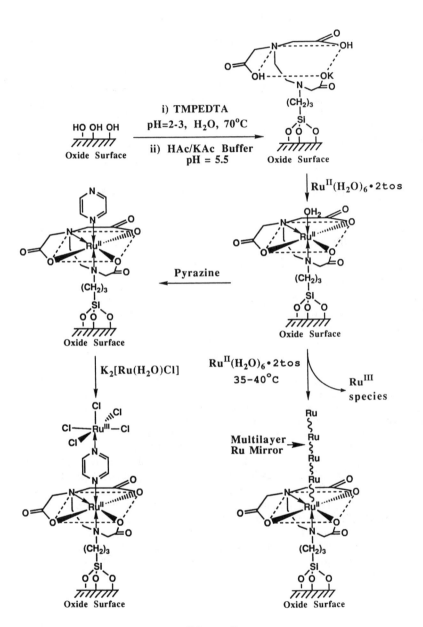

Scheme I

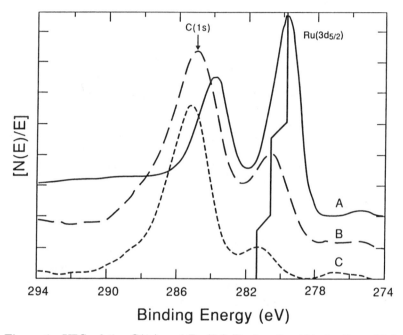

Figure 1. XPS of the C(1s) and Ru(3d) Region for (A) the "metallic" Ru Film (lightly sputtered to remove surface contamination), (B) the [TMPEDTA]Ru^{II}Pz powder and (C) the Covalently Attached [TMPEDTA]Ru^{II}Pz Complex.

negative ion static SIMS shows the presence of significant surface Cl. Ellipsometric measurements were also carried out on the thin films, Figure 2 (the ellipsometric parameters Δ and Ψ are plotted in degrees). Because Ψ is constant at about 16° there are no complicating strong absorption effects and, therefore, the 2-3 degree Δ change after each monolayer deposition can be interpreted as corresponding to a concomitant increase in film thickness ($\sim$10-15Å)(*15*)

When the TMPEDTA monolayer, buffered at pH $\approx$ 5.5, was heated to 35-40°C in a Ru^{II}(H$_2$O)$_6$•2tos solution, a shiny, conductive (ρ = 57.5 Ω/cm) film was formed on the TMPEDTA covalently attached adhesion layer. Scanning tunneling microscopy shows a fairly smooth surface with standard deviation of about 10 nm for a 1x1 μm^2 scan (Figure 3a) and 10-25 nm diameter domains were observed for higher resolution scans (Figure 3b). XPS data taken from the lightly sputtered cleaned film indicates that the Ru is predominately present in a metallic state (3d$^{5/2}$ transition at 279.7 eV). A significant concentration of O (0.08-0.18 a/o) was observed in the film from AES and SIMS depth profiles. A representative AES spectrum recorded in the interior of the film is shown in Figure 4, along with the depth profile (inset). Some indication that the TMPEDTA ligand is still present at the interface of the Ru film and the quartz substrate was obtained from positive ion SIMS depth profiles (e.g., increase in C and N positive ion yield seen when sputtering using high O$_2$ gas pressures (P$_{O_2}$ =3.3x10^{-7} mbar)).

The chemical composition, structure, and physical properties of the Ru films will be the subject of continuing investigations. Of particular interest is the characterization of the potentially mixed-valent [TMPEDTA]Ru^{II}PzRu^{III}(H$_2$O)$_5$ films and the nature of the conductive [TMPEDTA]Ru mirrors.

Mixed Inorganic-Organic Polymers

An alternate approach to the preparation of low-dimensional materials as described above is to develop methods for constructing mixed inorganic-organic polymers. Our initial investigations have focused on linking Ru$_2$(O$_2$CR)$_4$ centers with DMDCNQI bridging ligands in order to synthesize chains with extended overlap of π-symmetry orbitals. These materials should be complementary to the well known σ-systems, which include Wolfram's red salt derivatives (*16*) and halide bridged metal dimers (*17*), formed by σ-type interactions of M d$_{z^2}$ and halide p$_z$ orbitals. The Ru compounds, with the $\sigma^2\pi^4\delta^2\pi^{*2}\delta^{*2}$ electronic configuration (*18*), have partially filled π symmetry orbitals and are known to coordinate axial ligands with back donation into ligand π^* orbitals (*19*). Both of these factors are important with respect to setting up M-L π overlap and ultimately delocalized π-symmetry orbitals along the chain axis.

DMDCNQI

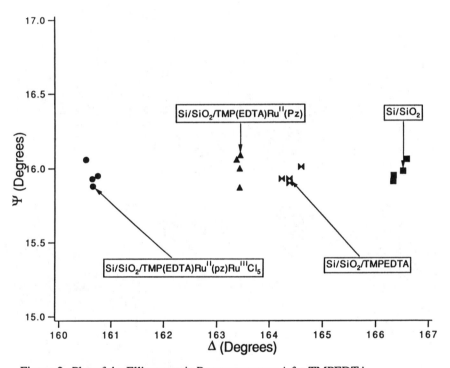

Figure 2. Plot of the Ellipsometric Parameters ψ vs Δ for TMPEDTA, [TMPEDTA]Ru^{II}Pz, and [TMPEDTA]Ru^{II}PzRu^{III}Cl$_5$ Films on SiO$_2$/Si Substrates.

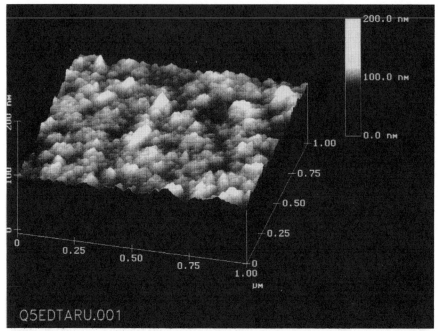

(a)

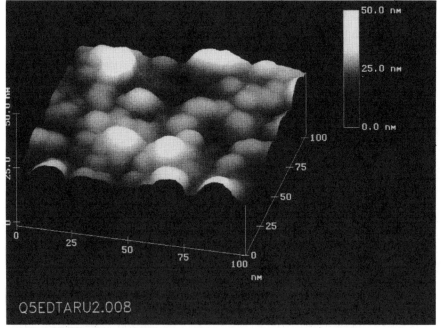

(b)

Figure 3. STM Images of Ru mirrors formed by self-assembled technique at (a) 1 x 1 μm^2 and (b) 100 x 100 nm^2 Scales.

The $Ru_2(O_2CR)_4$ entities are also easily oxidized (*20*). This then allows access to other oxidation states which would facilitate doping of the target one-dimensional materials.

Synthesis. Two routes can be used to obtain the $[Ru_2(O_2CR)_4 \cdot DMDCNQI]_X$ systems as shown in Figure 5 (Tol = $p\text{-}C_6H_4CH_3$, Mes = $2,4,6\text{-}C_6H_2(CH_3)_3$). However, the direct reaction of $Ru_2(O_2CR)_4$ and two equivalents of DMDCNQI produces purer products, as indicated by IR and elemental analysis, and this route is currently exclusively employed. As the yellow THF solution of DMDCNQI is dropped into the red-brown solution of $Ru_2(O_2CR)_4$ the color of the solution changes to an intense blue. The blue to purple/black products then precipitate from the solution. These observations indicate that: (1) the redox reaction between the Ru_2 center and DMDCNQI bridge has occurred (the Na radical anion salt of DMDCNQI has the same blue color as the reaction mixture) and (2) that the products are polymeric (they can not be redissolved in organic solvents including THF, Et_2O, or CH_2Cl_2).

Characterization. Preliminary X-ray powder diffraction results for $[Ru_2(O_2CMe)_4 \cdot DMDCNQI]_X$ show that the microcrystalline material does diffract and current effort is being directed towards obtaining structural information from this data.

We have formulated the $[Ru_2(O_2CR)_4 \cdot DMDCNQI]_X$ materials as chains of $Ru_2(O_2CR)_4^+$ centers bridged by radical anionic DMDCNQI ligands. This is based on the solubility properties of the materials, the color, optical spectroscopy , and elemental analysis data (*21*). The elemental analysis results show that the compounds contain a 1:1 stoichiometry of $Ru_2(O_2CR)_4$ to DMDCNQI. Several features contained in the infrared (IR) spectroscopic data shown in Table I are important to note. First, the position of the $v_{C \equiv N}$ band of the DMDCNQI ligand in the $[Ru_2(O_2CR)_4 \cdot DMDCNQI]_X$ compounds is shifted to lower frequency relative to the free ligands and the $Na[DMDCNQI]_2$ salt. This indicates coordination of the ligand to the metal center and formal reduction of the $C \equiv N$ bond order. Additionally, only one $v_{C \equiv N}$ band is observed, indicating that the DMDCNQI ligand is symmetrically bound. Even for the alkyl substituted carboxylates, bands in the aromatic region are observed; this is consistent with the formulation of the bridging DMDCNQI ligand as a radical anion which would have contributions from aromatic resonance structures.

Spectroscopic evidence for the oxidation state of the Ru_2 centers has been obtained by diffuse reflectance spectroscopy. For $[Ru_2(O_2CPh)_4 \cdot DMDCNQI]_X$, broad features are observed at ~600 and 950 nm, in addition to a weak band at 1158 nm. This is in the same position, ~1100 nm, where $\delta \rightarrow \delta^*$ transitions are observed for $Ru_2(O_2CR)_4Cl$ compounds (*22*) containing Ru_2^{5+} metal centers, suggesting that the Ru_2^{4+} center in the $Ru_2(O_2CR)_4$ starting material has been oxidized to Ru_2^{5+}.

Magnetic Measurements. Preliminary magnetic measurements for the $[Ru_2(O_2CR)_4 \cdot DMDCNQI]_X$ systems have been carried out. The $[Ru_2(O_2CTol)_4 \cdot DMDCNQI]_X$ exhibits apparent antiferromagnetic behavior with a T_{Neel} = 46K at H = 0.1 T, Figure 6, that does not change within our experimental uncertainty ($T_N \pm 1K$) for H = 1T. The magnetization exhibits strong field dependence which is consistent with the proposed one-dimensional nature of the material. At 10K, dM/dH is linear to ~2.6T and then increases more rapidly than linear. The inverse susceptibility is strongly nonlinear in temperature but for $T > T_N$ always extrapolates to a negative paramagnetic Curie temperature. On the basis of

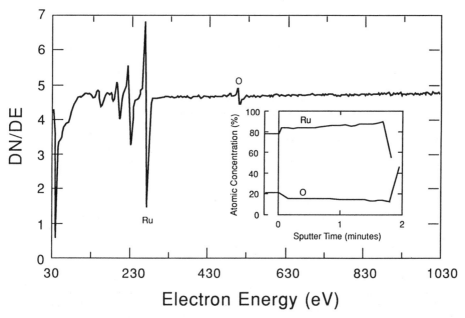

Figure 4. Representative AES Spectrum, dN/dE versus Electron Energy, Recorded in the Interior of the "Metallic" Ru Film. Inset: Depth Profile of the Same Surface.

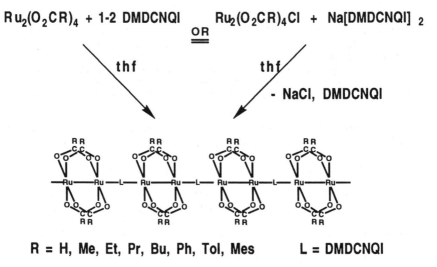

R = H, Me, Et, Pr, Bu, Ph, Tol, Mes   L = DMDCNQI

Figure 5. Synthetic Routes to $[Ru_2(O_2CR)_4 \cdot DMDCNQI]_x$.

Table I. IR Spectral Data for $[Ru_2(O_2CR)_4 \cdot DMDCNQI]_X$.

Compound	$v_{C \equiv N}$ (cm^{-1})
DMDCNQI	2166
Na[DMDCNQI]$_2$	2154
$[Ru_2(O_2CR)_4 \cdot DMDCNQI]_X$	
R = H	2098
R = Me	2132
R = Et	2116
R = Pr	2136
R = Ph	2132
R = Tol	2138

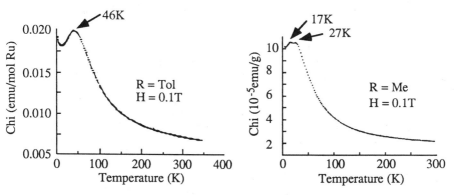

Figure 6. Magnetic Susceptibility of $[Ru_2(O_2CR)_4 \cdot DMDCNQI]_X$ (R=Tol, Me) at 0.1T.

these data it is postulated that either intrachain or interchain coupling mechanisms could be operative.

The $[Ru_2(O_2CMe)_4 \cdot DMDCNQI]_x$ compound appeared to show analogous behavior at H= 1T, T_{Neel} = 27K, but for H = 0.1T an additional magnetic transition was noted at 17K. Again the inverse susceptibility is nonlinear above 40K, but between 50 and 120K $1/\chi$ is approximately linear and extrapolates to a positive Curie temperature of ~20K, suggesting the presence of both antiferromagnetic (T>150K) and ferromagnetic-like (T<150K) correlations. The field dependent intensities of these transitions is surprising; one potential explanation is that they occur in two different dimensions, for example, in intrachain and interchain fashions. Concomitant intra- and interchain coupling has been proposed for other low-dimensional systems (23,24). That this behavior was not observed for $[Ru_2(O_2CTol)_4 \cdot DMDCNQI]_x$ measured at H= 1T, suggests that the R substituent plays a role in controlling these types of magnetic interactions. However, additional measurements at a variety of field strengths for the series of compounds are needed in order to obtain a clear understanding of the underlying interactions in these materials.

Conclusions. Reactions of $Ru_2(O_2CR)_4$ and DMDCNQI produce 1:1 adducts which are highly colored polymeric solids. Spectroscopic measurements indicate that the Ru_2^{4+} center has been oxidized to Ru_2^{5+} and a symmetrically bridging radical DMDCNQI anion has been formed. The magnetic properties of these materials are complex; continued effort will focus on obtaining an understanding of the tunability of these interactions.

Literature Cited.

1. Ferraro, J. R.; Williams, J. M. *Introduction to Synthetic Electrical Conductors*; Academic Press: New York, NY, 1987; p.6.
2. Marks, T. J. *Angew. Chem. Int. Ed. Eng.* **1990**, *29*, 857.
3. Li, D.; Ratner, M. A.; Marks, T. J.; Zhang,C.; Yang J. and Wong G. K. *J. Am. Chem. Soc.* **1990**, *112*, 7389-7390.
4. Lee, H.; Kepley, L. J.; Hong H. G.; and Mallouk, T. E. *J. Am. Chem. Soc.,* **1988**, *110*, 618-620.
5. Ulman A. and Tillman, N. *Langmuir,* **1989** ,*5*, 1418-1420.
6. Pomerantz, M.; Segmuller, A.; Netzer, L.; Sagiv, J. *Thin Solid Films,* **1985**, *132* , 153-162.
7. Putvinski, T. M.; Schilling, M. L.; Katz, H. E.; Chidsey, C. E. D.; Mujsce, A. M.; and Emerson, A. B. *Langmuir* **1990**, *6*, 1567-1571.
8. Allara,D. L.; Atre,S. V.; Elliger, C. A.; Snyder, R. G. *J. Am. Chem. Soc.* **1991**, *113*, 1852-1854.
9. Wasserman,S. R.; Tao, Y. T.; Whitesides, G. M. *Langmuir,* **1989**, *5*, 1074-1087.
10. Maoz, R.; Sagiv, J. *J. Colloid and Interface Science,* **1984**, *100(2)*, 465-496.
11. Dai, D.; Hubbard, M.; Li, D.; Park, J.; Ratner, M. A. ; Marks, T. J. ; Yang, J.; Wong, G. K. in *ACS Symp. Ser.* "New Materials for Nonlinear Optics," **1991**, 455, 226-249 (Boston, MA, April 22-27, 1990).
12. Li, D.; Marks, T. J.; Zhang, C.; Yang, J.; Wong, G. K. in *SPIE Proc.* "Nonlinear Optical Properties of Organic Materials III", **1990**, *1337*, 341-346
13. Bernhard,P.; Biner, M. and Ludi, A. *Polyhedraon*, **1990**, *9(8)*, 1095-1097
14. Bernhard, P.; Burgi, H.; Hauser, J.; Lehman, H. and Ludi, A. *Inorg. Chem.* **1982**, *21*, 3936-3941

15. The ellipsometrically measured parameters Δ and Ψ are related to the film thickness d and indexes of refraction n_{film}, n_{air}, and $n_{substrate}$ through a complex function ρ

$$\tan(\Psi)\exp(i\Delta) = \rho(\lambda,\theta,d,n_{film}, n_{air}, n_{substrate})$$

where λ and θ are the wavelength and angle of incidence used in the experiments. The explict form of ρ is given by Azzam, R. M. A.; Bashara, N. M. *Ellipsometry of Plarized Light*; North-Holland: Amsterdam, 1977; pp332-340

16. Miller, J. S.; Epstein, A.J.; *Prog. Inorg. Chem.* **1976**, *20*,1.

17. Che, C. M.; Herbstein, F. H.; Schaefer, W. P.; Marsh, R. E.; Gray, H. B.; *J. Am. Chem. Soc.* **1983**, *105*, 4604.

18. Clark, D. L.; Green, J. C.; Redfern, C. M.; Quelch, G. E.; Hillier, I. H.; Guest, M. F. *Chem. Phys. Lett.* **1989**, *154*, 326.

19. Lindsay, A. J.; Wilkinson, G. W.; Motevalli, M.; Hursthouse, M. B. *J. Chem. Soc., Dalton Trans.* **1987**, 2723.

20. Lindsay, A. J.; Wilkinson, G. W.; Motevalli, M.; Hursthouse, M. B. *J. Chem. Soc.; Dalton Trans.* **1985**, 2321.

21. Huckett, S. C.; Arrington, C. A.; Burns, C. J.; Clark, D. L.; Swanson, B. I. *Synthetic Metals* **1991**, *41-43*, 2769.

22. Miskowski, V. M.; Loehr, T. M.; Gray, H. B. *Inorg. Chem.* **1987**, *26*, 1098.

23. Kaisaki, D. A.; Chang, W.; Dougherty, D. A. *J. Am. Chem. Soc.* **1991**, *113*, 2764.

24. Caneschi, A.; Gatteschi, D.; Sessoli, R.; Rey, P. *Acc. Chem. Res.* **1989**, *22*, 392.

RECEIVED March 24, 1992

Chapter 5

Langmuir–Blodgett Films of Transition-Metal Complexes

R. E. Des Enfants II, Teddy A. Martinez, and M. Keith De Armond

Chemistry Department, New Mexico State University,
Las Cruces, NM 88003

Monolayer and multilayer Langmuir-Blodgett films of hydrophobic [Ru(t-mebpy)$_3$][ClO$_4$]$_2$ (t-mebpy is 4,4',-5,5'-tetramethyl-2,2'-bipyridyl, I) dissolved in stearic acid have been formed on various substrate materials. At a dilution ratio of 10:1 (acid to complex) stable films with up to 100 layers can be produced. Pressure-area isotherms have been determined for a number of mole ratios. Infrared and uv-vis absorption spectra have been obtained for the film samples as have room temperature emission data. These spectra suggest an ion pair mechanism for formation of the film.

Scanning tunneling microscopy (STM) for the monolayer and multilayer samples were obtained. Comparison of pure stearic acid films with mixed metal complex-stearic acid films imply that the films are homogeneous and that the metal complexes are dispersed in the stearic acid solution.

Typically Langmuir-Blodgett Films of transition metal complexes,(1-7) have incorporated long chain alkyl, ester, or acid functions to enhance their film forming capability. The early work on [Ru(bpy)$_2$R-bpy]$^{2+}$ (bpy is 2,2'-bipyridine and R-bpy is a 4,4'-diester-2,2'-bipyridine) by Whitten (1) and Gaines (2,3) illustrated the hydrolysis problems (and consequent film instability) that can occur for long chain ester functions on the chelate ligands. Japanese workers (4-7) solved this problem by replacing the reactive ester function with less reactive chemical functionalities.

More recently, electrochemical results from Bard (8-10) in which pure LB films [Ru(bpy)$_2$(bpy-R)]$^{2+}$ (R is a long alkyl chain) are coated on indium tin oxide substrates suggest two significant points:

(1)   the monolayers are not bound tightly and can be removed from the substrate in electrolyte solutions and

(2)   the Ru complexes tend to form "islands" of metal complexes on the surface.

Results for porphyrin derivatives (11,12) (metal substituted and/or containing aliphatic chains) show a range of behavior that can apparently be altered in a systematic fashion. For example, Mohwald (11) and coworkers can form homogeneous monolayers for pure porphyrins with aliphatic chains or by dilution

0097–6156/92/0499–0046$06.00/0

with phospholipids. Bocian *(12)* and coworkers observe heterogeneous films (domains of complexes) even for pure metal- containing porphryins with medium chain length substituents containing aromatic groups at the end of the chains.

The mixed layer strategy in which charge transfer complexes are dissolved in fatty-acid "solvents" to form stable monolayers was developed by Kuhn and Mobius*(13)*. Palacin and coworkers *(14,15)* have employed this general approach to produce LB films of water soluble phthalocyanine complexes with tricosenoic acid (24 carbons) by an ion-pair mechanism. Both in this case and with other amphiphilic Co (II) phthalocyanine complexes, the LB films formed (using fatty acid for mixed layers) result in homogeneous LB films with no evidence for domain structures.

Our interest in organized aggregates of Ru (II) diimine complexes derives from the unique charge localization properties of the lowest excited state (emitting state). Luminescence data *(16-18)* for the parent $[Ru(bpy)_3]^{2+}$ complex as well as a variety of spectroscopic data *(19-21)* for the reduced Ru (II) complexes, i.e. $[Ru(bpy)_2bpy^-]^{2+}$, demonstrate that the exciton (electron) is localized on a single chelate ring in the parent tris complex and hops from ring to ring in the molecule (Figure 1). Moreover this spatial isolation of the charge separation in the optical or electrochemically generated species can be readily verified with the synthesis of mixed ligand complexes such as $[Ru(bpy)_2L]^{2+}$ where L is another diimine ligand with different optical and redox properties. Therefore, by choice of L group the charge separation can be made "vectorial" either localizing on the Ru-L or Ru-bpy units. Such a property is useful in a range of applications from bimolecular electron transfer at interfaces, to molecular switching events and even nonlinear optical effects.

Thus, the general utility of the LB film technique for Ru (II) complexes became apparent to us. However, the recognition that aliphatic chains (R) on the π-chelate rings of the metal complex can diminish intermolecular interaction and, in particular, decrease intermolecular charge transfer, lowers the utility of these LB films. This and the apparent lack of stability of some of the monolayer and multilayer film structures as well as the suggestion that the films of at least one of these amphiphilic alkyl- substituted Ru (II) complexes give domain (heterogeneous) structures, motivated us to search for techniques that could produce stable homogeneous films of Ru (II) diimine complexes without long chain alkyl substituents. The desire for "homogeneous" films stems from the speculation that the more ordered homogeneous film would be better suited to device application. Consequently, the 1988 report *(22)* of the formation of stable LB films of a hydrophobic Ru (II) complex, $[Ru(dp-phen)_3]^{2+}$, (where dp-phen is 4,7-diphenyl-1,10 phenanthroline) in mixed layers of stearic acid was suggestive. Indeed, the authors report that LB films with up to 5:1 (stearic acid to Ru (II) complex) are stable and give homogeneous films (although the evidence for this conclusion is not strong). Subsequently, we surveyed some of the Ru (II) complexes available in our lab and determined that the $[Ru(t-mebpy)_3] [ClO_4]_2$ complex, I is insoluble in water and produces stable mixed mono- and multilayer films on various substrate materials.

This manuscript reports our characterization of these metal complex LB film materials. Simultaneously, we hope to develop a variety of spectroscopic techniques, most of which have been employed previously with other LB film materials and in some cases for Ru (II) complexes. For example, luminescence spectra and lifetimes have been measured for a number of the Ru (II) complexes. The report *(23)* of the use of the luminescence photoselection method to determine order in LB films of pyrene-lecithin mixed materials is pertinent since we have experience *(16-18)* in the use of the photoselection method with Ru (II) complexes. The grazing angle FT-IR method (Infrared Reflection Absorption Spectroscopy, RA) has been successfully utilized by the IBM group *(24-27)* as well as others *(28,29)* to examine mono- and multilayer LB films, primarily of molecular systems possessing long chain alkyl groups. This RA technique and polarized transmission spectra on the films can provide detailed information about the orientation of an amphiphilic molecule on the substrate surface. In at least one recent report, FT-IR results for mixed layers over a range of compositions were able to determine the occurrence of homogeneous and heterogeneous (domain) structures*(29)*.

Finally, the technique of scanning tunneling microscopy (STM) has been used to give angstrom resolution structural data for a variety of organic molecules *(30,31)* including fatty acids *(32-34)* on conducting substrates. The fact that a very recent study *(35)* of the graphite substrate places in doubt many of the DNA structures obtained with graphite requires caution with this substrate. However, other substrates such as silver give valid images for copper phthalocyanine *(36)*. Therefore, we have used other substrates such as Sn doped In_2O_3 (ITO) and Ag to ascertain if the STM technique can be used to characterize our mixed layer films to determine if the films are homogeneous and to learn if the metal complexes can be identified in the image. Evaporated films of metal complexes on ITO substrates appear to give images in which the π ring systems can be identified *(36)*, which lends promise to the use of the STM technique in the examination of mixed LB films of metal complexes.

Experimental Materials. [Ru(t-mebpy$_3$)] [ClO$_4$]$_2$ was available from a previous study and was synthesized using standard procedures. The stearic acid (99+%) and cadmium chloride (99.99%) were purchased from Aldrich. The subphase water was deionized water with a resistivity of 18 Mohm/cm at room temperature. HPLC grade chloroform (Aldrich) was used as a solvent for the spreading solutions.

The substrate plates for the LB films were 1 x 3 in glass or quartz microscope slides cleaned with hot HNO_3 and sonicated in $CHCl_3$ prior to deposition of a LB film or coating with a metal film. For the FT-IR and the STM measurements, these glass plates were first coated with a thin chromium film by evaporation. Subsequently, an Ag film of ~ 200 nm thickness was evaporated onto the chromium layer. The plates were soaked in $CHCl_3$ and dried in a stream of Grade V (99.999%) nitrogen gas prior to coating the LB film. For the STM measurements, the plate was cut to a convenient size for mounting on the STM head. The indium (tin) oxide (ITO) plates were 2.5 x 7.5 cm glass plates obtained from Delta Technologies, Ltd. (Stillwater, MN). They were cleaned with chloroform and dried prior to use. All substrate plates were kept in closed containers in the dark prior to preparation of the LB film.

Film Preparation. A NIMA alternate layer automated film balance was used to produce monolayer and multilayer films on one side of the substrate plate. The Ru (II) complex and stearic acid were dissolved together in chloroform typically at concentrations of 0.4 mM for the complex and 4 mM for the acid. This solution was added to the subphase surface with a glass syringe and allowed to spread and evaporate for 5 minutes prior to compressing the film. The monolayer films were prepared by compressing to a pressure of 15-30 mN/M (20°C) with the instrument feedback system controlling the pressure during the coating. The layers were produced with a dipping speed of 10 mm/min for the first layer and between 25 and 50mm min^{-1} for subsequent layers. Transfer ratios close to unity were obtained in all cases.

Measurements. Absorption spectra measurements on the film and solutions of the metal complexes were measured on a modified Cary 14 spectrometer. Luminescence spectra were recorded with a custom photon counting spectrometer or a PTI (Deer Park Drive, South Brunswick, NJ 08852) luminescence spectrometer system. The FT-IR spectra were recorded with a PE-1600 spectrometer or a Bio-Rad FTS-40 spectrometer (Professor R. Crooks, Chemistry Department, University of New Mexico). RA spectra for LB films were measured on the FTS-40 with a Harrick grazing angle attachment and MCT detection system. Typically 256 scans were required to obtain adequate intensity for the monolayer films. The scanning tunneling microscopy measurements employed a Nanoscope II system (Digital Instruments, Inc.). Etched Pt-Ir tips were used with the standard, 0.6 micron head. The system can be operated in either the constant current or constant height mode but the film contours were similar for either.

Typically, tunneling currents of +0.48 nA to + 2.0 nA were used with a negative (-200 mV) or positive (50 mV to 250 mV) bias potential, respectively. Monolayers were imaged in air at room temperature. The resolution obtained for these LB films and the evaporated metal complex film images was better than 2 angstroms. In all cases, a blank substrate sample (no film) was imaged prior to examination of the coated sample. Multilayer LB films with up to 10 layers could be imaged. The STM images of the LB films did not differ significantly, therefore the images presented here were with Ag substrates. To ascertain the effect of tip penetration into the LB film, the tunneling current was increased incrementally until penetration occurred. A similarly steady increase of the bias voltage was done until the tip penetrated. In both cases, penetration of the tip was recognized by the loss of the z direction structure. A rough estimate of the thickness of our film can be made from comparison to the ellipsometry measurement by Horber *(33)* and coworkers for a cadmium arachidate bilayer of 55Å thickness.

Results and Discussion

The chloroform solution of pure stearic acid was spread on the surface of a pH 5.5 water solution containing no additional ions. The pressure-area curve obtained is given in Figure 2. Figure 3 gives the π-A curves for the mixed stearic acid/I films at pH 5.5.

The surface areas were obtained for the various films assuming an area of 0.2 nm^2 for stearic acid and using the zero pressure extrapolation. Areas were assumed to be additive. The pressure area curves for the mixed films do not show plateau regions as observed by the Japanese workers with the mixed layers of $[Ru(dp\text{-}phen)_3]^{2+}$ and stearic acid. The area that we obtain for I is dependent on concentration unlike those reported by the Japanese workers with 95Å^2/molecule for the 50/1 film and 9Å^2/molecule for the 10/1 films. At ratios less than 10/1, the metal complex dissolves in the aqueous subphase at high pressure. The question of whether the complex is dispersed or clustered will be addressed in the STM data but these film data hint that a homogeneous film is produced only for low concentration of I. The temperature dependence of our 10:1 film was different from that observed by the Japanese for $[Ru(dp\text{-}phen)_3]^{2+}$. Indeed at none of our operating temperatures (18-25°C) do we observe a plateau region identified by the Japanese as a "liquid" state region. This suggests that the stearic acid or (stearate) ion) determines the film structure but that little perturbation upon the structure is seen even for the concentrated films (10:1 and 7:1).

The absorption and emission spectra for the mixed film (Fig. 4) are similar to that of the complex solution. Little change is seen between film and solution spectra which would suggest that the isolated molecule is little perturbed by the

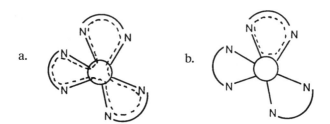

Figure 1. Spatial extent of the LUMO for d^6 tris(diimine) complexes: (a) multiring (D_3) delocalized; (b) single ring (C_2) localized.

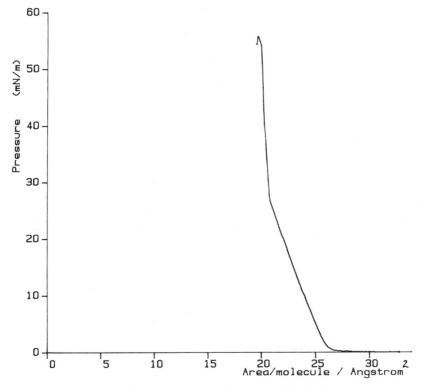

Figure 2. π-A isotherm for pure stearic acid.

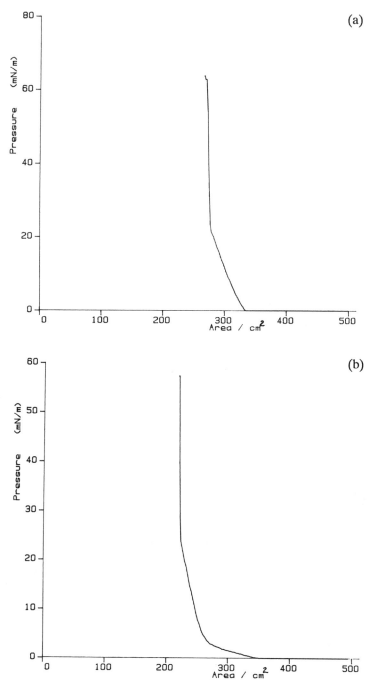

Figure 3. (a) π-A isotherm for 50/1 stearic acid/I. (b) π-A isotherm for 10/1 stearic acid/I.

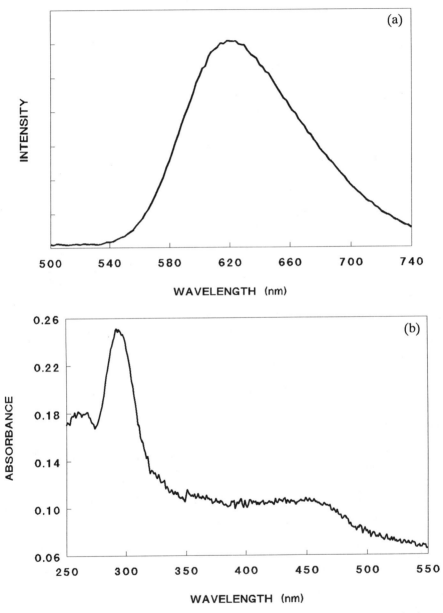

Figure 4. (a) Emission spectrum of 200 layers of 50/1 (SA/I). (b) Absorption spectrum of film in (a).

stearic acid. However, a more sensitive measure of the environmental perturbation in the film can be seen by comparing the lifetime data in solution and film. The decay data at room temperature for the aerated chloroform solution of I were obtained, but the films lacked sufficient intensity to be observed.

The FT-IR spectrum obtained from the grazing angle measurement for a 10 layer film of 10:1 stearic acid/I is given in Figure 5. In addition, IR data from KBr pellet samples are shown in Figures 6 and 7. Previous FT-IR results for stearic acid allow us to assign the CH_3 and CH_2 stretch bands in the 2800-3100 cm^{-1} region. Since surface molecules are excited almost exclusively by radiation polarized parallel (p-polarized) to the plane of incidence, molecular vibrations perpendicular to the substrate plate are most interesting. Comparison of the RA spectra with the isotropic (KBr) spectra and comparison with earlier published results enables assignment of the modes as well as estimates of the orientation of the stearate/stearic acid vibrations. The modes at 2918 and 2849 cm^{-1} can be assigned as asymmetric (γ_a) and symmetric modes (γ_s) of the CH_2 stretches and have opposite polarizations. The fact that both show intensity in the RA spectra indicates that the alkyl chains are not exactly perpendicular to the substrate plane.

The modes at 2964 and 2873 cm^{-1} can be assigned as CH_3 stretches (γ_a and γ_s, respectively). The CH_2 groups in the Ru (II) complex show strong bands at 2921 and 2849 cm^{-1} in the transmission spectra which would overlap with the stearic acid bands in the film even if they were observed. However, the D_3 symmetry of the metal complex precludes use the CH_2 or CH_3 stretching modes in the complex to determine the orientation of the metal complex in the film. Indeed, the effective spherical geometry of the metal complex means that the anisotropic orientation of the species is not possible. Much of the orientation information gleaned from the vibrational spectra will be that associated with the stearic acid and stearate C-O stretches at 1706, 1541, and 1433 cm^{-1}. Of greatest utility to ascertain the ion pair mechanism is the relatively strong ClO_4^- band at 1096 cm^{-1} (Figure 7). The disappearance of this band in the LB film spectra (Figure 5) suggests either that ClO_4^- is not present in the film or that the transition is forbidden. The symmetry of ClO_4^- implies that the vibrational modes should have some intensity in the RA spectra. Therefore, we conclude that the absence of the 1096 cm^{-1} band in the film implies that the ClO_4^- is not present in the film due to ion pairing of the Ru^{2+} complex with the stearate anion. The absence of significant band shifts between the isotropic pellet spectra and the RA film spectra and the similar line shapes imply that the metal complexes are dispersed.

A conclusive answer to the question of the structure of the metal complex stearic acid mixture may be possible from the STM images of the mixed monolayer films. The sequence of Figures 8 through 12 show successively the Ag substrate plate (side view), the stearic acid coated on the Ag plate, a large area scan of the substrate coated with the stearic acid/Ru complex (10:1) (Figure 10), "a blow up (insert Figure 10) showing individual stearic acid/Ru complex aggregates, (Fig. 11) and a cross section of the stearic acid/Ru complex aggregate (Figure 12). An average of 42 stearic acid/Ru complex aggregates per 10 nm^2, were counted by taking 3 diffferent 10x10 nm sections of Figure 10. Langmuir-Blodgett film data permit an estimated 46 stearic acid/Ru complex aggreegates per 10 nm^2. Figure 13 shows a spherical model for packing in the mixture using radii of 2.5Å for stearic acid and 5.5Å complex I. As shown, 10 molecules of stearic acid fit easily around each molecule of I. In Figure 12, the bumps outlined are thought to be the metal

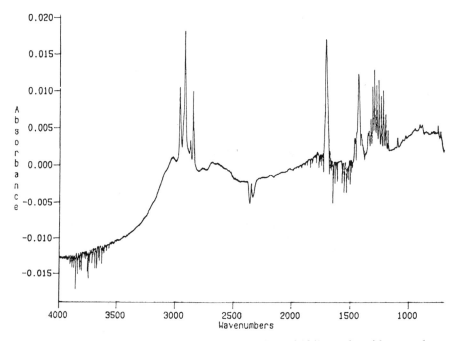

Figure 5. Grazing angle FT-IR for a 10 layer film of 10/1 stearic acid to metal complex.

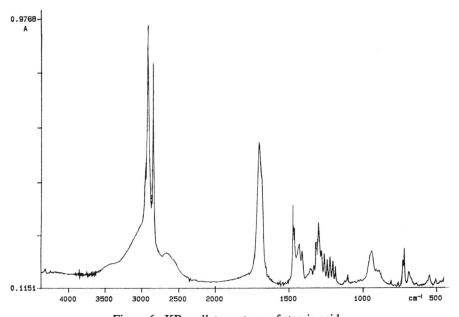

Figure 6. KBr pellet spectrum of stearic acid.

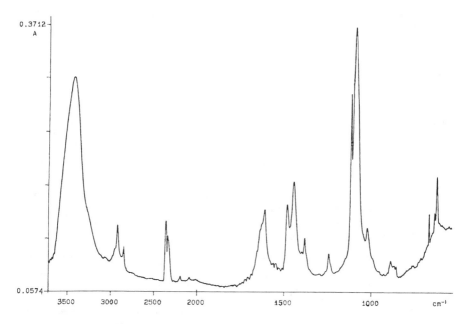

Figure 7. KBr pellet spectrum of metal complex.

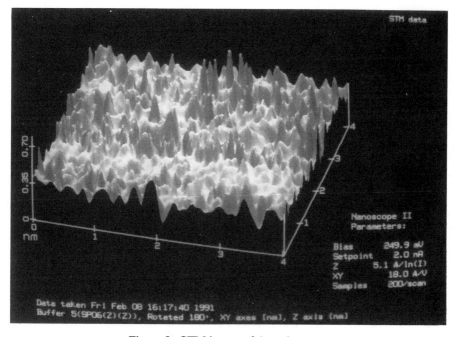

Figure 8. STM image of Ag substrate.

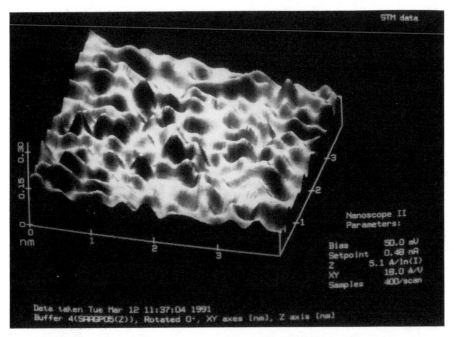

Figure 9. STM image of Ag substrate coated with stearic acid.

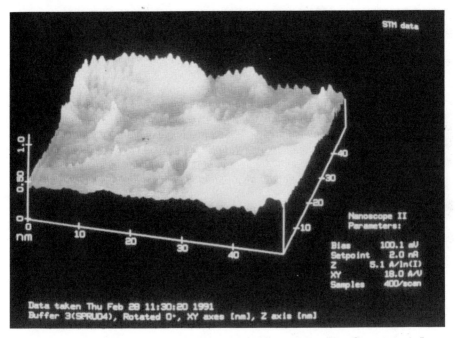

Figure 10. STM image of Ag substrate coated with Compound I dissolved in stearic acid (wide field).

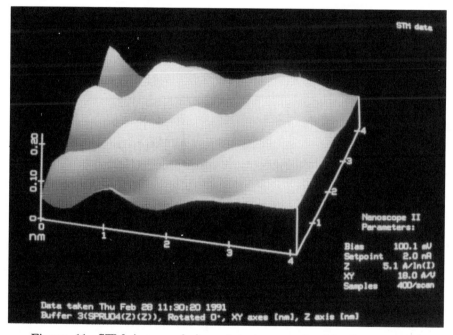

Figure 11. STM image of Ag substrate coated with Compound I dissolved in stearic acid (narrow field).

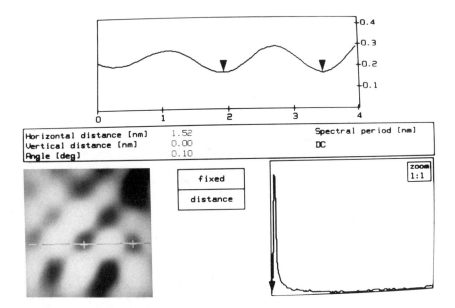

Horizontal distance [nm]	1.52	Spectral period [nm]
Vertical distance [nm]	0.00	DC
Angle [deg]	0.10	

Buffer 6(SPRUO4(Z)(Z)), Rotated 180°, XY axes [nm], Z axis [nm]

Figure 12. Top view of stearic acid-metal complex cluster.

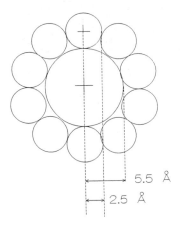

Figure 13. Spherical model for packing of stearic acid/I mixtures.

complex surrounded by the stearate and the stearic acid. The diameter calculated from pressure curves (16Å) of the stearic acid Ru complex aggregate correlates with that measured in Figure 11.

STM images of the stearic acid film (Figure 9) under high resolution suggest that the acid molecules are oriented nearly perpendicular (FIR indicates that the molecules are ~30° to the perpendicular) to the substrate material and are spaced by ~4.8 Å. The mixed layer film suggests that the spherical metal complexes are attached to the acid end of the stearic acid and that they are "dispersed" since evidence for aggregates of metal complexes with SA is seen when searching larger areas (0.18 μm x 0.18 μm) (Figure 10). Moreover, the clustering of the metal complex-SA aggregates is not apparent in these flat regions of image. Piling up of the Ag on the substrate can be found and this could obscure some heterogenous phase.

Since the 2+ charge of the metal complex can be readily compensated by the stearic acid at the ratio used, the basic stearic acid structure found for pure fatty acid structures is little altered. Thus the tilt angle of ~30° and the non-orthorhombic structure suggested by the FT-IR would be consistent with the STM images. Nevertheless, the likelihood is that STM will not be capable of providing the angstrom level resolution (complex on the head of the SA) required to elaborate the stearic acid-metal complex structure since the density of states for stearic acid would be primarily associated with a sigma bonding system. Perhaps X-ray may succeed in providing the resolution required to see the individual carbon atoms in the stearic acid and stearate as well as the Ru^{2+} ion in the complex.

Summary

The ion pair technique has been found to readily produce monolayer and multilayer films on a variety of substrate materials. The FT-IR technique was useful in verifying that the ion pairing of the metal complex and the stearic acid occurred with the ClO_4^- displaced by the stearate anion. The sequence of STM images are consistent with metal complex-stearic acid aggregates of 14.7Å diameter comparing well with the calculated aggregate diameter of 16Å. The ease of preparation and the ability to produce multilayer assemblies with up to 200 layers should be invaluable in arraying a wide range of hydrophobic Ru(II) diimine complexes. Subsequently, the unique charge localization and directional capability of the excited state and reduced species can be exploited to produce detectable switching states.

Acknowledgments

The support of the Army Research Office (Research Triangle Park, NC) is gratefully acknowledged.

Literature Cited

1. DeLaive, P.; Lee, J.R.; Sprintschnik, H.W.; Abruna, H.; Meyer, T.J.; Whitten D. *J. Am. Chem. Soc.* **1977**, *99*, 7094.
2. Gaines, G.L., Jr.; Behnken, P.E.; Valenty, L.J. *J. Am. Chem. Soc.* **1979**, *100*, 6549.
3. Valenty, S.J.; Behnken, D.E.; Gaines, G.L., Jr. *Inorg. Chem.* **1979**, *18*, 2160.
4. Daifuku, H.; Aoki, K.; Tokuda, K.; Matsuda, H. *J. Electroanal. Chem.* **1982**, *140*, 179.
5. Fujihira, M.; Nishiyama, K.; Yamada, H. *Thin Solid Films* **1985**, *132*, 77.

6.   Daifuku, H.; Aoki, K.; Tokuda, K.; Matsuda, H. *J. Electroanal. Chem.*
     1985, *183,* 1.
7.   Fujihira, M.; Nishiyama, K.; Aoki, K. *Thin Solid Films* **1988**, *160*, 317.
8.   Zhang, X.; Bard, A.J. *J. Phys. Chem.* **1988**, *92*, 5566.
9.   Zhang, X.; Bard, A.J. *J. Am. Chem. Soc.* **1989**, *111*, 8098.
10. Obeng, Y.S.; Bard, A.J. *Langmuir* **1991**, *7*, 191.
11. Möhwald, H.; Miller, A.; Stick, W.; Knoll, W.; Ruadel-Texier, A.;
     Lehmann, R.; Fuhrhop, J.H. *Thin Solid Films* **1986**, *141*, 261.
12. Schick, C.A.; Schreiman, E.C.; Wagner, R.W.; Lindsey, J.S.; Bocian,
     D.F. *J. Am. Chem. Soc.* **1989**, *111*, 1344.
13. Seefeld, K.P.; Möbius, D.; Kuhn, H. *Helv. Chim. Acta* **1977**, *60*, 2608.
14. Palacin, S.; Texier, A.R.; Barraud, A. *J. Phys. Chem.* **1989**, *93*, 7195.
15. Palacin, S.; Barraud, A. *J. Chem. Soc. Chem. Commun.* **1989**, 45.
16. Carlin, C.M.; De Armond, M.K. *Chem. Phys. Lett.* **1982**, *89*, 297.
17. Myrick, M.L.; Blakley, R.L.; De Armond, M.K.; Arthur, M.L. *J. Am.
     Chem. Soc.* **1988**, *110*, 1325.
18. De Armond, M.K.; Myrick, M.L. *Accounts Chem. Research* **1989**, *22,*
     364.
19. Motten, A.; Hanck, K.W.; De Armond, M.K. *Chem. Phys. Lett.* **1981**,
     79, 541.
20. Gex, J.N.; De Armond, M.K.; Hanck, K.W. *J. Phys. Chem.* **1987**, *91*,
     251.
21. Donohoe, R.J.; Tait, C.D.; De Armond, M.K.; Wertz, D.W. *Spectrochim.
     Acta* **1986**, *42A*, 233.
22. Murakata, T.; Miyashita, T.; Matsuda, M. *J. Phys. Chem.* **1988**, *92*,
     6040.
23. Yliperttula, M., Lemmetyinen, H.; Mikkola, J.; Virtanen, J.; Kinnunen, P.
     Chem. Phys. Letters **1988**, *152,* 61.
24. Allara, D.L.; Swalen, J.D. *J. Phys. Chem.* **1982**, *86*, 2700.
25. Rabolt, J.F.; Burns, F.C.; Schlotter, N.E.; Swalen, J.D. *J. Phys. Chem.*
     1983, *78*, 946.
26. Noselli, C.; Rabolt, J.F.; Swalen, J.D. *J. Chem. Phys.* **1985**, *82*, 2136.
27. Geddes, N.J.; Jurich, M.C.; Swalen, J.D.; Tweig, R.; Rabolt, J.F. *J.
     Chem. Phys.* **1991**, *94*, 1603.
28. Umemura, J.; Komata, T.; Kawai, T.; Takenada, T. *J. Phys. Chem.*
     1990, *94*, 62.
29. Watanabe, I.; Cheung, J.H.; Rubner, M.F. *J. Phys. Chem.* **1990**, *94*,
     8715.
30. Smith, D.P.E.; Horber, H.; Gerber, C.H.; Binnig, G. *Science* **1988**, *245,*
     43.
31. Caple, G.; Wheller, B.L.; Swift, R.; Porter, T.L.; Jeffers, S. *J. Phys.
     Chem.* **1990**, *94*, 5639.
32. Smith, D.P.E.; Bryant, A.; Quate, C.F.; Robe, J.P.; Gerber, C.; Swalen,
     J.D. *Proc. Natl. Acad. Sci. (USA)* **1987**, *84,* 969.
33. Horber, J.K.H.; Lang, C.A.; Hänsch, T.W.; Heckyl, W.M.; Möhwald, H.
     Chem. Phys. Letters **1988**, *145*, 151.
34. Wu, X.L.; Lieber, C.M. *J. Phys. Chem.* **1988**, *92*, 5556.
35. Clemmer, C.R.; Beebe, T.P. *Science* **1991**, *251*, 641.
36. Gimzewski, J.K.; Stoll, E.; Schlittler, R.R. *Surf Science*, **1987**, *181*, 267.

RECEIVED February 18, 1992

Chapter 6

Biomimetic Thin-Film Synthesis

Peter C. Rieke, Barbara J. Tarasevich, Susan B. Bentjen, Glen E. Fryxell, and Allison A. Campbell

Materials Sciences Department, Pacific Northwest Laboratory, Richland, WA 99352

Surfaces derivatized with organic functional groups were used to promote the deposition of thin films of inorganic minerals. These derivatized surfaces were designed to mimic the nucleation proteins that control mineral deposition during formation of bone, shell and other hard tissues in living organisms. By use of derivatized substrates, control was obtained over the phase of mineral deposited, the orientation of the crystal lattice, and the location of deposition. These features are of considerable importance in many technically important thin films, coatings, and composite materials. Methods of derivatizing surfaces are considered and examples of controlled mineral deposition are presented.

Biological hard tissues, e.g. bone and mollusk shell, are ceramic/polymer composites with remarkable mechanical properties (*1*). The composition is better than 90% mineral phase, and the remainder consists of structural proteins and polysaccharides. The choice of mineral phase is dictated by availability rather than intrinsic mechanical properties. The calcium phosphate and calcium carbonate found in bone and shell, respectively, would traditionally be considered very poor choices as structural materials. The impressive structural strength and fracture resistance arise from the organization of the two constituents; the result is greater than the sum of the parts. Practical application of biomineralization concepts requires a general understanding of the biology involved and identification of physical and chemical principles that can be applied to materials science. In this paper we briefly describe how biomineralization principles can be applied to control the deposition of thin film materials of technical importance. Specific examples of surface-controlled nucleation and growth are presented.

Because of its obvious importance to biology, medicine, and dentistry, the mechanism of biomineralization has been the subject of much research (*2-4*). This cannot be adequately reviewed here and only a simplified model is presented as background. Hard tissue formation begins with deposition of structural biopolymers.

0097–6156/92/0499–0061$06.00/0

In bone, collagens are the primary constituent; in mollusk shell, silk-fibroins and polysaccharide fibrils are utilized. The organization of these into a three dimensional, skeletal framework determines the form and function of the resulting hard tissue. Within the void spaces, nucleation proteins are deposited and the solution invading this network is brought to supersaturation with respect to the desired mineral phase. The nucleation proteins are thought to induce nucleation at the desired location. Crystal growth ensues and the void spaces are filled with mineral to form a fully dense ceramic/polymer composite.

As this process occurs in supersaturated solution, the problem consists of suppressing the kinetics of random homogeneous nucleation in the extrapallial fluid while enhancing the kinetics of oriented nucleation on the organic protein surface. Once nucleation on the surface is promoted, the problem shifts to controlling the kinetics and habit of crystal growth. In general, nucleation requires higher supersaturation than crystal growth, so when this process is initiated once per compartment crystal growth will dominate (5). The overall process is cyclic or incremental: a small fringe of unmineralized substrate forms, the mineral growth front advances, another small fringe of unmineralized substrate forms, ad infinitum.

A wide variety of minerals are deposited by organisms including metal oxides, sulfides, carbonates, and phosphates (6). Thus, biomineralization appears to be a general phenomenon the concepts of which should be applicable to deposition of minerals not found in nature. Furthermore, the mineral deposition process is not under direct cellular control and only the structure and composition of the nucleation protein appear to be controlling the mineralization. With this rationale, our researches have focused on understanding how organic substrates control nucleation and crystal growth of both biominerals and technically important minerals (7-11).

Three features are found in biomineralization that have important materials science implications: 1) the ability to control the phase of mineral deposited on the substrate, 2) the ability to control the orientation of the crystal lattice with respect to the surface, and 3) the ability to direct nucleation at sites specifically tailored for this purpose and to prevent nucleation either in solution or at random locations on the surface. These features can be of varying importance depending on the specific application in mind. The remarkable properties of natural composites arise from the sophisticated microstructures found in these materials. The microstructure is a direct result of the ability to direct nucleation and orient crystals within a polymer matrix.

Thin film ceramic materials with important magnetic, optical, electronic, and mechanical properties are often highly anisotropic. Thus, the ability to control orientation is critically important in thin film applications. For many of the oxide materials, as well as the ionic materials, aqueous solution or sol-gel routes are the most convenient or the only method of preparation. Examples of these include barium titanate ($BaTiO_3$) used in multilayer capacitors, lead-zirconate-titanate ($Pb(Zr,Ti)O_3$, "PZT") used as a piezoelectric material, and zinc oxide (ZnO) used in varistors. Thus, the use of substrates to control orientation can eliminate major problems in deposition of thin films. In some cases, e.g., the many magnetic and non-magnetic phases of iron oxide, the ability to control the phase formed is critical to production of the desired properties. While this can be controlled by solution conditions, the proper surface can add an additional and very effective mechanism of control.

In the previous discussion on biomineralization it was tacitly assumed that heterogeneous nucleation was the mechanism by which surfaces promote nucleation and crystal growth. Although this is clear concept in atom by atom vapor phase deposition of materials, the concept becomes muddled with increasingly complex solution chemistry. This problem is illustrated in Figure 1 using generic metal oxide formation. In sequence, the basic steps are ligand replacement, hydrolysis, condensation, polymerization, colloid formation, and phase transformation. A clear understanding of film growth translates to understanding and controlling the point in mineral formation at which the surface plays an important function.

At one end of the spectrum adhesion of relatively large crystalline colloidal substances is not expected to allow controlled film formation. On the other hand small, amorphous colloidal particles may undergo phase transformation and orientation at properly designed surfaces. This is a useful film formation mechanism although it may not be called "heterogeneous nucleation". Indeed it has been proposed that nucleation in bone occurs by transport of nuclei formed in vesicles to the desired location of bone deposition (*12*). At the other extreme is atom by atom deposition of the mineral, which many would accept as heterogenous nucleation. The extent of hydrolysis and condensation during these stages may be the critical question to address. In between these extremes, adsorption of condensation polymers of varying size, shape and charge may be prevalent.

The nucleation proteins have been isolated and partially characterized (*2,3,13-18*). Although considerable variation exists between organisms, some generalization can be made. The proteins are water soluble and contain many negatively charged functional groups. These may be carboxylate groups on the amino acid aspartic acid, phosphate groups on phosphoserine, or sulfate groups attached to polysaccharide residues. The choice of groups seems to be dominated by those having simple electrostatic interactions with cations rather than strong ligand binding characteristics. The close proximity of these groups, however, suggests cooperative interaction in binding cations.

The location of the nucleation proteins within the hard tissue has been well established and clearly associated with the induction of mineralization. For example, Greenfield et al. (*13*) have identified in the polymer matrix of mollusk shell nacre a site containing negatively charged groups. A single site is associated with each aragonite crystal and was shown to promote mineralization. In the case of bone, Stetler-Stevenson and Veis (*18*) have isolated a phosphoryn that is necessary for mineralization, strongly binds calcium, and preferentially adsorbs at sites in collagen fibers thought to be the loci of initial mineralization. Despite this considerable but circumstantial evidence, the ability of these proteins to actually induce heterogeneous nucleation *in vivo* has not been clearly demonstrated. Although the primary sequence structure of these proteins has been determined, very little has been done on the secondary structure of the proteins other than to propose the formation of ß-sheets (*19*). Hence almost nothing is known about the density or spatial arrangement of charged groups on the nucleation protein. The further understanding of the secondary and tertiary structure of the nucleation protein will be invaluable.

These characteristics suggest that the nucleation proteins function essentially as charged anionic surfaces that may or may not have long-range structural order. The emerging model is that the presence of nucleation proteins is all that is required to promote mineralization and that artificial surfaces can be designed which mimic their properties. This is a logical but still tenuous assumption and further biologically oriented studies on the exact role of nucleation proteins and related enzymes are needed. A few studies have been undertaken which utilize Langmuir-Blodgett films to promote mineralization. Zhao and Fendler (*20*) have prepared CdS and ZnS films by slow infusion of H_2S across L-B films of cadmium arachidate and zinc arachidate.

Landau et al. (*21*) have grown sodium chloride and glycine crystals at L-B films supported at the air-water interface. Mann and coworkers (*22,23*) have focused on calcium carbonate grown under a variety of L-B films under varying degrees of compression. Each of these works supports the premise that artificial surfaces may be tailored to promoted nucleation of minerals. Supported L-B films have limited durability and chemical head group variability. For thin film applications it would be desirable to functionalize a desired substrate with the necessary functional group required to promote mineralization. Addadi et al. (*19*) have used sulfonated polystyrene to which poly(aspartate) has been adsorbed to promote calcium carbonate nucleation. In this paper we describe some of our results using derivatized polymer substrates and covalently attached self-assembled monolayers to promote deposition of thin film materials.

Experimental

Sulfonated polystyrene was prepared from polystyrene samples cut from plastic petri dishes. Sulfonation was by either immersion in fuming sulfuric acid for time less than 60 seconds or by exposure to SO_3 gas for times up to 20 minutes. The fuming sulfuric acid method was difficult to control and severely etched the substrate surface. In the second method, polystyrene samples were placed in an dry glass chamber that was then evacuated. SO_3 gas was introduced by opening a stopcock connected to a round bottom flask containing a small amount of solid SO_3. Effusion of the gas over the polystyrene samples resulted in varying degrees of sulfonation depending on the time of exposure. With care this method allowed reproducible sulfonation.

Derivatized polyethylene was prepared by covalent attachment of primary amine to oxidized polyethylene via an amide linkage (see Figure 2). Details of preparation are described elsewhere (*11*). Low density polyethylene was first dip coated onto glass slides from an o-xylene solution. These films were then oxidized in chromic acid to introduce carboxylate groups. Treatment in a saturated solution of PCl_5 in ether gave the reactive acyl chloride. This was then immediately immersed in a solution of a primary amine dissolved in 50/50 volume percent triethylamine and pyridine to form the amide derivative. The primary amine used were n-propylamine, N',N' dimethylethylenediamine, β-alanineethylester, 2-aminoethanethiol, and 3-aminopropanol. The thiol and alcohol did not react significantly with the acyl chloride, and the amide was the predominant product. The thiol was further converted to the sulfonic acid by treatment with Cl_2 followed by hydrolysis. The alcohol was converted to the phosphate by treatment with phosphorous oxychloride. The films were characterized by XPS, IR, and advancing contact angle wetting.

Derivatized self-assembled monolayers were prepared from bromine terminated monolayers prepared from $Cl_3Si(CH_2)_{17}Br$. Silicon wafers, p-doped and polished along the 100 face, were cleaned in an air plasma unit (Harrick) for 10 minutes and broken into 2.5 x 1.0 cm samples. These were treated for 2 minutes in 0.1M KOH, 5 minutes in 0.1M HNO3, and rinsed in distilled water. The substrates were blown dry with clean nitrogen just prior to emersion for 30 minutes in a 1% solution by weight of the monomer in clean, dry cyclohexane. Upon removal, the substrates were rinsed twice in chloroform, and once in methanol. Conversion of the bromine monolayer to other derivatives followed the schemes shown in Figure 3. The sulfonic acid derivative was prepared from the thiocyanate or thiourea derivative by oxidation with sodium hypochlorite. The phosphonate was prepared by treating the phosphonate ester with trimethylsilane. A primary amine was prepared by reduction of the azide derivative in formic acid with a finely divided platinum metal catalyst. The films were

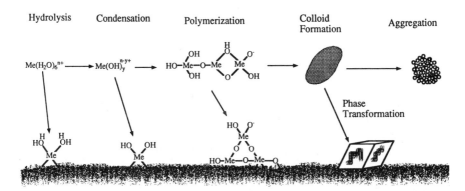

Figure 1. Schematic illustrating possible mechanisms of film growth.

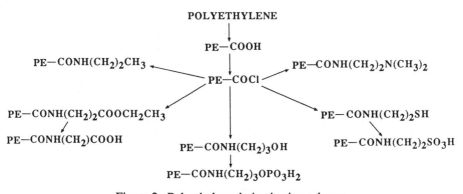

Figure 2. Polyethylene derivatization scheme.

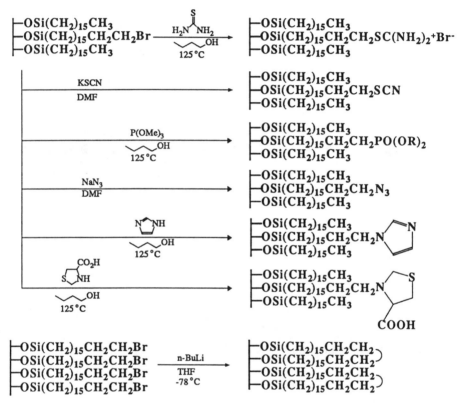

Figure 3. Self-assembled monolayer derivatization scheme.

characterized by XPS, IR, and contact angle wetting. Details of preparation and characterization will be presented in a forthcoming manuscript.

Calcium carbonate was deposited on substrates by the slow infusion of ammonium carbonate into calcium chloride solutions. Substrates were supported in 100 ml glass containers containing a 1.0 millimolar calcium chloride solution. A 0.5 mm hole was drilled into the lid of each container and covered with heavy piece of typing paper. A series of the containers were placed in a glass desiccator containing an open dish of fresh ammonium carbonate and left undisturbed for 1 to 7 days. The ammonium carbonate decomposed to ammonia and carbon dioxide, and these gasses slowly infused into the solutions via the paper-covered holes. The ammonium vapor buffered to solution to a pH of 9.2 and supersaturation was achieved by the continuous adsorption of CO_2 gas to form bicarbonate.

Iron oxide films were deposited from millimolar Fe(NO3)3 solutions adjusted to a pH of 2.0 to 4.0 with nitric acid. Hydrolysis and condensation was induced by heating to 90 °C with stirring. Cadmium sulfide films were prepared from 0.1M cadmium nitrate and 1.0M ammonium nitrate solutions at pH values ranging from 8.5 to 10.5. Sulfide was generated *in-situ* by base decomposition of thiourea at 0.05M concentration. The films were removed from solution at various time intervals up to the point at which a massive precipitate became visible in the bulk of solution.

Results and Discussion

The discussion of nucleation proteins suggests that a surface with high density of functional groups with strong acid/base character or ligand binding ability is the primary requirement for controlling mineralization. Artificial surfaces could be prepared with these groups in order to mimic the nucleation proteins. We have explored surface derivatization methods on polyethylene and self-assembled monolayers (SAM's) and developed methods for the introduction of hydroxy, phosphate, thiol, sulfonate, carboxylate, amine, and methyl groups.

It is desirable to have a single substrate upon which all types of groups may be attached. Unfortunately, most polymer substrates are limited in that the polymer is not easily derivatized or is soluble in the solvents required for the synthetic schemes. Polyethylene overcomes these problems, and recently a wide variety of synthetic schemes have been developed to attach a series of simple charged functional groups (9,11,24). Figure 2 illustrates the methods for introduction of various common organic functional groups. The basic method used was to form a strong amide linkage between a primary amine and a carboxylate group introduced to the surface of polyethylene by chromic acid oxidation.

An attractive alternative to polyethylene is to use derivatized self-assembled monolayers (SAM's) (25,26). For this work we have chosen long alkyl chains with a trichlorosilyl group on one end and a reactive group such as bromine on the other. The silyl group was used to covalently attach the alkyl chain to oxide substrates. This strong covalent linkage was an advantage over gold/thiol SAM system because the conditions required for mineral growth are often stringent. The dense layer of reactive end groups were then derivatized to obtain the desired organic functional groups. Figure 3 illustrates some of the reactions that we have achieved. By using mixed monolayers consisting of alkyl chains with and without the bromine groups, the average density of sites could controlled.

Experiments with various mineral systems and various surface systems have shown that artificial substrates can be used to control the phase of materials deposited, the orientation of the crystal lattice, and the location of deposition. Described below are selected systems which illustrate these points. For comparison purposes a dependence of mineral habit, phase, or orientation was considered relevant only when the conditions of deposition are identical.

Figure 4 is a series of electron micrographs of calcium carbonate on various surfaces. These were grown by slow infusion of ammonium carbonate vapor into millimolar calcium chloride solutions (7). The surfaces are from top to bottom: polystyrene, polystyrene sulfonated in fuming sulfuric acid (the striations arise from acid etching the polymer), and a clean, glass microscope slip cover. On polystyrene only the spherulitic form of vaterite occurred. On sulfonated polystyrene and on glass a majority of the crystals exhibited the clear cleavage faces of calcite; some spherulitic vaterite also was deposited on these surfaces. A striking difference in orientation of the calcite occurred on the two surfaces. The orientation on glass was with the 111 face (of the rhombohedral lattice) contacting the substrate; while a higher order face, estimated to be the 120 face from measurement of the crystal edge length, contacted the sulfonated polystyrene substrate. These experiments demonstrated that surface chemistry can be used to control the phase of materials deposited and also to control the orientation of the mineral.

The left side of Figure 5 shows a scanning electron micrograph of iron oxide deposited on sulfonated polystyrene; the right side of the figure shows a transmission electron micrograph of the film in cross section. The mineral was identified as goethite (α-FeOOH) from X-ray and electron diffraction studies with crystallites about 20-30 nm in diameter and 50-100 nm in length (8). The elongated axis of needle like goethite has previously been show to be the [001] axis.

This mineral phase, the highly anisotropic habit, the high crystallite density, and the orientation are ideally suited for formation of magnetic films. Typically maghemite particles are produced by reducing needle-like goethite to magnetite, followed by reoxidation to the more stable maghemite. Preliminary results on these iron oxide films indicate that conversion to the magnetic phase is possible by this processing method. The great density of crystallites and uniform orientation of these films hold promise for application as a very high density magnetic recording medium.

The kinetics of deposition depended strongly on the sulfonate site density, and high sulfonate densities showed rapid iron oxide deposition (27). Taken to extreme, this can be used to control the location of deposition. Shown in Figure 6 are photographs of polystyrene substrates that have been masked using tape. The masked substrates were sulfonated to prepare a surface with regions of bare polystyrene and highly sulfonated polystyrene. The masking was removed prior to oxide film deposition. In the lower portion of Figure 6 is an electron micrograph of the border between sulfonated and non-sulfonated areas. Iron oxide deposited in a uniform film with a very sharp break to a region without iron oxide deposit. Despite the crudeness of the masking experiment, it clearly shows that patterned derivatization can be used to pattern the film deposition. Based on the sharpness of the edge and the size of the individual crystallites, sub-micron resolution should be possible.

Cadmium sulfide films were deposited on substrates in basic cadmium nitrate solutions with cadmium complexed by excess ammonia. Slow decomposition of thiourea was used to generate the sulfide ion. Some initial studies of film growth on derivatized polyethylene suggested a dependence of film density surface on

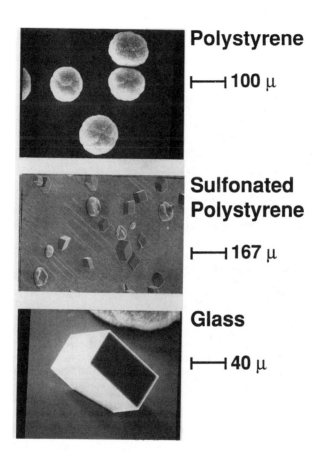

Figure 4. Electron micrographs of (top) vaterite on unmodified polystyrene, (middle) calcite with some vaterite on sulfonated polystyrene, and (bottom) calcite on glass.

Figure 5. Scanning electron micrograph (left) and cross-section transmission electron micrograph (Right) of iron oxide films deposited on sulfonated polystyrene.

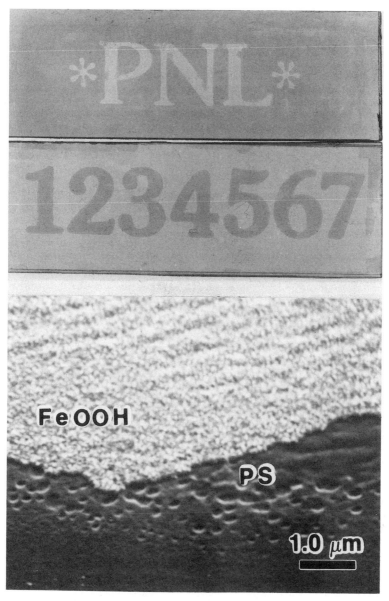

Figure 6. Photograph of patterned iron oxide films and electron micrograph near edge of pattern.

hydrophobicity (27). Hydrophobicity was measured by comparing the contact angle of water on the surfaces; oddly, film density increased with increasing hydrophobicity of the substrate.

To explore the anomalous result, films were grown on a non-wetting methyl terminated SAM surface, water contact angle > 105°, and bare silicon substrates with a 20 Å oxide layer, contact angle ≈ 20°. Figure 7 shows the film structure at various times of development for these two substrates. At 8 hours crystallites were visible on the SAM surface but not on the bare silicon/silicon dioxide surface. At later times, both surfaces showed evidence of mineralization but the SAM surface had more numerous crystallites of smaller size. Exact determination of crystallite density and size was difficult becaue of we are unable to clearly distinguish each crystallite. At 48 hours crystallite size was about 75 nm on the Si/SiO_2 substrate and less than 50 nm on the SAM substrate. The crystallites on Si/SiO_2 substrate were clearly defined and separated until high film densities but crystallites on the SAM substrate appear to be aggregating.

The important feature of this work was the ability to manipulate the particle size and density by proper choice of surface hydrophobicity. These results suggested more efficient nucleation on the hydrophobic substrate. This is in direct contradiction to the model of biomineralization discussed above which suggest that highly charged, cation adsorbing surfaces are needed to promote nucleation. The SAM surface most certainly does not bind cadmium but may, through van der Waals forces, attract neutral ion pair species or subnanometer colloidal particles.

Conclusions

In this article we have introduced the concepts of biomineralization and related how this information can be utilized in developing new materials processing methods. To realize this advancement requires development of new organic surface synthesis methods, extensive characterization of these surfaces, and a more detailed understanding of the mineral solution chemistry. With this knowledge, mineralization at surfaces can begin to be addressed in a systematic manner. However, we have just begun to perfect surface derivatization methods and need to expend more effort in characterizing the physical properties of these surfaces. The solution chemistry of some mineral systems, e.g. iron oxide, have been well characterized due to their technical importance. Most system have not benefitted from in-depth analysis and understanding film formation of these materials requires extensive analysis of the solution properties.

Despite these limitations we have demonstrated the feasibility of using the biomimetic method for controlling film deposition. In particular, we have shown that with specifically designed surfaces we can control the phase of the mineral deposited, the orientation of the mineral, and the location of the mineral. In addition, some control over the nucleation site density, and the habit of the mineral phase is possible. All of these characteristics are important in development of thin film devices with optimum properties. In the future, with close attention to surface chemistry and structure, thin film growth from solution may be comparable to vapor phase deposition methods in the ability to control film morphology and composition.

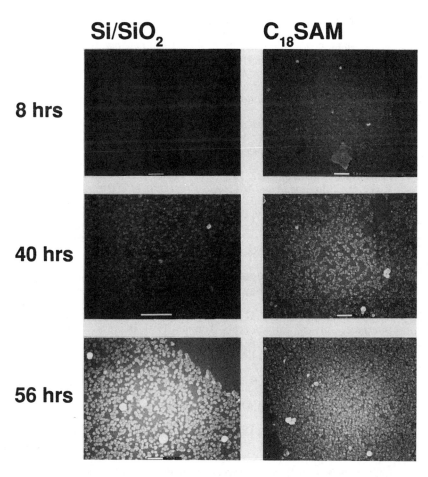

Figure 7. Electron micrographs of cadmium sulfide films deposited on a hydrophilic silicon substrate and on a hydrophobic SAM substrate at various times after initiation of film deposition.

Acknowledgments

This research was supported by the U.S Department of Energy, Office of Basic
Energy Sciences and the Office of Industrial Processes under contract DE-AC06-
76RLO 1830. Pacific Northwest Laboratory is operated for the U.S. Department of
Energy by Battelle Memorial Institute.

Literature Cited

1.   Wainwright, S. A.; Biggs, W. D.; Currey, J. D.; Gosline, J. M. *Mechanical
      Design in Organisms*; Princeton University Press: Princeton, NJ, 1976.
2.   Weiner, S.; *CRC Crit. Rev. Biochem.* 1986, *20*, 325.
3.   Veis, A.; Sabsay., B.; In *Insights into Mineralization, Biomineralization and
      Biological Metal Accumulation*; Westbroek, P.; de Jong, E. W.; Eds.; Reidel,
      Dordrecht, 1983, pp 273.
4.   *Biomineralization, Chemical and Biochemical Perspectives*; Mann, S.; Webb,
      J.; Williams, R. J. P., Eds.; VCH, Germany, 1989.
5.   Nielsen, A. E.; *Kinetics of Precipitation*; Pergamon Press, New York, NY,
      1964.
6.   Lowenstam, H. A.; *Science*, 1981, *211*, 1126.
7.   Rieke, P. C.; In *Atomic and Molecular Processing of Electronic and Ceramic
      Materials*; I. A. Aksay, G. L. McVay, T. G. Stoebe, and J. F. Wager, Eds.,
      Materials Research Society, Pittsburgh, PA, 1988
8.   Tarasevich, B.J.; Rieke, P.C.; In *Materials Synthesis Utilizing Biological
      Processes*; Rieke, P. C.; Calvert, P. D.; Alper, M., Eds.; Materials Research
      Society: Pittsburg, PA, 1988, Vol. 174; pp 51.
9.   Rieke, P. C.; Bentjen, S. B.; Tarasevich, B. J.; Autrey, T. S.; Nelson, D. A.;
      In *Materials Synthesis Utilizing Biological Processes*; Rieke, P. C.; Calvert, P.
      D.; Alper, M., Eds.; Materials Research Society: Pittsburg, PA, 1988, Vol.
      174; pp 69.
10.   Tarasevich, B.J.; Rieke, P. C.; McVay, G. L.; In *The Bone-Biomaterial
      Interface*; Univ. of Toronto Press, Toronto, 1991, in press.
11.   Bentjen, S. B.; Nelson, D. A.; Tarasevich, B. J.; Rieke, P. C.; *Journal of
      Applied Polymer Science*, 1992, *00*, 000.
12.   Boyan-Salyers, B. D.; In *The Chemistry and Biology of Mineralized
      Connective Tissues*; Veis, A., Ed.; Elsevier North Holland, New York, NY,
      1981.
13.   Greenfield, E. M.; Wilson, D. C.; Crenshaw, M. A.; *Amer. Zool.*, 1984, *24*,
      925.
14.   Butler, W. T.; Bhown, M.; DiMuzio, M. T.; Linde, A.; *Collagen Relats. Res.
      Clin. Exp.*; 1981, *1*, 187.
15.   Dimuzio, M. T.; Veis, A.; *Calcif. Tissue Res.*; 1978, *25* 6845.
16.   Stetler-Stevenson, W. G.; Veis, A.; *Biochemistry*; 1983, *22*, 4326.
17.   Takagi, Y.; Veis, A.; *Calcif. Tissue Int.*; 1984, *36*, 259.
18.   Stetler-Stevenson, W. G.; Veis, A.; *Calcif. Tis.*; 1986, *38*; 135.
19.   Addadi, L.; Moradian, J.; Shay, E.; Maroudas, N. G.; Weiner, S.; *Proc.
      Natl. Acad. Sci. USA*; 1987, *84*, 2732.
20.   Zhao, X. K.; Fendler, J. H.; *J. Phys. Chem*, 1991, *95*, 3716.
21.   Landau, E.M.; Levanon, M.; Leiserowitz, L.; Lahav, M.; Sagiv, J.; *Nature*,
      1985, *318*, 353.
22.   Rajam, S.; Heywood, B. R.; Walker, J. B. A.; Davey, R. J.; Birchall, J. D.;
      Mann, S.; *J. Chem. Soc., Faraday Trans. I*, 1990, *87*, 727.

23.   Heywood, B. R.;   Rajam, S.; Mann, S.; *J. Chem. Soc., Faraday Trans. I,* 1990, *87*, 735.
24.   Rasmussen, J. R.; Bergbreiter, D. E.; Whitesides, G. M.; *J. Am. Chem.Soc.,* 1977, *99*, 4746.
25.   Sagiv, J.; *J. Am. Chem. Soc.*; 1980, *102*, 92.
26.   Wasserman, S. R.; Tao, Y. T.; Whitesides, G. M.; *Langmuir*, 1989, *5*, 1074.
27.   Tarasevich, B. J.; Rieke, P. C.; Pacific Northwest Laboratory, unpublished data.

RECEIVED February 18, 1992

Chapter 7

Nanoscale Surface-Modification Techniques Using the Scanning Tunneling Microscope

Bruce Parkinson

Department of Chemistry, Colorado State University,
Fort Collins, CO 80523

The scanning tunneling microscope (STM) has proven to be a useful tool for the manipulation of matter on an extremely fine scale. The many new nanoscale surface modification techniques which have been developed for this instrument can be loosely divided into several categories based on the method of substrate modification. These categories include: 1. continuous and pulsed high voltage or high current applied to the tip, 2. chemical or electrochemical deposition or etching and 3. mechanical deformation or abrasion via direct tip surface interactions. A review of the literature in these areas will be presented with a few examples from my own work.

One of the first applications envisaged for the scanning tunneling microscope was its use to modify surfaces on an extremely fine scale. Small scale surface modifications would be useful for the production of tiny electronic devices which could exploit quantum transport and confinement effects or for ultra-high density information storage. Construction of tiny mechanical components or direct assembly or cleavage of individual molecules are other speculative applications for this relatively new technology. Manipulation of individual atoms has already been demonstrated but applications of this technology have yet to be developed. I will concentrate on the current state-of-the-art in this area and provide a few examples from my work and the work of others to illustrate the different experimental approaches to surface modification with the STM. I will not review the construction and operation of the STM since there are many excellent reviews already in the literature[1-3].

High Current or Voltage Techniques

The use of high voltages or currents pulses is by far the most common method used for surface modification with the STM. One of the initial studies, by Becker, Golovchenko and Swartzentruber[4], concerning the modification of a germanium (111) surface in UHV, already demonstrated the ultimate in miniaturization by creating single adatom sites via pulsing the tungsten tip to a bias of 4V. They reported some irreproducibility in the ability to produce the "atomic bits" and were

0097–6156/92/0499–0076$06.00/0

unable to produce similar adatoms on silicon (111) surfaces even at considerably higher biases (20 V). They speculate that the voltage pulse causes the tip to deposit a previously acquired germanium atom. Subsequent studies have used a variety of tip materials, substrates and voltage/current pulse programs to produce nanostructures. Many of these studies are summarized in Table I.

The modification of a favorite STM substrate, highly ordered pyrolytic graphite, is typified in the work of Albrecht et al [5]. They wrote very well defined 300Å tall letters each consisting of many 20 to 40 Å holes. They reported variations in success rates (up to 99.6%), voltage threshold and hole size with tip structure. A fortuitous observation of the removal of a graphite layer around some letters allowed them to observe that some damage was done to the graphite layer below by the voltage pulse. Another key observation in their report was the importance of water to the process. Almost no hole formation was observed in UHV, pure oxygen or nitrogen atmospheres even when voltage pulses as high as 10 V were applied. When the chamber was vented with 20 Torr of water vapor the writing process was observed to be as efficient as in air. The importance of water led them to propose a chemical mechanism for hole formation. More recent investigations by Lewis et al [6] under bulk water have shed more light on this process. They find that a modification of the surface can be observed at voltages below the threshold for hole formation. Processes such as field evaporation or local heating have also been proposed to explain feature formation upon application of voltage pulses. Calculations have demonstrated that the thermal evaporation is unlikely due to the small temperature rise expected in a well ordered solid[7], but melting of glassy materials, due to their hundred-fold lower thermal conductivity, has been proposed to explain the formation of cone-like features on the surface of these materials in response to current pulses[8,9].

Field evaporation has been proposed as a mechanism for deposition of Au islands from a Au tip onto a Au (111) substrate[10]. In this study a clear threshold of about 3.3 V was observed for the deposition of 10 to 20 nm diameter mounds that are 2 to 3 nm high. They support the field evaporation mechanism by pointing out that refractory metal tips, such as tungsten, most often produce pits in the Au substrate, rather than mounds, due to field evaporation now from the substrate to the tip. Other workers had also observed both hole and mound formation on gold and silver in response to voltage pulses applied to W, Pt or Pt/Ir tips (Table I). One disadvantage of modifying Au substrates is the high surface mobility of Au. Features have been observed to anneal out in times as short as a few minutes [11].

Although some impressive features have been fabricated, the mechanism for nanoscale surface modification via voltage/current pulses is not totally clear and may well vary with tip/substrate combinations as well as length, direction and magnitude of the voltage and current pulses. The presence or absence of air and other surface contaminants is also important to the process. In some cases different mechanisms have been proposed by different experimenters for the same conditions (Table I).

Chemical Methods

The distinction between chemical methods and the voltage pulse methods for STM nanoscale surface modification are quite blurred especially in light of the findings of several groups on the importance of water vapor to the previously discussed hole formation in graphite. Nevertheless there are several examples of a deliberate addition of a chemical species which reacts to the energetic voltage pulse to produce products which either deposit on or etch the substrate. An interesting example of this technique is given by McCord, Kern and Chang[12] who deposited lines and

Table I
Selected Voltage Pulse and Chemical Methods for STM Nanolithography

Substrate	Tip	Feature(s)	Reactant(s)	V(volts)	I	t	Mech*	Reference
Ag	W	holes, mounds		3-7	>1μA	0.25-2.5 ms	FE	26
Ag	Pt/Ir	holes, mounds		5.	4 nA	20-100 ns	FE	27
Au	Au	Au mounds		3.6-4.0	.1-1 nA	0.1-0.6 μs	FE	10
Au	Pt	holes, mounds	oil, WF$_6$	0.5-4.0	1 nA	0.1-10 ms	CR	28,29
Au	Pt/Ir	hole, mounds		0.7-3.0	--	s?	--	30,31
Au	W	lines	LB film	1-1000	10-500 nA	2-100 s	CR	32
Au	W	craters		2.7	90nA	>5 ms	TE	33
Au	W	hillocks		0.005	1 μA	s	TS	11
Ag$_x$Se	Pt	grooves		2-5	0.1-2nA	cw	IM	34
Graphite	W	holes (20-40 Å)	H$_2$O	3-8	--	1-100 μs	CR	5
Graphite	Pt/Ir	holes		2-4	0.5-2.0 nA	0.05-200 μs	--	35
Graphite	Pt	holes (100 Å)	H$_2$O	3.5	--	320 μs	CR	36
Graphite	W …	mounds, holes	DMP[α], decane	3-4.5	0.5-2.0 nA	0.2-2 μs	CR	37
Graphite	Pt/Ir	holes		4.0	0.5-1.0 nA	1-10s	--	38
Graphite	--	molecule	DMP[α],DEHP[α]	3.3-4.3	0.1-1.0 nA	100 ns	CR	39
Graphite	--	polymer	PODA‡	4-20	200 nA	100 ns	CR	40
a-Rh$_{25}$Zr$_{75}$	W …	cones		0.4-2.0	1nA-1μA	s?	LM	8,9
a-Pd$_{81}$Si$_{19}$	W	lines	hydrocarbon	0.1	10 nA	cw	CR	41

Si	Pt/Ir	lines, holes	O$_2$	1.7-3.5	3nA	cw	CR	42
Si	W	lines, dots	W(CO)$_6$, AuAc†	5-40	1-1000 nA	0.5 s -cw	CR	12
Si	W	lines	Cd(CH$_3$)$_2$	2-10	2 nA	10-400 ns	CR	43,44
Si	--	dots, lines	P4BCMU#	5.0-30	0.5 nA	cw	CR	45,46
Si	--	mounds, holes	WF$_6$	15-20	---	100 ns	CR	47
SiO$_2$	--	work function		6	1 mA	35 µs	LM,DB	48
Cr, SiO$_x$	--	lines	PMMA°, PVC°	20-30	10 pA - 1nA	cw	CR	49
WSe$_2$	W	depression		1.5-6	3 na	2-3 ms	NC	50-52

* Mechanism as proposed by the authors: FE = field evaporation, TE = thermal evaporation, IM = ion migration, CR = chemical reaction, LM = local melting, DB = dielectric breakdown, TS = tip to surface transfer, NC = not clear

□DMP = dimetyl phthalate, DEHP = di(2-ethylhexyl) phthalate # P4BCMU = polydiacetylene urethane resist polymer,
° PMMA = polymethyl methacrylate, PVC = poly(vinyl cinnamate) ‡PODA = poly (octadecylacrylate) †AuAc = dimethyl-Au-trifluoro acetylacetonate

dots of gold and tungsten on a silicon substrate by decomposition of 5 to 20 mTorr of gaseous dimethyl-Au-trifluoro acetylacetonate or tungsten hexacarbonyl. Large voltages (up to 40 V) were used to effect the decomposition of the gases and produce features in the 10 to 30 nm range. The deposited W lines were rather conductive but they contained substantial amounts of carbon (40%) and oxygen (12%), as determined by Auger electron spectroscopy. High aspect ratio W structures (25 nm diameter and 280 nm high) were prepared and the authors point out the STM's suitability for preparing such structures due to the fact that the tip will be withdrawn as the features grow due to the tunneling current feedback. Modification of the W tip, from a pointed to a rounded structure, was also observed after deposition of surface features. Table I also contains other similar examples from the literature.

Another twist on chemical methods for nanofabrication, the scanning electrochemical microscope (SECM), was pioneered by Bard[13-16] . The SECM uses feedback from the Faradaic current of electrochemical processes to sense the surface of either insulators or conductors in an electrochemical environment (electrolyte or ion conducting polymer). The nice feature of this technique is that it accesses all the redox chemistry known to electrochemists for the generation of deposits and etchants. Bard's group deposited silver lines as narrow as 300 nm from a Nafion film with the SECM. They also demonstrated local etching of semiconductors and metals, such as GaAs, CdTe and Cu, by electrogeneration of an etchant at the tip while the tip was held very near the surface to be etched. Unlike STM the modification of insulating materials is possible with this technology. To date the resolution of SECM has not approached that of the STM but progress is being made on reducing the size of the tip/microelectrode.

Mechanical Techniques

A straightforward method for modifying a surface with the STM tip is to simply extend the tip towards the substrate, via application of a voltage to the z piezoelectric element, and physically indent the surface. An elegant manifestation of this technique was demonstrated by workers at Phillips[17,18] for Si(110) and Si(001) surfaces in UHV. They were able to produce stable indentations on the 2 to 10 nm scale by moving the tip down a distance of about 2 nm and then returning the tip to its original position. This process was completed in about 8 ms. They observed that the atomic structure in the region of the indentation remained relatively undisturbed and that the quality of the tip for atomic scale imaging was not degraded by the contacts with the surface, in fact it was often observed to improve. The process was proven to not be due to local heating from the relatively large increase in tunneling currents observed on tip contact (30 nA) by repeating the experiment with the tunneling voltage set to zero. Similar experiments reported by other authors on graphite, even with much larger z excursions, do not result in modification of the surface presumably due to the ease of compressing graphite.

A very different etching technique has been reported by several workers who have imaged materials with layered structures. Garfunkel et al [19,20] have shown that features, in the shape of the raster pattern of the microscope, can be fabricated on the conducting molybdenum bronzes, $Rb_{0.3}MoO_3$ and $Na_{0.9}Mo_6O_{17}$. They observe nucleation and "liquid like" hole growth of pits on some surfaces and use voltage excursions of several seconds to create hole nuclei on less defective surfaces. In the latter case on the Na bronze they observe faceting along crystallographic axes as the pits grow. They have discussed thermal and abrasive mechanisms for the etching process but tend to favor an abrasive mechanism for atom removal from tip contact with step and kink sites.

In my lab we have observed similar behavior with the layered metal dichalcogenide materials[21]. We are able to remove individual molecular layers (0.5 -0.7 nm thick) from the region of the crystal rastered by the STM. The substrates we have been able to pattern include among others ZrS_2, SnS_2, $SnSe_2$, MoS_2, $TiSe_2$, $NbSe_2$, TaS_2. The mechanism again involves nucleation and growth of holes on a single layer, which then repeats on subsequent layers to allow the controlled removal of many layers. Too high a nucleation density results in ragged holes many layers deep because lower layers have nucleated and begun to etch before the higher layers are completely removed. When nucleation density is low we can fabricate structures at least 20 layers deep (Figure 1). The accuracy of the crystallographic c-axis measurements also gives us confidence that we know the depth of our etched features to very high accuracy if we keep track of the number of layers we have removed.

We also observe faceting during etching of certain materials. Figure 2 shows the remarkable triangular morphology we have observed in the etching of a natural crystal of MoS_2 (we have also seen the triangular etching pattern on $NbSe_2$). The stacking polytype of the material is revealed when the triangles on the next layer appear rotated 180° with respect to the top layer as would be expected for the most common MoS_2 polytype (2H). In this case the etching of the material is very slow, taking over 5 hours of continuous scanning to remove a 0.25 μm^2 area. A similar region of $SnSe_2$ would be removed in about 20 minutes but the nucleation density on the $SnSe_2$ is much higher than the few (5) nucleai observed on the MoS_2 surface in Figure 2.

Harmer and Parkinson[22] have extended this method to several high T_c superconducting materials. For example lines and square holes could be etched into single crystals and laser ablated thin films of $YBa_2Cu_3O_{7-x}$ by continuously rastering the tip over a given region. The layers are again removed in multiples of the c-axis length in this material (1.2 nm).

The exact mechanism of the layer-by-layer etching process is also still not established as is the range of materials which exhibit this effect. It is clear that all reports of etching layered type materials have established the importance of nucleation of the holes. Recent results in our laboratory have shown no or little etching on sulfur vacancy free crystals of SnS_2, demonstrating the importance of chalcogenide vacancies for the nucleation of the holes which grow and result in the removal of an entire layer. An experiment in C. Liebers lab at Columbia University[23] established that $SnSe_2$ does not etch in an UHV environment. This result suggests that the interaction of the tip with the surface is important. We also have recent results showing that the atomic force microscope (AFM) etches layered compound crystals in the same manner as the STM. Many researchers have shown that the water and hydrocarbon layers present on surfaces exposed to air will wet the STM or AFM tip resulting in capillary forces which keep the tip and the substrate in contact. Transfer of energy from the piezoelectric translator to the substrate via frictional forces could provide the necessary activation to remove atoms with their nearest neighbor interactions disrupted. Whether this process involves a non-local transfer of energy via phonon coupling or a local bonding of dangling bonds from coordinately unsaturated atoms to the tip is still an open question. The materials which have been studied to date all have anisotropy associated with their structures which may interfere with the conduction of frictional energy away from the interface. However the observation of etching may simply be due to the fact that these materials have large regions of atomic flatness where removal of small amounts of material is very noticeable.

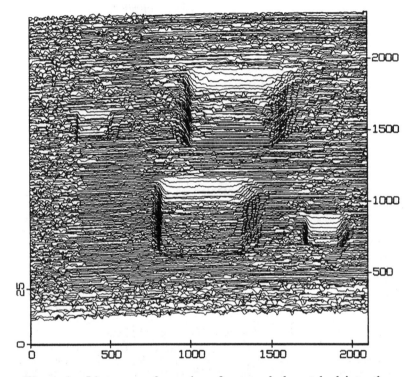

Figure 1 - Line scan of a series of square holes etched into the surface of SnSe₂. Two 500 nm square holes are shown, one of which is 20 layers deep (123 Å) and the other which is 10 layers deep (61.5 Å). Several 200 nm squares are also shown, one of which is ten layers deep and the other (upper left) which is 5 layers deep (30.7 Å). For details about the etching process see reference 21. The distance scale is in nanometers.

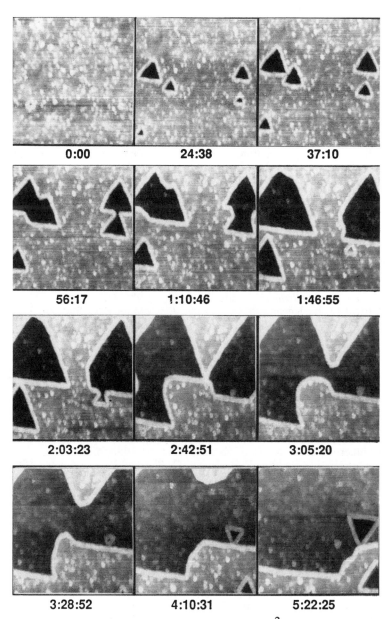

Figure 2. A series of images showing a $0.5\mu m^2$ region of a MoS_2 crystal which was continuously rastered with the STM. The appearance and growth of triangular nucleation sites is apparent. After several hours the triangular regions grow and merge, eventually resulting in removal of an entire MoS_2 layer (3.16 Å). Nucleation sites then appear on the next layer rotated 180° with respect to the layer above. The rotation is indicative of the 2H polytype of MoS_2.

The Ultimate is Already Here

No review of nanoscale lithography would be complete without a description of the recent reports of very controlled manipulation of individual atoms. Eigler and Schweizer [24] have been able to move individual Xe atoms on the nickel (110) surface by lowering the tip (by increasing the tunneling current) above a given atom to increase the tip-atom interaction. The atom is then dragged to the desired location and the tunneling current decreased to release the atom by raising the tip. The letters IBM were spelled out in 5 nm letters. The drawback is that relying on the van der Waals interactions between the Xe atoms and the Ni surface requires working at very low temperatures, and taking heroic measures for thermal, electrical and vibration isolation. A more recent announcement from Hitachi reported writing less than 2 nm high letters at room temperature by removing individual sulfur atoms from the surface of an MoS_2 crystal with voltage pulses[25]. Their message "PEACE '91 HCRL" (Hitachi Central Research Laboratory), a nice thought with which to end this review.

Literature Cited

(1)   Avouris, P. *J. Phys. Chem.,* **1990**, *94*, 2246-2256.
(2)   Denley, D. R. *Ultramicroscopy,* **1990**, *33*, 83-92.
(3)   Griffith, J. E.; Kochanski, G. P. *Annu. Rev. Mater. Sci.,***1990**, *20*, 219-244.
(4)   Becker, R. S.; Golovchenko, J. A.; Swartzentruber, B. S. *Nature,* **1987**, *325*, 419-421.
(5)   Albrecht, T. R.; Dovek, M. M.; Kirk, M. D.; Lang, C. A.; Quate, C. F.; Smith, D. P. E. *Appl. Phys. Lett.,***1989**, *55*, 1727-1729.
(6)   Lewis, N. *et al, Appl. Phys.Lett.,* in press
(7)   Marella, P. F.; Pease, R. F. *Appl. Phys. Lett.,***1989**, *55*, 2366-2367.
(8)   Staufer, U.; Scandella, L.; Wiesendanger, R. *Z. Phys. B-Condensed Matter,* **1989**, *77*, 281-286.
(9)   Staufer, U.; Wiesendanger, R.; Eng, L.; Rosenthaler, L.; Hidber, H. R.; Guntherodt, H. J.; N., G. *J. Vac. Sci. Tech. A,* **1988**, *6*, 537-539.
(10) Mamin, H. J.; Guethner, P. H.; Rugar, D. *Physical Review Letters,* **1990**, *65*, 2418-2421.
(11) Abraham, D. W.; Mamin, H. J.; Ganz, E.; Clarke, J. *IBM J. Res. Develop.,* **1986**, *30*, 492-499.
(12) McCord, M. A.; Kern, D. P.; Chang, T. H. P. *J. Vac. Sci. Technol. B,* **1988**, *6*, 1877-1880.
(13) Mandler, D.; Bard, A. J. *J. Electrochem. Soc.,* **1989**, *136*, 3143-3144.
(14) Bard, A. J.; Denuault, G.; Lee, C.; Mandler, D.; Wipf, D. O. *Acc. Chem. Res.,* **1990**, *23*, 357-363.
(15) Mandler, D.; Bard, A. J. *Langmuir,* **1990**, *6*, 1489-1494.
(16) Liu, C.; Bard, A. J. *Chemical Physics Letters,* **1990**, *174*, 162-166.
(17) van Loenen, E. J.; Dijkkamp, D.; Hoeven, A. J.; Lenssinck, J. M.; Dieleman, J. *Appl. Phys. Lett.,* **1989**, *55*, 1312-1314.
(18) van Loenen, E. J.; Dijkkamp, D.; Hoeven, A. J.; Lenssinck, J. M.; Dieleman, J. *J. Vac. Sci. Technol. A,* **1990**, *8*, 574-576.
(19) Garfunkel, E.; Rudd, G.; Novak, D.; Wang, S.; Ebert, G.; Greenblatt, M.; Gustafsson, T.; Garofalini, S. H. *Science,* **1989**, *246*, 99-100.
(20) Saulys, D.; Rudd, G.; Garfunkel, E. *J. Appl. Phys.,* **1991**, *69*, 1707-1711.

(21) Parkinson, B. *J. Am. Chem. Soc.,* **1990**, *112*, 7498-7502.
(22) Harmer, M. A.; Parkinson, B. A.; Fincher, C. *J. Applied Physics,* submitted
(23) C. Lieber, private communication
(24) Eigler, D. M.; Schweizer, E. K. *Nature,* **1990**, *344*, 524-526.
(25) Clery, D. *New Scientist,* **1991**, 31.
(26) McBride, S. E.; Wetsel, G. C., Jr. *Appl. Phys. Lett.,* **1990**, *57*, 2782-2784.
(27) Rabe, J. P.; Buchholz, S. *Appl. Phys. Lett.,* **1991**, *58*, 702-704.
(28) Baba, M.; Matsui, S. *Japanese Journal of Applied Physics,* **1990**, *29*, 2854-2857.
(29) Matsui, S.; Ichihashi, T.; Baba, M.; Satoh, A. *Superlattices and Microstructures,* **1990**, *7*, 295-301.
(30) Schneir, J.; Sonnenfeld, R.; Marti, O.; Hansma, P. K.; Demuth, J. E.; Hamers, R. J. *J. Appl. Phys.,* **1988**, *63*, 717-721.
(31) Schneir, J.; Hansma, P. K. *Langmuir,* **1987**, *3*, 1025-1027.
(32) McCord, M. A.; Pease, R. F. W. *J. Vac. Sci. Technol. B,* **1986**, *4*, 86-88.
(33) Li, Y. Z.; Vazquez, L.; Piner, R.; Andres, R. P.; Reifenberger, R. *Appl. Phys. Lett.,* **1989**, *54*, 1424-1426.
(34) Utsugi, Y. *Nature,* **1990**, *347*, 747-749.
(35) Miller, J. A.; Hocken, R. J. *J. Appl. Phys.,* **1990**, *68*, 905-907.
(36) Mizutani, W.; Inukai, J.; Ono, M. *Japanese Journal of Applied Physics,* **1990**, *29*, L815-L817.
(37) Bernhardt, R. H.; McGonigal, G. C.; Schneider, R.; Thomson, D. J. *J. Vac. Sci. Technol. A,* **1990**, *8*, 667-671.
(38) Terashima, K.; Kondoh, M.; Yoshida, T. *J. Vac. Sci. Technol. A,* **1990**, *8*, 581-584.
(39) Foster, J. S.; Frommer, J. E.; Arnett, P. C. *Nature,* **1988**, *331*, 324-326.
(40) Albrecht, T. R.; Dovek, M. M.; Lang, C. A.; Grutter, P.; Quate, C. F.; Kuan, S. W. J.; Frank, C. W.; Pease, R. F. W. *J. Appl. Phys.,* **1988**, *64*, 1178-1184.
(41) Ringger, M.; Hidber, H. R.; Schlogl, R.; Oelhafen, P.; Guntherodt, H. J. *Appl. Phys. Lett.,* **1985**, *46*, 832-834.
(42) Dagata, J. A.; Schneir, J.; Harary, H. H.; Evans, C. J.; Postek, M. T.; Bennett, J. *Appl. Phys. Lett.,* **1990**, *56*, 2001-2003.
(43) Ehrichs, E. E.; Yoon, S.; de Lozanne, A. L. *Appl. Phys. Lett.,* **1988**, *53*, 2287-2289.
(44) Silver, R. M.; Ehrichs, E. E.; de Lozanne, A. L. *Appl. Phys. Lett.,* **1987**, *51*, 247-249.
(45) Dobisz, E. A.; Marrian, C. R. K.; Coton, R. J. *J. Vac. Sci. Technol. B,* **1990**, *8*, 1754-1758.
(46) Marrian, C. R. K.; Colton, R. J. *Appl. Phys. Lett.,* **1990**, *56*, 755-757.
(47) Ehrichs, E. E.; de Lozanne, A. L. *J. Vac. Sci. Technol. A,* **1990**, *8*, 571-573.
(48) Jahanmir, J.; West, P. E.; Colter, P. C. *J. Appl. Phys.,* **1990**, *67*, 7144-7146.
(49) Zhang, H.; Hordon, L. S.; Kuan, S. W. J.; P., M.; Pease, R. F. W. *J.Vac.Sci.Technol. B,* **1989**, *7*, 1717-1722.
(50) Schimmel, T.; Fuchs, H.; Akari, S.; Dransfeld, K. *Appl. Phys. Lett.,* **1991**, *58*, 1039-1041.
(51) Fuchs, H.; Laschinski, R.; Schimmel, T. *Europhysics Letters,* **1990**, *13*, 307-311.
(52) Fuchs, H.; Schimmel, T. *Adv. Mater.,* **1991**, *3*, 112-113.

RECEIVED January 16, 1992

LAYERED AND LOW-DIMENSIONAL STRUCTURES

Chapter 8

Design and Chemical Reactivity of Low-Dimensional Solids

Some Soft Chemistry Routes to New Solids

Jean Rouxel

Institute des Matériaux de Nantes, Centre National de la Recherche Scientifique UMR No. 110, Université de Nantes, 2 rue de la Houssinière, 44072 Nantes Cedex, France

Low dimensional solids can be regarded as built up from a stacking of slabs or from a juxtaposition of fibers. These are infinite two dimensional or one-dimensional giant molecules weakly bonded to form a three dimensional architecture. General considerations about the stability of such arrangements will be followed by a critical discussion of various approaches introduced in the form of particular examples. Low dimensional solids have a fascinating chemical reactivity. It is largely a low-temperature "soft-chemistry" mostly based on redox processes or acido-basic reactions, that are reversible. It is therefore a powerful tool for the preparation of new solids that can be low dimensional compounds but also large open frameworks with empty cages or tunnels. This contribution will be primarily devoted to these subjects.

To a large extent low-dimensional solids represent by themselves supermolecular architectures in which slabs or fibers, that can be regarded as giant two- or one-dimensional molecules, are weakly bound together to build a three dimensional arrangement.

These solids show very specific physical properties (1-4) (charge density waves (CDW) instabilities, low-dimensional magnetism...). They also have a fascinating chemical reactivity (5) largely associated with the so-called "chimie douce", i.e, soft chemistry done close to room temperature that has three main components (i) redox intercalation/deintercalation processes, (ii) acido-basic reactions followed by structural recondensations, (iii) grafting reactions in which the van der Waals gaps between slabs or fibers are considered as many internal surfaces of these solids. The reversibility attached to some of these processes makes them useful for the design of new low-

0097–6156/92/0499–0088$07.50/0

dimensional solids. All are at the origin of new synthesis routes in solid state chemistry.

I - General outlines for the crystal chemistry of low-dimensional solids.

The so-called van der Waals gap which separates slabs or fibers is bound on each side by atomic layers of the same nature, usually anionic layers (like in the transition metal dichalcogenides MX_2 where the slabs can be regarded as [XMX] sandwiches). This has two major consequences :
(i) the surrounding of the layers or rows of metal atoms by the non-metal element implies a chemical formula MX_n, with a rather high non-metal content, the value of n increasing when going from two-dimensional (2D) to one dimensional (1D) arrangements.
(ii) there is a repulsion between these similar anionic layers on each side of the van der Waals gap. This partially fixes the slab to slab separation and consequently the low-dimensional character of the structure. The more ionic the materials, the stronger the tendency to exhibit low-dimensionality through anion- anion repulsion with separation of the sheets or fibers, but also the less stable the structure. The limit of the low-dimensionality character is fixed by the intrinsic instability that it generates. Therein lies the difficulty of synthesis of these phases, which can rely only on qualitative approaches.

To counter the slab to slab repulsion, the bonding through the interslab or the interfiber spacings must be stronger than the simple van der Waals interaction. It has been recently shown that an interaction between the $3p_z$ orbitals of sulfur atoms of one slab and the empty states provided by the metal atoms in the neighbouring slab, is to some extend responsible for the structural cohesion of layered chalcogenides and for the existence of polytypes *(6)*.

In the case of more ionic oxides, the repulsion would be stronger and very destabilizing. It is observed that the MO_2 oxides have the rutile or fluorite structures and not the layered structures of the parent dichalcogenides. Low-dimensional oxide structures have to be stabilized by separating the slabs or fibers by counter-ions like in layered Na_xCoO_2 phases *(7)* or pseudo one-dimensional $K_{0.30}MoO_3$ blue bronzes of molybdenum *(8)*. Stability is also achieved by reducing the formal charge of the anions through a protonation of oxide ions (to give layered hydroxides like brucite $Mg(OH)_2$), or with transition metals in a high oxidation state (MoO_3, V_2O_5). In the latter case there is a strong polarization of the electrons of the anions towards the interior of the slabs. Finally, layered silicates represent some compromise among all of the above situations. Extra cations are often located between slabs that can be partially protonated. In addition, one remarks that the tetrahedra are always at the external part of a composite slab, the heart of the slab being made from the octahedra. The tetrahedra, with the shortest anion-cation distances, represent another approach to achieve a strong polarization of the electronic density and a stabilization of the structure. Such a situation is obtained in the recently prepared layered phosphatoantimonic acids and salts with slabs built up from internal $[SbO_6]$ octahedra and external $[PO_4]$ tetrahedra *(9-10)*.

With the exception of these particular situations which allow the stabilization of a low-dimensional structure in the case of oxides, a convenient way to get layered or one dimensional arrangements is to decrease the ionicity of the bonds. A low-dimensional arrangement may then become stable (sulfides as compared to oxides). However, at the same time the ionicity of the bonds is decreased, there will be a loss in the low-dimensionality character of the structure as mentioned above. A layered framework is more easily stabilized in the case of selenides than that of sulfides but on the other hand the low-dimensional character of the properties is less evident.

The stabilization of low-dimensional arrangements when going down a column in the periodic table may also reach a limit with the heaviest elements (tellurium for example), for which the loss of directionality of the bonds may lead to three dimensional metals. Between the extreme cases of ionic oxides and quasi-metallic, but mostly 3D, tellurides, sulfides and selenides represent the most favorable domain for the design of low-dimensional solids. Similar conclusions can be drawn for neighbouring columns.

However structures involving sulfides and selenides may result in additional problems that arise from a redox competition, now possible, between d cationic levels and sp anionic ones. This can be easily understood when looking at the classical band scheme for transition metal chalcogenides (figure 1a,b,c). Between a valence band which is essentially sp anionic in character and antibonding levels mostly issued from the corresponding cationic levels, the d orbitals of the metal, split by the crystal field, play an essential role in the physical properties. But they also govern the stability of the layered arrangements and even their stoichiometries. With sulfur, a maximum cationic oxidation state of four is observed in the case of IVB elements leading to a d^0 configuration and no electronic conductivity. ZrS_2 with a CdI_2 structural type and an octahedral coordination of the metal is a semi-conductor. It shows a broad and empty t_{2g} band. A d^1 configuration in the case of its neighbour Nb, lowers the symmetry which becomes trigonal prismatic with a half filled d_{z^2} (a'_1) band, and a metallic behaviour. Immediately after with a d^2 configuration and the same symmetry, MoS_2 and WS_2 are diamagnetic semiconductors. Going further to the right, one would expect to come back to an octahedral symmetry for a layered MnS_2 for example (no stabilization through a distortion for a d^3 configuration). But that phase cannot be obtained. Indeed, when going to the right of the periodic table the d levels progressively decrease in energy and may enter the sp valence band. If such a situation occurs for an empty d level

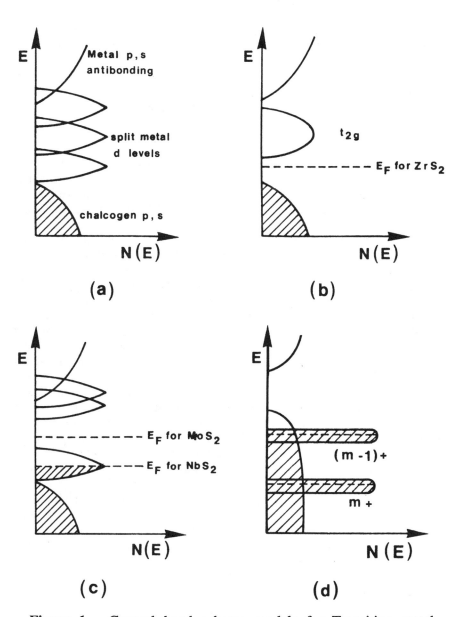

Figure 1 - General band scheme models for Transition metal chalcogenides (a). Particular cases of ZrS_2 (b), NbS_2 and MoS_2 (c), d-cationic levels and sp anionic band at the end of a period (d).

it will be filled at the expense of the valence band, at the top of which holes will appear *(11)* (figure 1d). Chemically speaking this means that the cation is reduced and that the anion is oxidized via the formation of anionic pairs. Following that scheme, one goes from layered structures, such as TiS_2 with Ti^{4+} and $2S^{2-}$ to pyrites and marcasites with Fe^{2+} and $(S_2)^{2-}$ for example. Selenium being less electronegative than sulfur, the top of the sp valence band is situated at a higher energy, and such a transition is observed earlier in selenides as compared to sulfides.

II - Dimensionality and compositional variations

Niobium triselenide was obtained during a careful study of the Nb-Se system on the selenium rich side *(12)*. The increased Se/Nb ratio resulted in a chain like arrangement of $[NbSe_6]$ trigonal prisms that are stacked on top of each other instead of being arranged in layers like in $NbSe_2$ (figure 2). The chains are irregular due to the presence of a $(Se_2)^{--}$ pair. However this structure is very complicated due to complex couplings between chains which modify the simplest d cationic - sp anionic redox competition expected to give $Nb^{4+}Se^{--}(Se_2)^{--}$. One is left with three chains (d Se-Se values are 2.37, 2.48 and 2.91 Å, Fig. 2c). The chalcogen pair behaves as an electron reservoir. According to the population of its antibonding level (which governs its length), it takes or gives less or more electrons to the adjacent metallic chain. The two chains with the shortest intra-chain Se-Se distances are conducting and show charge density waves respectively below 59 and 140 K. The third chain is an insulating one.

As compared to $NbSe_2$, the increase of the Se/Nb ratio, in $NbSe_3$, is accommodated by the structure through a diminution of its dimensionality. This a consequence of the possible condensation modes of polyhedra in order to form slabs or chains with an increasing number of unshared corners. Let us consider now the MPS_3 phases. They are layered materials with slabs built up from sharing edges of octahedra, like in TiS_2, except that the octahedra occupancy is 2/3 by M^{2+} cations and 1/3 by (P-P) pairs. Once again a study of the chalcogen rich side of M-P-S systems leads to series of new low-dimensional compounds, most of them presenting one-dimensional arrangements. Extended studies have been done in the case of the P-V-S and P-Nb-S systems. Although presenting either a dimensionality or a chemical composition different from each other, the compounds derive from the same basic building unit which is a tetracapped biprism $[M_2S_{12}]$ (figure 3a). All these biprisms are bonded together to form infinite puckered $[M_2S_9]_\infty$ chains (figure 3b). Then the bonding between these chains takes place through $[PS_4]$ tetrahedra and the way such bondings are made determines the dimensionality of the phases which is for example :
(i) 1D in PV_2S_{10} *(13)* when the M_2S_9 lines are linked two by two (figure 3c)

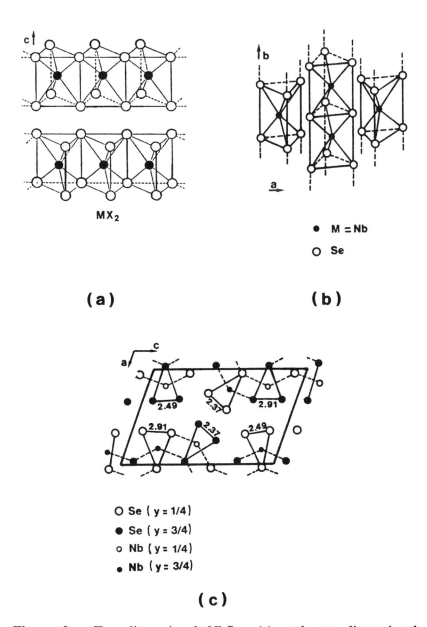

Figure 2 - Two-dimensional NbSe$_2$ (a) and one dimensional NbSe$_3$ (along the chains (b) and perdicular to them (c))

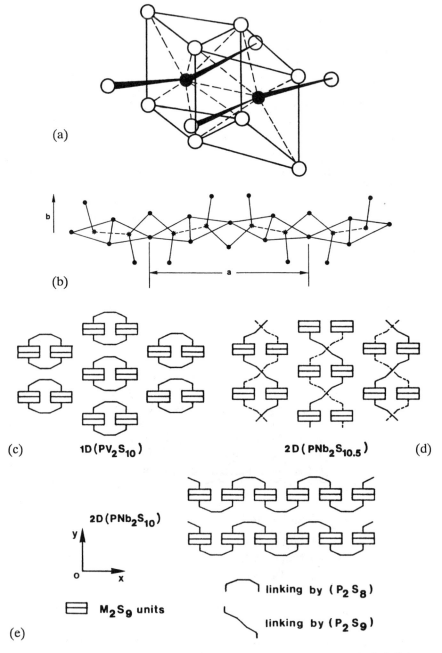

Figure 3 - M_2S_{12} bicapped biprism (a), formation of $(M_2S_9)_\infty$ chains from M_2S_{12} biprisms (b). Association of these chains by $[P_2S_8]$ and $[P_2S_9]$ groups in 1D PV_2S_{10} (c), 2D $PNb_2S_{10.5}$ (d) and 2D PNb_2S_{10} (e). The little rectangles represent a cross section of the $[M_2S_9]$ chains.

(ii) 2D in PNb_2S_{10} *(14)* when two PS_4 groups unite the chains on both their sides (figure 3e)

(iii) 2D in $PNb_2S_{10.5}$ *(15)* when longer $[PS_4-S-PS_4]$ connecting groups link the chains above and sideways to each other (figure 3d).

These are supermolecular architectures based on different connections of similar chains. Many other structures based on 3D arranged bicapped biprisms have been found in these series. However the most fascinating compound is $Ta_4P_4S_{29}$ *(16)*.

It is built up from Ta_2S_{12} capped biprisms, as above, but these are connected by PS_4 tetrahedra which associate an upper corner of one biprism to a lower corner of the adjacent biprism (figure 4). This leads to a helix which has four biprisms per repeat distance. Such a biprism helix can be either right handed or left handed. It contains an other helix which is a sulfur helix, a S_{10} chain, also right handed. In the unit cell there are two big intertwined right handed helices and two small left handed ones which give much smaller tunnels that are empty. The space group is $P4_32_12$. But there is another form with space group $P4_12_12$ which is left handed for the big helices and right handed for the small ones. $P_2Nb_2S_8$, $Ta_2P_2S_1$, and $V_2P_4S_{13}$ are other members of the series with related open structure arrangements *(17)*. In situ polymerizations are certainly possible in a similar way as in some zeolites.

III - Tuning with the ionicity of bonds.

$NbSe_3$ is not a true one dimensional conductor. It remains metallic below the C.D.W. transitions which means that the Fermi surface is not completely destroyed (inperfect nesting condition). Due to the importance of this compound (first observation of the depinning of a C.D.W.) it was of great interest to increase the low-dimensional character. As seen above this is basically a problem of increasing the ionicity of the chemical bonds in the corresponding framework. One has to change the niobium for the more electropositive tantalum, and the selenium for the more electronegative sulfur. TaS_3, which was already known with an orthorhombic unit cell *(18)*, demonstrates effectively to the increased 1D character in three ways :

(i) a transition at 210 K which goes as far as a metal to semi-conductor transition *(19)*. The superstructure is commensurate with the initial lattice.

(ii) important pretransitional effects manifested by diffuse lines which condense at the transition into spots between the main Bragg rows of spots. These pretransitional effects imply that the C.D.W. exists dynamically above the transition but without phase coherence from one chain to another.

(iii) the existence of a polytype. Like in the case of most low-dimensional compounds polytypism is expected in trichalcogenides. It expresses the fact that slabs or fibers weakly bound to each other can

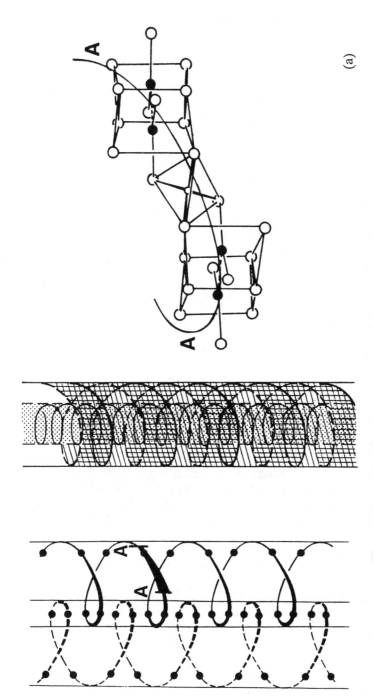

Figure 4(a) - M_2S_{12} biprisms are connected by PS_4 tetrahedra (right) to form intertwinned helix (left), each of them containing an other helix of sulfur (centre).

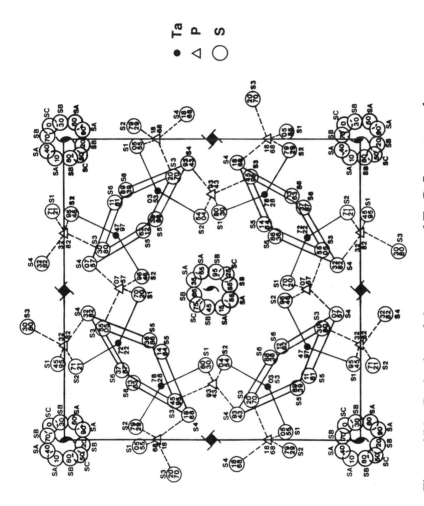

Figure 4(b) - Projection of the structure of $Ta_4P_4S_{29}$ onto a plane perpendicular to the direction of the helix. It shows clearly the sulfur helix inside the bigger one.

(b)

behave as independant units in a true low-dimensional structure. Polytypism usually associates the same chains or slightly different ones in new arrangements. No polytype was observed in the case of $NbSe_3$. However trials were successful for TaS_3. This is once more in agreement with the increased low-dimensional character of TaS_3 as compared to $NbSe_3$. In addition to the orthorhombic form, a monoclinic TaS_3 was found (20). The two structures are built from groups of four TaS_3 chains (21) as shown on figure 5 (chains with 2.835, 2.068 and 2.105 Å S-S distances are present). The b monoclinc axis is equal to the c orthorhombic axis. This corresponds to the height of one $[TaS_6]$ prism. The monoclinic form shows also a metal to semi-conductor transition at 240 K (22), but the superstructure is now incommensurate with the initial sublattice.

    A still more critical interplay with the ionicity of bonds concerns the tetrachalcogenides. The mineral patronite VS_4 presents a 1D organization with S_4 rectangles built up from true $(S_2)^{2--}$ pairs, rotating around a -V-V- chain and giving a rectangular antiprismatic coordination to the metal atoms. Vanadium is in the 4^+ oxidation state but there is a d^1-d^1 pairing which leads to a diamagnetic semi-conducting situation. Long and short V-V distances alternate along the chain with a doubling in that direction of the lattice period which can be taken as a classical illustration of a Peierls distortion. If tantalum and niobium homologues of VS_4 could be prepared, an increased 1D character is to be expected due to the more electropositive character of the metal.

    In addition more expanded d orbitals could favor an electronic delocalization along the chains. Direct combinations of the elements in evacuated silica tubes at various temperatures between 400 and 600°C have been unsuccessful. Even with the help of pressure it has not been possible, up to now, to prepare NbS_4 and TaS_4.

    The impossibility to realize such a synthesis is certainly due to the increased ionicity of the bonds which destabilizes the one-dimensional structural arrangement. We are beyond the stability limit. The solution is either to decrease the ionicities by changing sulfur by selenium or tellurium, or to separate the chains by counter ions. Working with selenium is not sufficient. One has to go to tellurium to stabilize a binary tetrachalcogenide of niobium and tantalum (these tetratellurides have a complicated incommensurate structure (23-24). In the case of selenium a counter ion is needed to stabilize the structural arrangement. This is the situation of the tetrachalcogenoiodides of niobium and tantalum (25). There is a complete series of $(MSe_4)_nI$ phases with M = Nb, Ta and n = 1/3, 1/2, 10/3... In the common structural type $[MSe_4]$ chains are separated by columns of I^- anions, both running parallel to the c axis of unit cells which are generally tetragonal (figure 6). Large distances (~6.70 Å) separate the MX_4 chains. The various compounds differ not only by the amount of iodine but also by the sequence of long and short metal-metal distances along the chains, and by the way Se_4 rectangles

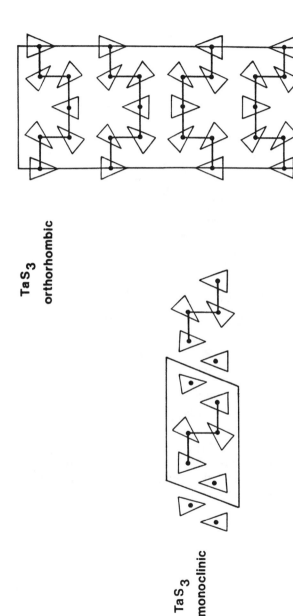

TaS₃
orthorhombic

TaS₃
monoclinic

Figure 5 - The two TaS₃ polytypes are built up from similar groups of TaS₃ chains (sections perpendicular to the chains direction). The solid lines figure the different associations of similar TaS₃ chains in the two structures (they do not represent metal-metal bonds).

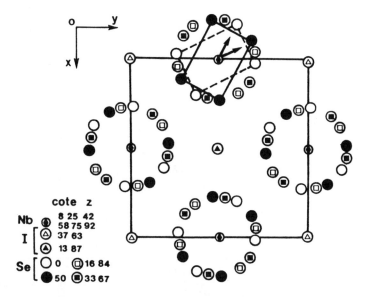

Figure 6 - Niobium (and tantalum) halogeno-tetrachalcogenides show rectangular antiprismatic MX_4 chains separated by I^- columns. The arrows indicate the relative orientations of two successive X_4 planes above, and below, a metal atom.

rotate around the chains to form the rectangular antiprismatic coordination of the metal. Every two adjacent units $[Se_4]^{4-}$ have a dihedral angle of 45° measured between the arrows that characterize their orientation (minimization of the repulsion between their π and π^* orbitals along with obtaining the best set of symmetry adapted orbitals to interact with the metal orbitals). But the relative displacement of such successive rectangles can refer to a right - or left - handed rotation. Thus $(NbSe_4)_3I$ is a $(1\ 2\ 3\ 4\ 3\ 2)_n$ tetrachalcogenide (rotation in the same way between rectangles 1, 2, 3, 4, then inverse rotation back to 3 and 2 before coming back to 1 and starting a new repeat unit) $(TaSe_4)_2I$ and $(NbSe_4)_{10}I_3$ are $(1\ 2\ 3\ 4)_n$ and $(1\ 2\ 3\ 4\ 5\ 6\ 5\ 4\ 3\ 2)_n$ chalcogenides, respectively. In the M-Se combination all the metal orbitals are involved, except d_{z^2} which remains isolated and builds a d_{z^2} band along the chain. This band governs the 1D electronic properties of the system *(26)*. Assuming that each iodine takes one electron, the average number of d electrons per metal is $(n-1)/n$ with a corresponding filling of the d_{z^2} band of $f = (n-1)/2n$, possibly leading to a distortion increasing the repeat distance by a factor $1/f$. As n increases the d_{z^2} band of $(MX_4)_nI$ phases becomes nearly half-filled. Indeed $(MX_4)_nI$ phases are CDW materials. Among them $(TaSe_4)_2I$ and $(NbSe_4)_{10}I_3$ have shown interesting features, particularly a memory effect in the CDW depinning experiments in the latter case *(27)*. Vanadium tetrasulfide constitutes finally an example of the n -> ∞ limit. It has consequently the 2c insulating Peierls distortion.

IV - Counter ions and counter chains.

The tetraselenoiodide phases just described above represent a successful attempt to stabilize chains by counter ions. That structure also recalls the organisation of Krogman's salts in coordination chemistry, or organic conductors like tetraselenotetracene I_x phases *(28)*. Most of the time the counter ion is a cation. From that point of view the so-called blue bronzes of molybdenum, known for a long time represent a symmetrical situation with respect to the $(MX_4)_nI$ derivatives, the counter ion being a cation instead of I^- species. With a general formulation $A_{0.30}MoO_3$ (A = K, Rb, Tl) they show complex chains built from groups of $[MoO_6]$ distorted octahedra *(8, 29)*.
    It is worth noticing that most of these structure are stabilized by big counter ions which have the strongest favorable influence on the Madelung energies. Most of the time the corresponding arrangements cannot be obtained for small counter ions (Cl^- in the case of tetraselenides, Li^+ in the blue bronzes).
Instead of being stabilized by counter ions chain-like arrangements could probably become stable when different chains requiring opposite conditions for a stabilization are associated in the same structure. For example $[NbSe_4]$ chains are stabilized by negative I^- columns in

$(NbSe_4)_nI$ phases. But, in $A_2Mo_6Se_6$ compounds *(30, 31)*, Mo_6Se_6 chains described below are stabilized by positive A^+ cations (A = alkalimetal or thallium). Attempts have been made to associate in the same structure chains having opposite stability requirements. They have been essentially unsuccessful. However, even when "clever" approaches have not yet given the expected result, the concept is correct as proved by the recent characterization of a crystal with a mixed chain structure in a batch corresponding to preparations made in the Nb-Se-Br system. The formulation is $Nb_3Se_{10}Br_2$ *(32)* and the structure is built from an alternation of $[NbSe_4]$ chains centered along a two-fold axis and $[NbSe_3Br]$ chains running a 2_1 axis parallel to the former one (figure 7). Se-Br exchanges on certain sites explain the rather complicated formulation.

V - Redox soft-chemistry preparations.

Cationic intercalation chemistry is based on coupled ion-electron transfer reactions.

$$x\,(A^+ + e^-) + Host \quad \rightarrow \quad A_x\,Host^{x-}$$

These reactions represent an important part of the chemical reactivity of low-dimensional solids. They are reversible. In that case A^+ ions are extracted and the host is reoxydized. A scale of reagents able to intercalate, or deintercalate, a given alkalimetal (lithium) in or from various host structures has been proposed *(33)*.

Interesting is the fact that the deintercalation reaction can be the base of a real concept to define a new and very important synthetic route in solid state chemistry. The idea is to consider a ternary compound $A_xM_yY_z$ (usually Y = O, S, Se) that can prepared by classical solid state chemistry as the intercalation of A^+ ions in a hypothetical M_yY_z host structure. Removing A^+ ions by electrochemistry or by using chemical reagents such as iodine in acetonitrile solutions in the case of lithium, yields M_yY_z which can be a new compound or a new structural form of an already known phase. For example one can get VS_2 *(34)* from $LiVS_2$ or a new form of TiS_2 (named cubic TiS_2) from the spinel $CuTi_2S_4$ *(35, 36)*.

A critical discussion of this approach is possible under the light of the general considerations of section I. VS_2 and TiS_2 are favorable cases from the point of view of both the ionicity of bonds and the d-sp redox competition.

But let us consider oxides. $NaTiO_2$ presents the $\alpha NaFeO_2$ structure, like $NaTiS_2$. It can be regarded as an occupancy of octahedral sites by Na^+ ions between pseudo $[TiO_2]$ slabs similar to those of TiS_2. Removing A^+ ions is possible. At the beginning the process is reversible. It suggests that one could end up with a layered TiO_2 similar to TiS_2. But this is not the case. After a critical extent in deintercalation titanium moves irreversibly from the $[TiO_2]$ "slabs" to the van der Waals gap

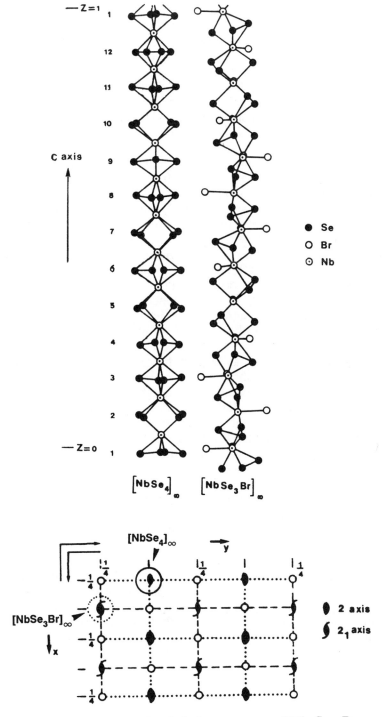

Figure 7 - The mixed-chains structure of $Nb_3Se_{10}Br_2$.

(37). Beyond that point the deintercalation can be continued but at the end the process does not lead to a layered oxide. This is a situation which has been reproduced in various AMO_2 systems. Very likely the threshold value for which the cationic displacement occurs is simply related to the electropositivity of the M element. M cations must replace Na^+ ions in maintaining the stability of the structure. One system, Li_xCoO_2, behaves differently *(38)*. Actually $LiCoO_2$ is a ferroelectric compound, and a cooperative ferroelectric displacement of the cobalt ions from the center of the octahedra in the slabs in the direction of the c axis, i.e. perpendicular to the slabs, is probably responsible for dipole-dipole couplings between CoO_2 slabs, which stabilizes the structure.

Now, in the case of a ternary sulfide involving an element of the second part of transition metal rows, the d-sp redox competition will interfere. When deintercalating Li_2FeS_2 *(39)* a new form of FeS_2 is obtained in which both iron and (S^{--}) anions are oxidized as shown by the formulation $Fe^{3+} S^{--} (S_2^{--})_{0.5}$. This is reflected in structural changes, for example, a shifting of the Fe^{+++} from octahedral to tetrahedral sites.

The modifications affecting site symmetries are those expected when considering ion sizes, crystal field stabilization energies, or band structures. An interesting example concerns the preparation of an "octahedral" MoS_2. In paragraph I we have discussed the fact that a layered MnS_2, which would have been similar to TiS_2 with octahedral coordination of the metal and a half filled t_{2g} band, can not be prepared. However such a situation has been indirectly achieved by adding one electron to MoS_2 through intercalation of lithium. A phase transition is observed which leads to "octahedral" molybdenum in $LiMoS_2$ *(40)*. It corresponds to a simple translation of a sulfide layer relative to the other one limiting the [SMoS] slab. Figure 8 shows the two corresponding band schemes from which the electronic stabilization associated with the transition clearly appears. A minimum intercalation rate is certainly needed to initiate the transition in order that the gain in electronic energy overcomes the cost in elastic energy associated with the structural modification. Indeed intercalation usually starts at a critical non-zero value each time a structural modification is involved, even when it is a slight alteration like a simple shifting of slabs relatively to each other (compare Li_xZrS_2 with a modification of the stacking of slabs which starts at x $\simeq$ 0.25 and $LiTiS_2$ with a progressive filling of octahedral sites between slabs which starts from x = 0). The driving force for intercalation is to be found in that favorable competition between a cost of elastic energy to modify the host and a gain of electronic energy.

In deintercalation processes, as well as in intercalation reactions, it is absolutely necessary to check carefully whether the host is, or is not, a rigid framework.

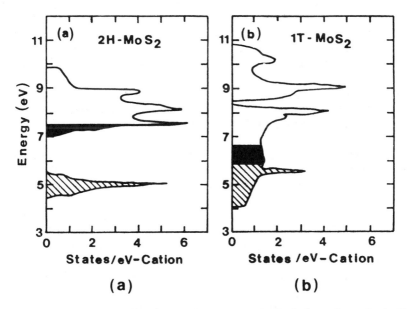

Figure 8 - Intercalating in a "trigonal-prismatic" or "octahedral" MoS$_2$. The figure shows a comparison of d-band density of states for 2H-MoS$_2$ (trigonal prismatic) and hypothetical 1T-MoS$_2$ (octahedral). The hatched region corresponds to occupied states. The blackened region shows the additional states that are filled in 2H and 1T forms when intercalating.

VI - Acido-basic soft chemistry from low-dimensional solids.

The acido-basic behaviour of layered materials is a pleasing aspect of their reactivity. It has lead to spectacular new syntheses such as new forms of TiO_2 *(41)* or WO_3 *(42)*. The method is well illustrated by considering TiO_2. Figure 9 shows the structure of potassium tetratitanate $K_2Ti_4O_9$ with K^+ ions between slabs made up of zig-zag ribbons of four $[TiO_6]$ octahedra. The K^+ ions are exhanged by $(H_3O)^+$ when washing by 3N nitric acid at room temperature. At 60°C $(H_3O)^+$ decomposes with a loss of water and primarily a protonation of the angular oxygen atoms indicated by arrows. Now we have -(OH) groups facing each other. Heating at higher temperature, one gets a further elimination of water and a sealing of the two blocks. This is a new titanium dioxyde, named TiO_2 (B) because it retains the framework of the sodium bronzes of titanium.

It is important to discuss this reaction which is more or less a transposition to solid state chemistry of mechanisms which are known to occur in solution, for example in the process of formation of hetero- or homo poly ions.

The principle of the mechanism is based on a selective protonation of sites according to a gradation of their basicity. This basicity depends on the geometry of the O^{--} position, the nature of the cation, the metal oxygen distance.

Here there is one type of polyhedra which are all centered with the same cation. The angular oxygen atoms represent the most basic sites because their electronic density is the less attracted by the cations that are behind. They represent active sites to be protonated. Now if we could change Ti^{4+} by Nb^{5+}, keeping the same geometry, the basicity of oxygen will considerably decrease. Eventually the corresponding structure will much more easily intercalate ammonia and amines in the starting layered oxides (this is indeed the case of titano-niobates *(43)*.

If tetrahedra are present, the shortening of the metal-oxygen distance, will result in less basic sites as compared to those provided by octahedra. This point sheds some light on the results obtained by Poeppelmeier *(44)* when trying to exchange and compensate Li^+ ions in $LiAlO_2$ by a protonation. $LiAlO_2$ has three structural forms, $\alpha LiAlO_2$ with aluminium octahedra and β and $\gamma LiAlO_2$ built from tetrahedra. Lithium can be exchanged for protons in α, but not in β and γ.

VII - Exfoliation reactions.

Exfoliation represents the extreme limit of deintercalation : when the counter ions stabilizing certain low-dimensional structures are removed one gets a separation of the slabs or fibers which, in a solvent, can lead to a dissolution or to colloïdal suspensions. It is now possible to

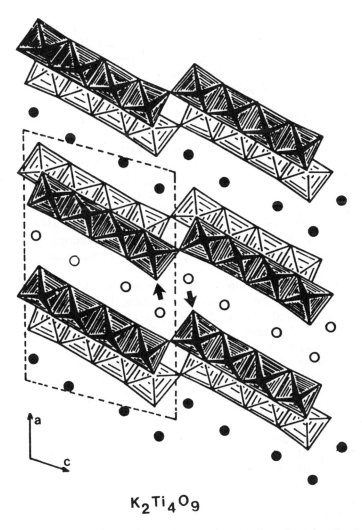

$$K_2Ti_4O_9$$

Figure 9 - Potassium titanate, precursor for the synthesis of TiO_2 (B).

recognize two approaches among the various reactions that have been proposed.

In a first group the stabilizing energy brought by the counter ion is overcome by a solvation energy. Polar solvents such as dimethylsulfoxide (DMSO), N-methylformamide (NMF), propylene carbonate (PC) are strong enough to solvate lithium and cause the swelling of Mo_6Se_6 chains in $Li_2Mo_6Se_6$ (45). Lithium represents the most favorable case from a double point of view. Lithium is small and it is the chain to chain repulsion that largely fixes the chain to chain separation as can be seen by extrapolating from the interchain distances of the other alkalimetal compounds $A_2Mo_6Se_6$. Lithium is "floating" between the chains (figure 10 represents the $A_2Mo_6Se_6$ structural type). On the other hand the smaller size of lithium results in a high polarizing power which leads to high solvation energies. There is no swelling in case of the big counter ions K^+, Rb^+ and Cs^+. $Na_2Mo_6Se_6$ is a borderline for which the presence of macroscopic rod-like particles in contrast to $Li_2Mo_6Se_6$ is remarkable. When a solution is formed it has been proposed that negatively charged $[Mo_6Se_6]$ chains would be surrounded by a cloud of solvated Li^+ ions. Stabillity against aggregation is then a consequence of both the strong repulsive interactions between the external layers that are similarly electrically charged and the cation-solvent affinity.

In the second group an intermediate acido-basic reaction is considered and the process yields neutral slabs. This is well illustrated by the perovskite related layered phases $A[Ca_2Na_{n-3}Nb_nO_{3n+1}]$ (46). These phases (figure 11) show a rather high layer charge ($0.147e^-/nm^2$). Exfoliation of such layered phases with high layer charges is not so easy as in the case of chalcogenides. In addition there is no lithium phase allowing an easy extraction of the cation. Exfoliation is generally achieved by preintercalation of a molecule which increases the interlayer space. The A^+ cations are first replaced by protons in an acidic solution. This leads to solid acids which react with organic bases to form intercalation compounds with large interlayer expansions (47). Individual crystallites begin to exfoliate. The process can be enhanced by using special surfactant molecules having an amine head group and a hydrophilic polyether tail (48). The amine head is protonated while the hydrophilic tail attracts solvent molecules and enhances the interlayer space. The last step is to wash with water (or acetone). The role of water can be envisioned as an hydrolysis reaction in which a protonation of the surface of the slabs leads to neutral slabs (49). The intercalated molecules are washed out as $R\ NH_3OH$.

Conclusion

From observations on already known low dimensional solids a few general rules can be pointed out concerning their stability. These rules

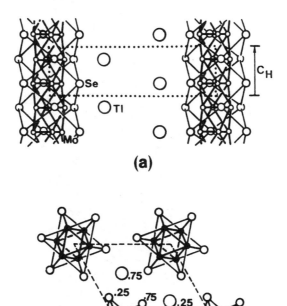

(a)

(b)

Figure 10 - The $A_2Mo_6Se_6$ structural type with A^+ ions between Mo_6Se_6 chains (a) representation along the chain, (b) perpendicular to the chains.

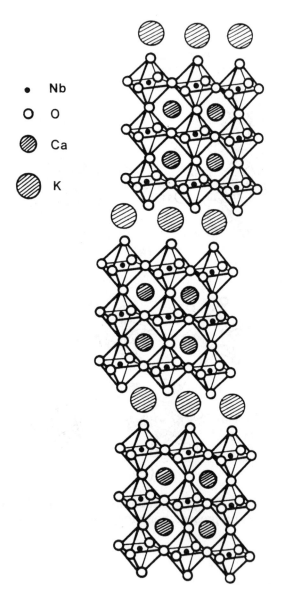

Figure 11 - Layered perovskites with potassium cations stabilizing thick perovskite blocks.

suggest a few synthesis approaches and indicate the best domains in inorganic chemistry where these compounds are to be expected. One cannot predict of course the structural types that will be obtained, but it is possible to a large extent to tune the properties within a given framework.

Soft chemistry routes represent a very powerful tool and the relevant concepts can now be presented. It is largely preoriented by solid state chemistry and leads essentially to metastable phases. However it represents an important thrust in the direction of innovative design of new compouds.

References

(1) L.P. Gorkov, G. Grüner, "Charge Density Waves in Solids", North Holland, 1989.

(2) J. Rouxel, "Crystal Chemistry and Properties of Materials with quasi 1D-structures", D. Reidel, 1986.

(3) P. Monceau, "Electronic Properties of Inorganic quasi 1D compounds", D. Reidel, 1985.

(4) Organic Inorganic L.D. Materials P. Delhaes, M. Drillon Ed. Nato Series B, Vol. 168, Plenum N.Y. 1987.

(5) J. Rouxel, Chem. Scripta, **28**, 33, (1988).

(6) E. Sandré, R. Brec and J. Rouxel, J. Phys. Chem. Solids, **50**(8), 801, (1989).

(7) C. Fouassier, C. Delmas and P. Hagenmuller, Mat. Sci. Eng., **31**, 97, (1977).

(8) A. Wold, W. Kunmann, R.J. Arnott and A. Feretti, Inorg. Chem., **3**, 545, (1964).

(9) Y. Piffard, A. Verbaere, A. Lachgar, S. Deniard and M. Tournoux, Rev. Chim. Min., **23**, 766, (1986).

(10) M.P. Crosnier, D. Guyomard, A. Verbaere and Y. Piffard, Eur. J. Solid State and Inorg. Chem., **26**, 529, (1989).

(11) F. Jellinek, Inorg. Sulfur Chemistry, G. Nickless Ed., Elsevier, Amsterdam (1968).

(12) A. Meerschaut and J. Rouxel, J. Less-Common Met., **39**, 197, (1975).

(13) R. Brec, G. Ouvrard, M. Evain, P. Grenouilleau and J. Rouxel, J. Sol. State Chem., 7, 174, (1983).

(14) R. Brec, P. Grenouilleau, M. Evain and J. Rouxel, Rev. Chim. Miner., **20**, 295, (1983).

(15) R. Brec, M. Evain, P. Grenouilleau and J. Rouxel, Rev. Chim. Miner., **20**, 283, (1983).

(16) M. Evain, M. Queignec, R. Brec and J. Rouxel, J. Sol. State Chem., **56**, 148, (1985)

(17) M. Evain, R. Brec and M.H. Whangbo, J. Sol. State Chem., **71**, 244, (1987).

(18) E. Bjerkelund and A. Kjekshus, Z. Anorg. Allg. Chem., B 328, 235, (1964).

(19) T. Sambongi, K. Tsutsumi, Y. Shiozaki, K. Yamaya and Y. Abe, Sol. State Comm., 22, 729, (1977).

(20) A. Meerschaut, L. Guémas and J. Rouxel, J. Sol. State Chem., 36, 118, (1981).

(21) A. Meerschaut and J. Rouxel in Crystal Chemistry and Properties of materials with quasi 1D structures, cited in (2).

(22) C. Roucau, R. Ayroles, P. Monceau, L. Guémas, A. Meerschaut and J. Rouxel, Phys. Stat. Sol., 62, 483, (1980).

(23) S. Van Smaalen and K.D. Bronsema, Acta Cryst., B42, 43, (1986).

(24) H. Böhn and H.G. Von Schnering, Z. Krist., 162, 26, (1983).

(25) A. Meerschaut, P. Palvadeau and J. Rouxel, J. Sol. State Chem., 20, 21, (1977).
and P. Gressier, A. Meerschaut, L. Guémas, J. Rouxel and P. Monceau, J. Sol. State Chem., 51, 141, (1984)

(26) P. Gressier, M.H. Whangbo, A. Meerschaut and J. Rouxel, Inorg. Chem., 23, 1221, (1984).

(27) see A. Meerschaut and J. Rouxel in (2).

(28) P. Delhaes, C. Coulon, S. Flandrois, B. Hilti, C.W. Mayer, C. Rihs and J. Rivory, J. Chem. Phys, 73, 1452, (1980).

(29) G.A. Bouchard, J.H. Perlstein and M.J. Sienko, Inorg. Chem., 6, 1682, (1987) ;
C. Schlenler in Proc. Int. Conf. on the Physics and Chimistry of Low-dimensional Synthetic Metals, vol. 121, Mol. Cryst. and Liq. Cryst., 1985 ;
R.J. Cava, R.M. Fleming, P. Littlewood, E.A. Rietman, L.F. Schneemeyer and R.F. Dunn, Phys. Rev., B30, 3228, (1984) ;
M. Ganne, A. Boumaza and M. Dion, Mat. Res. Bull., 20, 1297, (1985).

(30) M. Potel, R. Chevrel and M. Sergent, Acta Cryst., B36, 1545, (1980).

(31) W. Houle, H.G. Von Schnering, A. Lipka and K. Yvon, J. Less Common. Met., 71, 135, (1980).

(32) A. Meerschaut, P. Grenouilleau, L. Guémas and J. Rouxel, J. Sol. State Chem., 70, 36, (1987).

(33) D.W. Murphy and P.A. Christian, Science, 205, 651, (1979).

(34) D.W. Murphy, J.N. Carides, F.J. di Salvo, C. Cros, J.V. Waszczak, Mmat. Res. Bull., 12, 827, (1977).

(35) R. Schöllhorn and A. Payer, Angew. Chem. Int. Ed. Eng., 24, 67, (1985).

(36) S. Sinha and D.W. Murphy, Sol. State Ionics, 20, 81, (1986).

(37) C. Delmas in "Chem. Phys. of Intercalation", p. 209, A.P. Legrand and S. Flandrois Ed., Plenum Press, N.Y., (1988).

(38) M.G. Thomas, P.G. Bruce and J.B. Goodenough, Sol. State Ionics, 17, 13, (1985).

(39) L. Blandeau, G. Ouvrard, Y. Calage, R. Brec and J. Rouxel, J. Phys. C, **20**, 4271, (1987).

(40) M A. Py and R.R. Haering, Can. J. Phys., **61**, 76, (1983).

(41) R. Marchand, L. Brohan and M. Tournoux, Mat. Res. Bull., **15**, 1129, (1980).

(42) B. Gérand, G. Nowogrocki, J. Guenot and M. Figlarz, J. Sol. State Chem., **29**, 429, (1979).

(43) H. Rabbah, G. Desjardins and B. Raveau, Mat. Res. Bull., **14**, 1125, (1979).

(44) K.R. Poeppelmeier and S.J. Hwu, Inorg. Chem., **26**, 3297 (1987).

(45) J.M. Tarascon, F.J. di Salvo, C.H. Chen, P.J. Carroll, M. Walsh and L. Rupp, J. Sol. State Chem., **58**, 290, (1985).

(46) M. Dion, M. Ganne and M. Tournoux, Mat. Res. Bull., **16**, 1425, (1985);

(47) A.J. Jacobson, J.W. Johnson, J.T. Lewandowski, Mat. Res. Bull., **22**, 45, (1987).

(48) M.M.J. Treacy, S.B. Rice, A.J. Jacobson and J.T. Lewandowski, Chem. Mat., **2**, 279, (1990).

(49) M. Tournoux, personal communication.

RECEIVED January 16, 1992

Chapter 9

Iron Oxychloride as a Precursor to Lamellar Iron Phosphonates by Soft Chemistry

P. Palvadeau[1], Jean Rouxel[1], M. Queignec[1], and B. Bujoli[2]

[1]Institut des Matériaux des Nantes, Centre National de la Recherche Scientifique UMR No. 110, and [2]Laboratoire de Synthèse Organique, Centre National de la Recherche Scientifique UA No. 475, Université de Nantes, 2 rue de la Houssinière, 44072 Nantes Cedex, France

The behaviour of phosphonic acids toward the lamellar oxyhalide FeOCl is presented as a new route for the synthesis of lamellar iron phosphonates. The action of methyl, ethyl and phenyl phosphonic acids is compared in order to get a better understanding of the different steps involved. In all cases, kinetic compounds [I] $HFe(RPO_3H)_4$ have been isolated that gradually convert to compounds [II] $HFe(RPO_3)_2 \cdot H_2O$. An additional evolution is observed in the case of $HFe(C_2H_5PO_3)_2 \cdot H_2O$, with removal of one ethylphosphonic acid molecule and a corresponding reduction of Fe^{3+} to Fe^{2+}, yielding a new lamellar compound $Fe(C_2H_5PO_3) \cdot H_2O$.
The reactivity of phosphonic acids bearing functional groups with FeOCl was also investigated. The first results obtained for the action of 2-carboxyethylphosphonic acid are presented here.

A great number of investigations concerning the lamellar iron oxychloride FeOCl have been reported (see for example Ref. *1-7*). FeOCl is built up of two corrugated metal-oxygen sheets sandwiched between two layers of labile chlorine atoms (Figure 1). A very interesting feature of this compound is the possibility to realize not only intercalations but also grafting and pillaring via topochemical reactions (*8-10*). Our approach has been to prepare lamellar iron phosphonates by anchoring organophosphonates onto the FeO layers.

It is well known that metal phosphonates have applications in various fields : agriculture (*11*), catalysis (*12*), sorption of heavy metals (*13*) etc. Special interest has been given to layered phosphonates with alternating organic and inorganic layers because of their sorptive and catalytic properties (*14*). Good examples are the intensively studied zirconium phosphates and organophosphonates (*15*). Also vanadium forms a family of layered vanadyl phosphonates (*16,17*). Recently new metal organophosphonates have been obtained $M(II)(RPO_3) \cdot H_2O$ (*18-19-20*).

0097–6156/92/0499–0114$06.00/0

I - Preparation and kinetic relations

Iron oxychloride was prepared from Fe_2O_3 and $FeCl_3$ by the usual sealed-tubes technique. All iron phosphonates were obtained according to the following chemical process : n mmoles of phosphonic acid react with one mmole of FeOCl in 5 mL of solvent in sealed Pyrex tubes. Some reactions have been also realized in round-bottom flasks under nitrogen.

The results indicate that chemical reactions are governed by two main factors : the nature of the solvent and the phosphonic acid / FeOCl molar ratio.

For all three acids RPO_3H_2 (methyl, ethyl and phenyl phosphonic acids), the rate for the consumption of FeOCl increased in the order water < acetone < toluene, dichloromethane. Since these acids are much less soluble in dichloromethane and toluene than in acetone and water, the observed reaction rates can be rationalized by decreased association with the solvent, resulting in increased accessibility to its reactive groups (phosphoryl group P=O and acidic OH), and increased rate of complex formation. On the other hand, the reaction rate (in acetone) according to the R radical increased in the order $C_6H_5 < C_2H_5 < CH_3$. A bulky radical borne by the phosphonic acid is unfavourable to good reactivity, taking into account the probable preliminar intercalation step. Phenyl phosphonic acid has been selected to carry out a systematic study of the influence of different solvents.

From these results, the solvents were classified into three groups that exhibited different rates for the consumption of FeOCl, which can be easily observed due to color change in the starting material from violet to white (Table I).

Table I - Influence of the solvent on the reaction of phenylphosphonic acid with FeOCl

Group	Solvent	progress of the reaction
1	Dichloromethane toluene hexane	very rapid kinetics complete consumption of FeOCl in 30 minutes for CH_2Cl_2, 5 hours for toluene and hexane
2	Acetone Acetonitrile Tetrahydrofuran	slow kinetics, complete consumption of FeOCl in 24 to 48 hours
3	water ethyl alcohol	very slow kinetics incomplete reactions even after 2 weeks

When the molar ratio (S) of phosphonic acid to FeOCl varied with all other parameters fixed, two types of compounds were obtained $HFe(RPO_3H)_4$, [I] and $HFe(RPO_3)_2 \cdot xH_2O$ [II].

In the studied solvents, when S was close to 2, only phases of type [II] were produced whereas the reaction yielded [I] when S was greater than or equal to 4 (Equations 1 and 2).

$$S = 2 \quad FeOCl + 2\,RPO_3H_2 \; \rightarrow \; HFe(RPO_3)_2 \cdot xH_2O \; [II] + HCl \qquad (1)$$

$$S \geq 4 \quad FeOCl + 4\,RPO_3H_2 \; \rightarrow \; HFe(RPO_3H)_4 \; [I] + H_2O + HCl \qquad (2)$$

In the intermediate cases, $2 < S < 4$, a competition between the formation of both species was observed and well followed by room-temperature Mossbauer spectra. The isomer shift value, close to 0.45 $mm.s^{-1}$, is in agreement with the classical results for iron Fe^{3+} in high spin state for both [I] and [II] derivatives.

When the reaction time was prolonged beyond the duration necessary for the complete consumption of FeOCl $(t_2 > t_1)$, $HFe(RPO_3H)_4$ evolved, leading to a compound of type [II]. The transition from type [I] to type [II] (Equation 3) is favoured as the phosphonic acid is soluble in the reaction medium (case of acetone).

$$HFe(RPO_3H)_4 + xH_2O \; \rightarrow \; HFe(RPO_3)_2 \cdot xH_2O \; + 2\,RPO_3H_2 \qquad (3)$$

Consequently, compounds [I] appear to be kinetic products and [II] seem to be thermodynamically stable phases. It is noteworthy that an additional conversion of $HFe(C_2H_5PO_3)_2 \cdot H_2O$ is observed, $(t_3 > t_2)$ leading to a new lamellar derivative $Fe(C_2H_5PO_3) \cdot H_2O$ with a correlative reduction from Fe^{3+} to Fe^{2+} (Equation 4)

$$HFe^{3+}(C_2H_5PO_3)_2 \cdot H_2O \; \rightarrow \; Fe^{2+}(C_2H_5PO_3) \cdot H_2O \qquad (4)$$

All these reactions are summarized Figure 2.

The preparation of our different compounds have been described elsewhere $(R = C_6H_5 \; (21), \; R = CH_3, \; C_2H_5 \; (22), \; R = C_2H_4CO_2H \; (23))$. The synthesis of $Fe(II)(C_2H_5PO_3) \cdot H_2O$ is given as example. The reaction procedure consisted of sealing 107 mg (1 mmol) of FeOCl, 220 mg (2 mmol) of ethylphosphonic acid, and 5 mL of dry acetone in a Pyrex tube and heating the mixture at 80°C for one month. The white crystalline product was filtered off with suction, rinsed with acetone, and dried at room temperature under vacuum. Yield 50%.

II - Compounds of type [I]

Suitable single crystals have been obtained in the case of the phenylphosphonic compound $HFe(C_6H_5PO_3H)_4$. This structure has been solved in the P1 space group (triclinic) and appeared as a quasi one-dimensional compound along the b axis. The FeO_6 octahedra (average Fe-O bond length = 1.99 Å) and $PO_3C_6H_5$ tetrahedra (average P-O bond length = 1.55 Å) are arranged in infinite chains running along

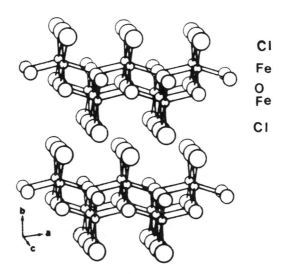

Cl
Fe
O
Fe

Cl

Figure 1 - The FeOCl structural type.

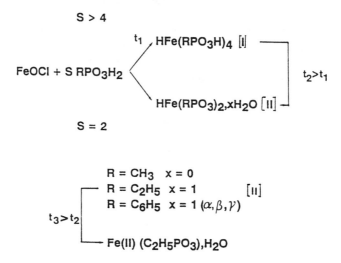

Figure 2 - Kinetic relations between iron phophonates.

the b axis. Each FeO_6 octahedron is connected to six different phosphonate groups. Fig 3 shows that the chains form layers in which the metal atoms are coplanar.

The distance between two metals along the b axis is 5.36 Å, close to the value observed in αZrP for two zirconium atoms linked by O-P-O bridges (5.29 Å) (24). The phenyl groups make Van der Waals contacts along the a axis.

II - Compounds [II] $HFe(RPO_3)_2 \cdot xH_2O$

Trivalent iron phosphonates previously prepared, show the general formula $Fe(RPO_2OR')_3 \cdot xH_2O$ (25-28) with a P/Fe ratio different from the one we have obtained. Two differences appear between methyl, ethyl and phenyl phosphonates $HFe(RPO_3)_2 \cdot xH_2O$. Only one crystallographic form of methyl and ethyl was obtained. $HFe(C_6H_5PO_3)_2 \cdot xH_2O$ was detected as three structural forms α, β, γ. These three phases were obtained in a precise order, only depending on the reaction time. An other hand, there is no water molecule in the methyl compound $HFe(CH_3PO_3)_2$.

Unfortunately no suitable crystals were isolated. However, based on X-ray powder data, infrared and Mössbauer spectroscopies, it is possible to propose a structural model close to the one observed in zirconium phosphate.

In all cases, the white semicrystalline solids gave X-ray powder patterns exhibiting prominent series of h00 diffraction lines, indicative of layered structures in which a is the stacking axis [CH_3: a = 12.09 Å, C_6H_5: a = 15.0 Å (α), 14.66 Å (β), 15.11 Å (γ)]. Some parameters are given Table II.

These results are consistent with the values found in many compounds whose structure is close to that of αZrP. For example, in $Zr(C_6H_5PO_3)_2$ the a parameter value is 15 Å (29).

Moreover, we have shown by kinetic studies, that compounds [I] (21) were precursors to compounds [II] (Figure 4). If the layers are shifted in the (b.c) plane along a gliding vector v with elimination of monodentate phosphonate groups, a structure close to that of αZrP is obtained.

Table II. Crystallographic parameters of iron phosphonates

$HFe(C_6H_5PO_3H)_4$	$Fe(C_2H_5PO_3) \cdot H_2O$	$Fe(HPO_3C_2H_4CO_2H)_2$
Triclinic	monoclinic	orthorhombic
a = 14.968_9Å	a = 4.856_1Å	a = 5.142_1Å
b = 5.36_1Å	b = 10.33_2Å	b = 15.667_1Å
c = 8.678_6Å	c = 5.744_2Å	c = 15.432_1Å
α = 88.5°		
β = 86.41°	β = 91°	
γ = 89.6°		
V = 694Å^3	V = 288Å^3	V = 1243Å^3
Z = 1	Z = 2	Z = 4

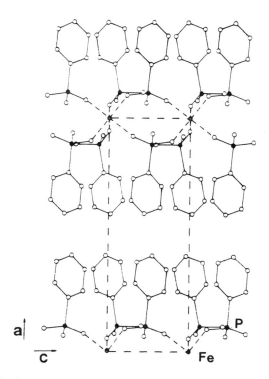

Figure 3 - Structure of HFe(C$_6$H$_5$PO$_3$H)$_4$ viewed down the *c* axis.

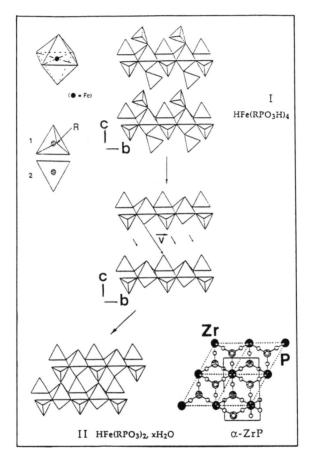

Figure 4 - Schematic illustration of the possible structural change upon transition from [I] to [II].

It seems likely that compounds [II] have layered structures. In the ethyl and phenyl derivatives, two phosphonate anions are probably connected by a weak hydrogen bond as in $H(NdPO_3H)_2 \cdot 2H_2O$ (30).

If in the three phases α, β, γ, the same chemical formula has been established by chemical analysis, differences clearly appear in the Mössbauer and infrared spectra (Figure 5) (Bujoli B. *Eur. J. Sol. State Inorg. Chem.*, in press). The three structures would then differ probably in their interlayer structural arrangement, that is in the location of the water molecule. A confirmation of this hypothesis has been given by a thermogravimetric study. The same compound $HFe(C_6H_5PO_3)_2$ is obtained after the first loss of water at 200°C, characterized by same infrared, X-Ray and Mössbauer spectra, independent of the starting material (α, β, γ).

$Fe(C_2H_5PO_3) \cdot H_2O$. As mentioned above, $HFe(C_2H_5PO_3)_2 \cdot H_2O$ evolved in the reaction medium to give a new phosphonate $Fe(C_2H_5PO_3) \cdot H_2O$. The structure solved in the monoclinic symmetry shows a lamellar character. (Figure 6). The sheets are built up with FeO_6 octahedra. The Fe-O chains are similar to those observed in FeOCl with the same distances. The most interesting feature of this compound is its magnetic behaviour. Figure 7 shows the susceptibility curve $\chi = f(T)$ recorded with different applied magnetic fields using a Faraday balance. Extended magnetic measurements are currently in progress to explain this puzzling evolution. The first results obtained with a SQUID magnetometer are indicative of a metamagnetic state between 25 and 10 K. (31)

Substituted iron phosphonates, the layered $Fe(HPO_3C_2H_4CO_2H)_2$

The previous results led us to investigate the reactivity of other phosphonic acids bearing functional groups. It thus seemed interesting to study the behaviour of the carboxylic function in the reaction process and to know if these groups were accessible to organic compounds for heterogeneous acid catalyzed reactions such as deacetalization. The first results have been obtained with the 2-carboxyethylphosphonic acid.

The structure of $Fe(HPO_3C_2H_4CO_2H)_2$ has been solved in the space group Pbca (orthorhombic symmetry). Figures 8 shows that this phosphonate is layered and that the metal atoms are coplanar. The linkage is such that four O-P-O bridges are in the *a.b* plane, forming a corner-sharing of octahedra and tetrahedra in infinite chains running parallel to the *a* axis. The structure is very closed to that of zirconium phosphate. A similar compound has been recently prepared by Mallouk et al. (20).

Conclusion

These results suggest that the chemical process of the reactions between FeOCl and phosphonic acids consists of a first intercalation step increasing the Van der Waals gap (also observed with other phosphonic acids actually studied). The following

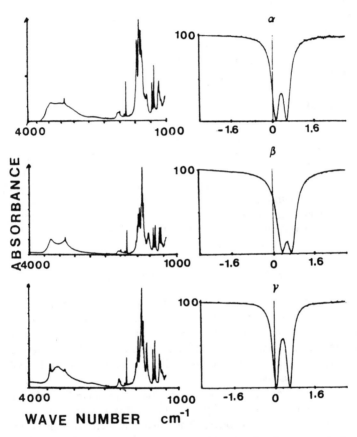

Figure 5 - Room temperature Mössbauer (relative absorption versus velocity en mm.s^{-1}) and infrared spectra of the three phases α, β, and γ HFe(C$_6$H$_5$PO$_3$)$_2$·H$_2$O.

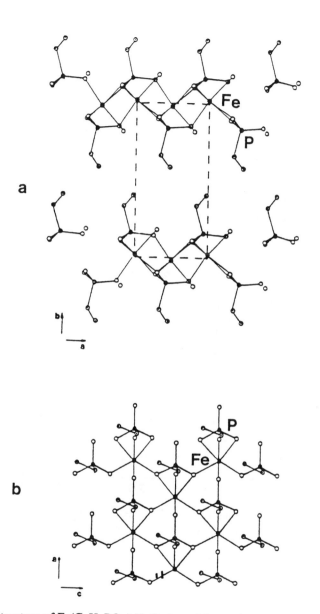

Figure 6 - Structure of $Fe(C_2H_5PO_3) \cdot H_2O$ viewed down the c axis (a) and the a (stacking) axis. (b)

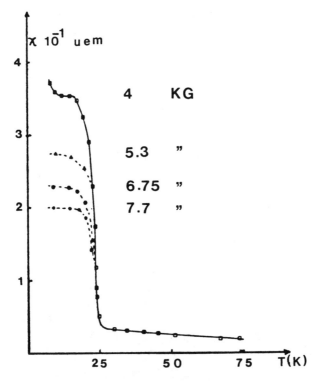

Figure 7 - Magnetic susceptibility of $Fe(C_2H_5PO_3) \cdot H_2O$ between 4 K and 75 K. Curves were recorded at different magnetic fields (in KG).

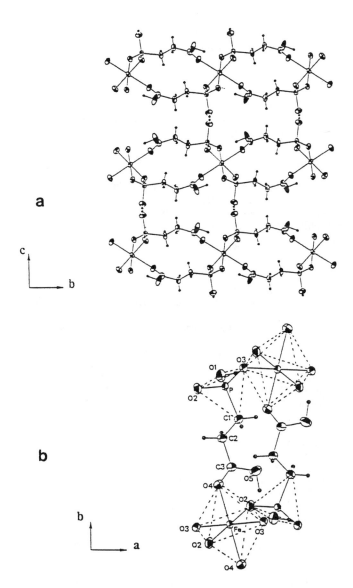

Figure 8 - Fe(HPO$_3$C$_2$H$_4$CO$_2$H)$_2$ (a) Structure viewed down the *a* axis. - (b) Molecular representation and atom numbering scheme

step is a substitution of the chlorine atoms via the phosphoryl group with a breakdown of the Fe-O-Fe network.

We have tried to obtain the same complexes starting from other materials : such as Fe_2O_3, $FeCl_3$, $Fe_2(SO_4)_3$. Following the same procedures, reactions give the same compounds, but are slower than in the case of FeOCl. Probably the layered structure and the labile character of the Fe-Cl bond play a great role in the kinetics.

Acknowledgments

We acknowledge the Commission of European Communities for support of this work under grant MA1E-0030-C.

Literature cited

(1) - Goldstaub, S. *Bull. Soc. Franc. Miner.* **1935**, *58*, 49
(2) - Lind, M.D. *Acta Cryst.*, **1970**, *B 26*, 1058
(3) - Kanamaru, F.; Yamanaka, S.; Koizumi, M.; *Chem. Letters* **1974**, 374
(4) - Hagenmuller, P.; Portier,J.; Barbe, B.; Bouclier, P. *Z. Anorg. Allg. Chemie* **1967**, *355*, 209
(5) - Rouxel, J.; Palvadeau, P. *Rev. Chim. Miner.* **1982**, *19*, 317
(6) - Herber, R. H.; Maeda, Y. *Physica (B+C)* **1981**, *105*, 243
(7) - Kanatzidis, M.G.; Wu, C.G.; Marcy, H.O.; DeGroot, D.C.; Kannewurf, C.R.; Kostikas, A.; Papaefthymiou, V. *Adv. Materials.* **1990**, *2(8)*, 364
(8) - M. Armand, M.; Coic, L.; Palvadeau, P.; Rouxel, J. *J. Power Sources* **1978**, *3*, 137
(9) - Villieras, J.; Chiron, R.; Palvadeau, P.; Rouxel, J. *Rev. Chim. Min.* **1985**, *22*, 209
(10) - Kikkawa, S.; Kanamaru, F.; Koizumi, M. *Inorg. Chem.* **1980**, *19*, 259
(11) - Chalandon, A.; Crisinel, P.; Horriere, D.; Beach. D. *Proc. Br. Crop. Prot. Conf. Pest. Dis.* **1979**, *2*, 347
(12) - DiGiacomo, P.M.; Dines, M.D. *Polyhedron*, **1982**, *1*, 61
(13) - Tsvetaeva, N.E.; Rudaya, D. Y.; Fedosseev, D.A.; Ivanova, L.A.; Shapiro, K.Y. *USSR patent* N° 623392
(14) - Johnson, J.W.; Jacobson, A.J.; Brady, J.F.; Lewandowski, J.I. *Inorg. Chem.* **1984**, *23*, 3842, and references therein
(15) - Alberti, G.; Constantina, U.; Alluli, S.; Tomassini, J. *J. Inorg. Nucl. Chem.* **1978**, *40*, 1173
(16) - Huan, G.; Jacobson, A.J.; Johnson, J.W.; Corcoran Jr., E.W. *Chem. Mat.* **1990**, *2*, 91
(17) - Johnson, J.W.; Brody, J.F.; Alexander, R.M.; Pilarski, B.; Katritsky, A.R. *Chem. Mat.* **1990**, *2*, 198
(18) - Cao, G; Lee, H.; Lynch, V.M.; Mallouk, T.E.; *Inorg. Chem.* **1988**, *28*, 2781
(19) - Ortiz-Avila, Y.; Rudolf, P.; Clearfield, A. *Inorg. Chem.* **1989**, *27*, 2137
(20) - Cao, G.; Rabenberg, L.K.; Numm, C.N.; Mallouk, T.E. *Chem. Mat.* **1991**, *3*, 149
(21) - Bujoli, B.; Palvadeau, P.; Rouxel, J. *Chem. Mat.* **1990**, *2(5)*, 582
(22) - Bujoli, B.; Palvadeau, P.; Rouxel, J. *C.R. Acad. Sci. Paris* ;**1990**, *310 (série II)*, 1213
(23) - Bujoli, B.; Courilleau, A.; Palvadeau, P.; Rouxel, J. *Europ. J. Solid State and Inorg. Chem* (accepted) **1991**, 9
(24) - Clearfield, A.; Smith, G.D. *Inorg. Chem.* **1969**, *8*, 341

(25) - Mikukski, C.M.; Karyannis, N.M.; Minkiewicz, J.V.; Labes, M.M.;
Pytlewski, L.L. *Inorg. Chim. Acta*, **1969**, *3*, 523
(26) - Kodama, Y.; Kodama, T.; Nakabayashi, M.; Semoura, M.;. Kiba, Y. *Japan
Patent N°770324* (cl co7f9/40), appl 75/79351, **1975**
(27) - Klein, J.M.; Kokalas, J.J. *Spectroscop. Lett.* **1972**, *5*, 497
(28) - Fedoseev, D.A.; Nefedov, V.F.; Shesterikov, E.G.; Teterin, N.N.; Filin, V.M.
Zh. Neorg. Khim. **1979**, *24*, 3023
(29) - Dines, M.B.; DiGiacomo, P.M. *Inorg. Chem.* **1981**, *20*, 92
(30) - Loukili, M.; Durand, J.; Cot, L.; Rafiq, M. *Acta Crystal.* **1988**, *C44*, 6
(31) - Palvadeau, P.; Bujoli, B.; Pena, O. *Sol. State Com.* (to be published)

RECEIVED January 16, 1992

Chapter 10

Pillared Layered Materials

Abraham Clearfield and Mark Kuchenmeister

Department of Chemistry, Texas A&M University,
College Station, TX 77843

The structure of several types of layered compounds which behave as
ion exchangers will be described. These include clays, group 4 and 14
phosphates, titanates and antimonates. The techniques of pillaring used
for each type will be presented along with a description of the structure
and porous nature of the products. Potential applications as catalysts,
sorbants and ion exchangers will be discussed and predictions for the
future of this field of research will be presented.

A great deal of progress has been made in the field of pillared layered materials since
our last review article *(1)*. Two monographs have appeared *(2-3)* as well as many
notes and papers. However, since this is by way of a tutorial, a brief introduction to
the field and review of earlier work is in order.

Along with the oil embargo in 1973 the United States experienced a sharp rise
in oil prices. This price increase acted as a stimulus to develop catalytic materials
which would process heavy crude oils. Normal processing of these crudes, such as
coking and hydroprocessing, is relatively expensive but the cost could be
considerably reduced using conventional fluid catalytic cracking (FCC). Zeolites are
normally used as FCC catalysts, but these materials have relatively small pore
openings (~8 Å for Zeolite Y) and are unsuitable without expensive preprocessing of
the feed.

Clays and layered silicate minerals have a long history of use as petroleum
cracking catalysts *(4)*. However, only the outer surface of the clays are useful in
catalysis as the molecules cannot penetrate between the layers at the temperatures
used for cracking. It has also been known that certain classes of organic molecules,
such as amines and alkylammonium ions, readily intercalate between clay layers, *(5)*
and that porous structures can be formed from such host-guest complexes *(6)*. These
organically pillared structures suffer from the thermal instability of the organic
component. In 1974 Lussier, McGee and Vaughan *(7)* initiated work leading to the
synthesis of inorganically pillared clays *(8)*. The general procedure is to incorporate
a large inorganic cation between the layers to prop them open *(8-9)*. The props or
pillars are then thermally cross-linked to the layers. This propping makes the inner
surfaces of the clay available for catalytic purposes.

0097–6156/92/0499–0128$06.00/0

Before discussing these pillaring reactions, we will examine some aspects of the clays themselves.

Description of Smectite Clays

The principal class of swelling clays is termed smectites.   These clays characteristically have alumina octahedra sandwiched between silica tetrahedra as shown in Figure 1. The smectites are distinguished by the type and location of cations in the layered framework. In a unit cell formed from twenty oxygens and four hydroxyl groups, there are eight tetrahedral sites and six octahedral sites. Idealized formulas are given in Table I.

Table I.   Idealized Structural Formulae for Principal Smectite Clays

Name	Formula
Montmorillonite:	$M_{X/N}^{N+}$ $[Al_{4-X}Mg_X]$ (Si_8) O_{20} $(OH)_4$ • yH_2O
Hectorite:	$M_{X/N}^{N+}$ $[Mg_{6-X}Li_X]$ (Si_8) O_{20} $(OH)_4$ • yH_2O *
Beidellite:	$M_{X/N}^{N+}$ $[Al_4]$ $(Si_{8-X}Al_X)$ O_{20} $(OH)_4$ • yH_2O
Saponite:	$M_{X/N}^{N+}$ $[Mg_6]$ $(Si_{8-X}Al_X)$ O_{20} $(OH)_4$ • yH_2O

*Some F^- may substitute for $(OH)^-$

When all the octahedral sites of hectorite are filled with Mg^{2+}, a magnesium talc of ideal formula $[Mg_6](Si_8)(OH)_4O_{20}$ is the result. Pyrophyllite is like talc except that 2/3 of the octahedral sites are filled by Al^{3+}. In smectites, substitutions of Mg^{2+} for Al^{3+} and Li^+ for Mg^{2+} can take place in the octahedral sites, or M^{3+} ions can substitute for Si^{4+} in the tetrahedral sites. The deficiency of positive charge is compensated by the presence of hydrated cations between the layers *(10)*. Typically, the positive charge deficiency in smectites ranges from 0.4 to 1.2 e per Si_8O_{20} unit *(11)*. If the charge deficiency arises from octahedral substitution, then this excess negative charge is distributed over all the oxygens in the framework. These clays tend to be <u>turbostratic,</u> that is, the layers are randomly rotated about an axis perpendicular to the layers. However should the smectite have been formed by substitution at tetrahedral sites, then the extra charge is more localized and the resultant clay tends to exhibit greater three dimensional order *(12)*.

The remarkable properties of the smectite clays are due to their relatively low layer charge. The interlayer cations are loosely held and therefore swelling via solvent sorption, ion exchange and intercalation of a variety of species is possible.

Ion Exchange Behavior of Smectite Clays

The term "intercalation" should be reserved for the incorporation of neutral species between the layers and the term "ion exchange" when there is a one for one exchange of species based on charge. The ion exchange capacity (IEC) of smectite clays ranges from about 0.5 to 1.5 meq/g. These capacities are low in comparison with those of sulfonated styrene-divinylbenzene resins (IEC ~ 5 meq/g) or a-zirconium phosphate, $Zr(HPO_4)_2 \cdot H_2O$ (6.6 meq/g) *(13)*. Given the fact that the interlamellar surface is about 750 m^2/g, univalent ions would be about 8.3 Å apart in the interlayer space at an IEC of 1 meq/g.

An important property of the smectites is their ability to swell by intercalation of water or alcohols. The extent of the swelling depends upon the layer charge, the interlayer cation and the nature of the swelling agent *(14)*. When the hydration forces of the interlayer cation are strong, and the layer charge low, the clay is most susceptible to swelling in water. Under the right conditions, the layers can be separated by hundreds of Angstroms. If such a dispersion is concentrated, gelation usually occurs. The gel is believed to have a "house-of-cards" structure (Figure 2) resulting from layer edge to face interactions *(15)*. This swelling behavior allows large complex ions, which can be used for catalytic purposes, to be exchanged into the interlamellar spaces. For additional information on complexes in clays the reader is referred to a review article by Pinnavaia *(16)*.

Pillared Clays (PILCS)

As was mentioned in the introduction, the driving force for the preparation of PILCS is the desire to synthesize temperature-stable porous materials with large pores. For this purpose, inorganic polymeric species were chosen since they would form oxide particles on heating, which would act as props for the clay layers, as shown in Figure 3.

The initial work was done with either aluminum or zirconium polymers *(8-9,18-19)* . In the case of the aluminum pillaring, two types of solutions were used: an aluminum chloride solution to which was added up to 2.33 mol of NaOH per mol of Al (9), and a commercially available (Reheis Chemical Co., Chlorhydrol·) solution in which Al(OH)$_3$ or Al metal is dissolved in aluminum trichloride solutions, until the OH/Al ratio is equal to or slightly less than 2.5. Such solutions have been reported to contain polynuclear species with 6 to 400 aluminum atoms per structural unit.[20,21] However, the exact form of these species is unknown.[22] It has now been shown (see below) that in both types of solutions a major polymeric species is the Keggin ion, $[Al_{13}O_4(OH)_{24} \cdot 12H_2O]^{7+}$, *(23)*. However, NMR data have revealed some interesting differences between the two types of solution *(24)*. An aluminum chloride solution initially exhibits a single resonance characteristic of the $Al(H_2O)_6^{3+}$ ion *(25)*. As base is added to this solution, a resonance at 62.8 ppm increases at the expense of a resonance at near zero ppm indicative of the trivalent aluminum species. The new resonance is attributed to the Al_{13} Keggin ion, and when the OH/Al ratio is 2.42, it is the only species observed by NMR. In contrast, the Chlorhydrol· solution (OH/Al = 2.50), which was diluted from a 6.2 **M** solution, exhibited a small peak for the Keggin ion, a large resonance for $Al(H_2O)_6^{3+}$ and a broad resonance at 10.8 ppm attributed to species of higher nuclearity.

In all of the early pillaring studies irrespective of the type of aluminum solution used, the interlayer spacing of the smectite increased from approximately 9.3 Å to 18-19 Å *(8-9,26)*. This result is consistent with the incorporation of the Al_{13} Keggin ion, which has the shape of a prolate spheroid with a long axis of about 9.5 Å. The pillaring is illustrated (somewhat naively) in Figure 3. Hydrolysis may change the charge on the Al_{13} Keggin ion. This in turn increases the loading of the pillar into the clay. Evidence for this charge lowering was obtained by the analysis of products obtained in the pillaring of a Bentonite as a function of pH. Based on the analytical data a plot of Al_{13} charge variation with pH of the solution was developed (Figure 4) *(7)*. However, it appears that no systematic study of PILC properties as a function of pillar content has been carried out.

On heating, the Keggin ion loses protons to the layers to balance the negative layer charge as the pillar is transformed into an oxide particle *(8,21)*. Surface areas of

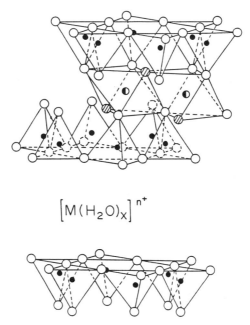

$$\left[M(H_2O)_x\right]^{n+}$$

Figure 1. Schematic drawing of smectite clay: ◑: octahedrally coordinated ion; ●: tetrahedrally coordinated ion; O: oxygen, ∅: OH group.

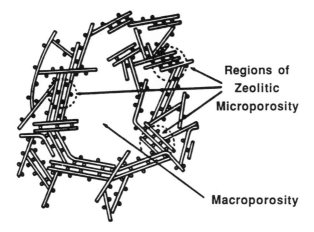

Figure 2. Schematic representation of a "house of cards" arrangement of a pillared clay. Ideal drying conditions would have all the layers lying parallel. (Reproduced with permission from ref. 17. Copyright 1984 T. J. Pinnavaia.)

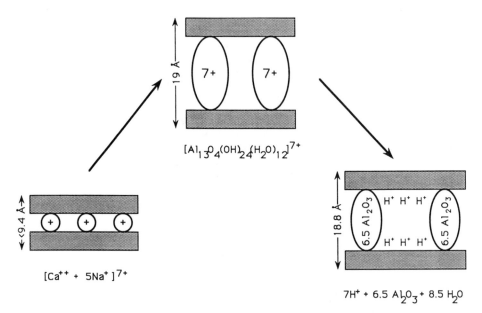

Figure 3. Schematic representation of the formation of a pillared clay. (Reproduced with permission from ref. 26. Copyright 1980 Wiley).

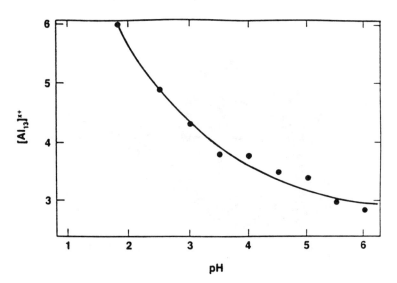

Figure 4. Hydrolysis of $[Al_{13}]$ as a function of pH to give lower charge polymers. (x^+) was determined from chemical analysis of the final PILC after the exchange reaction. (Reproduced with permission from ref. 7. Copyright 1988 Elsevier Science Publishers.)

200-300 m^2/g have been observed for the PILCS *(7,24)*. Simultaneous thermal gravimetric analysis and surface area measurements show that about 50% of the water is lost at temperatures below 175 °C; this is considered to be surface water. As the temperature is increased to 500 °C, there is very little loss of surface area. At higher temperature, there is rapid loss of surface area accompanied by about a 1-Å loss in interlayer spacing, and at 700 °C there is a total collapse of the structure *(35)*.

Vaughan et al. *(8,26)* reported pore size distributions for two clays pillared by aluminum oxycations. Their results are shown in Table II.

Table II. Pore Size and Surface Area Distribution for Pillared Clays and Zeolites

Pore size Range, Å	% N$_2$ Surface Area			Amorphous Catalyst
	PILC	PILC	NaY	
> 100	1.1	0.6	2.4	1.1
100 - 80	0.5	0.2	0	3.0
80 - 60	1.0	0.4	0	19.3
60 - 40	16.5	19.0	0	47.5
40 - 30	4.3	7.5	0	23.1
30 - 20	0.7	0	0	6.0
20 -14	14.1	18.3	0	0
< 14	61.9	64.0	97.6	0
Average pore diameter, Å	23.6	21.9	18.2	51.4
Total surface area, m^2/g	477	373	860	575

Adapted from ref. 8.

In both examples, the majority of pores were less than 14 Å in diameter. However, there were significant numbers of pores between 14 and 20 Å and between 40 and 60 Å. The larger pores indicate some similarity to the amorphous catalyst. However, at low sorption pressures negligible amounts of large molecules are sorbed (Table III) *(26,27)*.

Table III. Molecular Sieve Properties of PILC

Probe Molecule	Size (Å)	Sorption Pressure (torr)	Amount Sorbed(wt.%)
n-Butane	4.6	600	8.0
Cyclohexane	6.1	60	8.4
Carbon Tetrachloride	6.9	106	11.5
1,3,5-trimethylbenzene	7.6	9.4	5.3
1,2,3,5-tetramethylbenzene	8.0	6.9	0
Perfluorotributylamine	10.4	31.0	0

Adapted from ref. 26.

Sorption data obtained by Occelli et al. *(28,29)* are shown in Table IV. Although the amount of straight chain hydrocarbons sorbed decreases with increasing chain length, the total volume occupied by the sorbed molecules is roughly constant and amounts to 60-70% of the total pore volume.

Table IV. Saturation Loadings on ACH-bentonite

Sorbate	Liquid densities (g cm^{-3})	Liquid sorbed (cm^3 g^{-1})
n-Hexane	0.6532	0.1099
n-Heptane	0.6795	0.1048
n-Octane	0.6966	0.1072
n-Nonane	0.7114	0.0981
n-Decane	0.7077	0.0806

Reproduced from ref. 28

One would imagine that the pore size of a particular pillared clay might depend upon the initial layer charge of the unpillared clay. Further, it is known that the charge distribution in smectite clays is highly irregular, *(30,31)* which increases the expectation of non-uniform pillaring. However, it has been demonstrated that the pillars fill each interlayer region to essentially the same extent regardless of the layer charge, and the exchanging polynuclear hydroxo species apparently adjusts its charge by hydrolysis *(17,24)*. This finding would appear to be at odds with the results obtained by Vaughan *(7)* indicating a highly variable uptake of pillars.

NMR Studies. Only aluminum pillared clays have been examined to any extent by NMR techniques *(32)*. Two separate cases need to be considered: clays in which the ion of lower charge substitutes isomorphously in the octahedral layer and those clays where the substitution occurs in the tetrahedral layer. Only the $^{27}Al^{IV}$ resonance is observed for the Al_{13} Keggin ion in solution at +62 ppm (with respect to $Al(H_2O)_6^{3+}$). The octahedral Al^{VI} resonance is not observed. Within the clay layers the Al^{VI} resonance of the Keggin ion was clearly seen at ~6-7 ppm as a result of immobilization of the Al_{13} species. The ratio Al^{IV}/Al^{VI} indicated that the pillars were indeed the Al_{13} Keggin ion.

Upon calcination in the temperature range where the pillars dehydroxylate (300-400 °C), they are not transformed into spinel like particles *(32)*. In fact, up to 400 °C no drastic changes are observed in the ^{29}Si or ^{27}Al NMR spectra of pillared montmorillonites. However, the pillars do become fixed and can no longer be exchanged out. Protons that split out from the pillars behave as dictated by the electrostatic forces of the layer which drive them towards the sites of negative charge density. For pillared beidellite, in which Al^{3+} substitutes in the tetrahedral sites, calcination at temperatures up to 350 °C produced a drastic change in the sharpness and intensity of the ^{27}Al resonance for the tetrahedrally coordinated aluminums both in the clay layers, Al_s, and the pillar, Al_p. This spectral change, shown in Figure 5, is interpreted in terms of a linkage between the layer and the pillar *(33)*. Since the Al_{13} Keggin ion is acidic, it transfers protons to the layers upon calcination. The most likely point of attack is at an Al-O-Si bridge, as shown in eq (1), with formation of a silanol group.

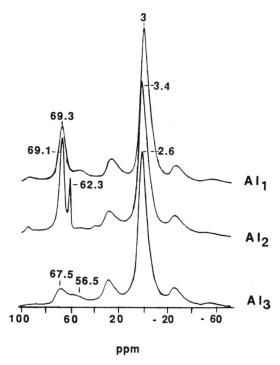

Figure 5. 11.7-T 27Al MAS spectra, corresponding to a resonance frequency of 130.3 MHz, of beidellite (B; Al_1), pillared beidellite (PB; Al_2), and calcined pillared beidellite (CPB; Al_3). (Adapted with permission from ref. 32. Copyright 1988 Elsevier Science Publishers.)

(1)

Subsequent reaction with the pillar could lead to either

where the Roman numerals denote the coordination of the Al and the subscripts s and p their location in the clay or pillar, respectively. The formation of silanol or Al-OH groups in pillared beidellite makes it a much stronger acid than corresponding montmorillonites and accounts for its catalytic behavior in cracking *(35)* and hydroisomerization reactions *(33)*.

The NMR spectral changes in fluorohectorite also indicate that a cross-linking reaction has occurred *(34)*. In both beidelite and fluorohectorite it is suggested *(33,34)* that the cross-linking operates through inversion of some clay lattice tetrahedra as shown in Figure 6. In beidelite the center of the inverted tetrahedron would be Al^{3+} since the negative charge is concentrated there and $\overset{O}{\underset{Al \quad Al}{/\backslash}}$ cross-linking occurs. The inversion in fluorohectorite requires some explanation since isomorphous substitution of Li^+ for Mg^{2+} occurs in the octahedral clay layer. Here it is expected that the charge would be distributed evenly over the oxygens along the periphery of clay layer. It has been suggested that since lithium ion in fluorohectorite is surrounded by 2 F^- and 4 O^{2-}, the two most electronegative elements, that the Li^+ is highly ionic and thus the charge imbalance is concentrated near the lithium ions. This charge concentration is responsible for the inversion of Si tetrahedra near the pillars and subsequent $\overset{O}{\underset{Si \quad Al}{/\backslash}}$ cross-linking.

Thomas et al. *(36,37)* have also examined both pillared and unpillared clays by a variety of physical methods including NMR, FTIR and variable temperature powder X-ray diffraction. The clays studied were gilwhite, a naturally occurring montmorillonite, and hectorite. In the unpillared clays it was shown that small cations such as Li^+, or cations which generate protons at high temperature (NH_4^+, $Al(H_2O)_x^{3+}$), show structural changes on heating to 500 °C which are attributed to the diffusion of H^+ or Li^+ into the sheets. This effectively lowers the charge in the interlayer, resulting in reduced Bronsted acidity. At 700 °C octahedrally coordinated aluminum is converted to 5-coordinate aluminum. Similarly, in the pillared clays the Bronsted acidity is also drastically reduced by diffusion of protons, generated from hydrolysis of the pillars, into the octahedral layer. At approximately 500 °C, hydroxyl groups on the pillars condense with lattice hydroxyls on the clay sheets. The oxide pillars then become directly linked via oxygen to the aluminum and magnesium atoms in the octahedral layer.

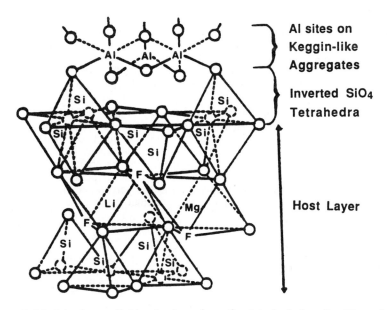

Figure 6. Model of cross-linking by inversion of a tetrahedral unit. (Reproduced with permission from ref. 32. Copyright 1988 Elsevier Science Publishers.)

Pillaring by Cations Other Than Aluminum

A variety of polynuclear cations other than the Al_{13} Keggin ion have been used to pillar clays. Among them, we cite the zirconium tetramer *(8,19,27,38,39)*, chromia polymers *(40)*, bismuth *(41)*, silicon *(42)*, and niobium and tantalum halide clusters *(43)*. Of these, the most thoroughly studied have been zirconium polymers. The cation species in the salt $ZrOCl_2 \cdot 8H_2O$ is actually a tetramer, $[Zr(OH)_2 \cdot 4H_2O]_4^{8+}$. The zirconium atoms form a square and are connected by double -ol (hydroxyl) bridges *(44)*. The same species has been shown to exist in solution in a more hydrolyzed form *(45)*. Boiling of zirconyl chloride solutions leads to the formation of colloidal, crystalline oxide particles *(46)*. These particles grow by -ol bridging between tetramers *(47-49)*, forming a sheet-like structure *(50)*. Boiling for shorter periods yields a polydisperse system with average degrees of polymerization of 30-40 *(51)*.

Recent discussions of polymeric species of Zr(IV) solutions as well as Al(III) solutions are available *(52,53)*. Burch et al. *(38,39)* prepared zirconium pillared montmorillonite. They obtained three different pillared products with gallery heights of 3-4 Å, 6.3 - 7.4 Å and 9.6 - 10.5 Å. By gallery height we mean the increase in interlayer spacing resulting from pillaring. Refluxing the Zr(IV) solution and dispersing the clay prior to pillaring produced PILCS with larger interplanar spacings, higher surface areas and greater stability. Surface areas were generally 200-300 m^2/g. On the basis of the small increase in interlayer spacing (3-4 Å), it was proposed that the tetramers lie parallel to the clay layers in these PILCS. This result contrasts with the results of Yamanaka and Brindley *(19)*, who obtained an interlayer increase of 10 Å with about 10% Zr uptake. From this, Burch and Warbutron *(38)* concluded that the tetramer species, the dimensions of which are 4.6 Å x 10 Å x 10 Å, must stand perpendicular to the layers in the Yamanaka preparations. The free space in such pillared structures would arise only when the pillars are dehydrated *(54)*. In any event whether the pillars are parallel or perpendicular, the zirconium PILCS appear to have higher thermal and hydrothermal stability than the aluminum PILCS *(54)*.

By careful control of the pillaring conditions, Farfan-Torres and Grange *(55)* were able to obtain highly stable (to 700 °C) zirconia pillared clays with surface areas as high as 280 m^2/g. Chromium pillared clays represent an interesting case: Products with interlayer spacings of 20-27 Å were obtained by the use of chromium nitrate solutions which had been heated for different lengths of time at OH/Cr = 2 *(40)*. Surface areas ranged from 353-433 M^2/g. All of these PILCS showed reduced basal spacings on heating in air and total collapse between 300 and 400 °C. However, by heating in vacuum their stability was maintained to above 500 °C. This added stability is attributed to the non-oxidizing conditions at reduced pressure *(56)*. Appreciable amounts of perfluorotributylamine, kinetic diameter = 10.4 Å, were sorbed by the chromia PILCS.

Layered Double Hydroxides. Layered double hydroxides (LDH) are complex hydroxides of general formula $M_n^{2+}M^{3+}(OH)_{2+2n}X$, where n = 2-5 and X is a charge balancing anion. The anions are located in the interlamellar space but do not fill it. Variable amounts of water reside in the pores or cavities created by the anion positioning. The LDH may be considered as the inverse of the swelling smectite clays because the layers are positively charged, balanced by anions, whereas the clays have negatively charged layers balanced by cations in the interlamellar region. Only in the case of Mg-Al and Ni-Al LDH can the full range of n values be attained. In most other systems n = 2 or in some cases 3 also. There are two attractive reasons for wanting to pillar the layered double hydroxides. First, they have already been shown to possess interesting catalytic behavior *(57,58)* and second, there are an enormous number of large anions which can be used for pillaring. These include the isopoly- and heteropolyanions *(59)*. Drezdon *(60)* reported pillaring $Mg_2Al(OH)_6X^-$ by first replacing the halide ion by phthalate ion. He was then able to exchange out the organic anion with $V_{10}O_{28}^{6-}$. Shortly thereafter Pinnavaia et al. *(61)* were able to show that the LDH $Zn_2Al(OH)_6Cl$ was able to directly intercalate the $V_{10}O_{28}^{6-}$ ion. Further studies by these workers demonstrated that the same LDH was able to exchange heteropoly anions such as $[H_2W_{12}O_{40}]^{6-}$ and a-$[SiV_3W_9O_{40}]^{7-}$ but not polyoxymetallates of lower charge *(62)*. Surface areas of 63-155 m^2/g with pore volumes of 0.023 - 0.061 cm^3/g were reported.

In a most recent communication Dimotakis and Pinnavaia *(63)* reported a new route for the pillaring of LDH. $Mg_3Al(OH)_8(CO_3)_{0.5}\cdot2H_2O$ was first heated at 500 °C to obtain the mixed oxides. The oxide was in turn stirred in water for 30 h to reconstitute the layered double hydroxide with the hydroxide ion as the anion. By adding glycerol to the mix, the layers swell and are able to incorporate large organic acids as anions. These in turn were exchanged out for polyoxometallate anions. The field of pillared LDH would appear to hold great promise for future applications *(57,64)* however, a more general method of pillaring them is required.

Layered Phosphates. The phosphates of groups 4 and 14, with the a-structure, are layered *(65)* and bear a resemblance to clay layers. The general formula is $M(IV)(HPO_4)_2\cdot H_2O$ and the layers consist of a central octahedrally coordinated metal layer sandwiched between tetrahedral phosphate layers. The major difference from the clays is the fact that the phosphate groups are inverted. Three oxygens of each phosphate group bridge across metal atoms and the fourth points into the interlamellar space. The ion exchange capacity of these solids depends upon the mass of the metal. For example, the exchange capacity of the Zr compound is 6.6 meq./g, a value 4-12 times that of a clay. All of this charge is concentrated in the oxygens not bonded to metal, which results in covalent bonding with the protons *(66)*. Exchange of Na^+ for the protons occurs with some expansion of the layers, but in general no swelling occurs for the crystalline phases *(67)*. Because of the lack of swelling of this class of compounds, they did not lend themselves to pillaring reactions. However, it was demonstrated *(68)* that large charged species or complexes could be exchanged into zirconium phosphate if first the layers were spread apart by amine intercalation. Shortly thereafter, we succeeded in pillaring both zirconium and titanium phosphates with the Al_{13} Keggin ion, via the amine intercalation reaction *(69)*.

Strangely, titanium phosphate (TiP) yielded porous products with surface areas of 35-237 m^2/g while the zirconium phosphate (ZrP) was non-porous. Sorption of molecules with increasing kinetic diameters demonstrated the presence of micropores of ~ 7 Å diameter. TiP was found to lose phosphate ion during pillaring which lowers the layer charge and allows less pillar to be inserted. Tin phosphate has recently been pillared with chromium polymers *(70)*. Here also it is suspected that phosphate hydrolysis may be responsible for the porous nature of the products. MacLachlan and Bibby *(71)* have pillared a-zirconium phosphate by first preparing the half-exchanged sodium phase. This phase has an expanded interlayer spacing of 11.8 Å and in this condition could be pillared by a basic chromium acetate polymer. As with pillaring by the aluminum Keggin ion, the interlamellar space was stuffed, i.e., non-porous. However, in our laboratory we were able to pillar amine intercalated ZrP with chromium acetate *(72)* and achieve a surface area of 119 m^2/g.

A general discussion of pillaring of phosphates has been published *(73)* as well as descriptions of organically pillared materials *(1,73,74)*. Very recently we have pillared antimony phosphates of composition $HSb(PO_4)_2$ *(75)* and $H_3Sb_3P_2O_{14}$ *(76)*. The former compound has a structure that is isomorphous with ZrP but contains half the hydrogen ion. It was pillared with the aluminum Keggin ion so as to satisfy the total layer charge and also approximately half the charge, i.e., $H_{0.5}Sb(PO_4)_2[Al_{13}O_4(OH)_{24}(H_2O)_{12}]_{1/14}$. Interlayer spacings of 17-18 Å were observed as compared to an initial interlayer spacing of ~8 Å for the anhydrous proton phase. Surface area and pore size measurements are in progress *(77)*.

Pillaring of Oxides.   Cheng and Wang *(78)* were able to pillar $H_2Ti_4O_9$ by a variant of the Clearfield method.   After intercalation of hexylamine, the hexylammonium ion was exchanged by tetramethylammonium ion. In this condition, the Al_{13} species could be inserted between the layers. Although high surface areas were observed, no pore size measurements were obtained. However, the pores were estimated (from N_2 sorption isotherms) to be ~10 nm in average diameter. It is altogether possible that these authors missed the presence of micropores in their products.

Perovskite Layered Compounds.   Recently, Dion et al. *(79)* synthesized a new family of layered oxides with the general formula $ACa_2Nb_3O_{10}$ (A = K, Rb, Cs, Tl). The layers consist of perovskite slabs separated by A^+ ions as shown in Figure 7. Subsequently, Jacobson et al. *(80)* were able to increase the layer thickness by substitutions of the type $K[Ca_2Na_{n-3}Nb_nO_{3n+1}]$, $3 \leq n \leq 7$. The layer thickness increases by approximately 3.9 Å for each integral increase in the value of n. These compounds were converted to solid acids in 6M HCl and subsequently shown to intercalate amines *(81,82)*. This behavior makes them excellent candidates for pillaring because the layers are expected to be relatively rigid or stiff and robust as compared to clays or phosphates. Furthermore, the proton may be highly acidic because of the high charge on Nb^{5+}. We prepared the following new compounds in addition to $KCa_2Nb_3O_{10}$: $KCa_3Nb_3TiO_{13}$, $KCa_4Nb_3Ti_2O_{16}$, $KCa_2Sr_{0.5}Nb_3Ti_{0.5}O_{11.5}$ and $KCa_2Sr_{0.25}Nb_3Ti_{0.25}O_{10.75}$. The first two compounds were prepared by incorporating one and two moles of $CaTiO_3$, respectively, into the parent potassium calcium niobate *(82)*. All of the compounds were converted to the acid form in 6 M HCl and subsequently intercalated with either butylamine or hexylamine. Exchange of the alkyl ammonium salt for the aluminum Keggin ion $[Al_{13}O_4(OH)_{24}(H_2O)_{12}]^{7+}$ was carried out using chlorohydrol solutions at 50 °C. These compounds gave surface areas of about 8 m^2/g indicating a lack of pores. In all of the above cases, the amount of Al_{13} Keggin ion incorporated was close to $\frac{1}{7}$ of the charge on the host compound. Calculations show that the interlayer space is crowded and a reduction in the amount of pillar incorporated is required.

The foregoing examples are sufficient to show that a broad range of layered compounds are amenable to pillaring. Thus, a vast new group of materials can be synthesized which show promise for use as catalysts, sorbants, ion exchangers, chemical sensors and host of other uses.[3] Before this promise becomes a reality, we need to learn how to control the reactions to obtain pores of specific size and stability and relate structure to the observed properties of the pillared layered materials.

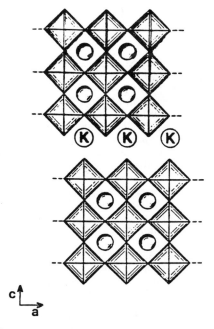

Figure 7. Schematic representation of the structure of $KCa_2Nb_3O_{10}$. (Reproduced from ref. 80. Copyright 1985 American Chemical Society.)

Acknowledgment. Our work on pillared layered materials has been supported by NSF grant DMR-88012283 and the State of Texas through the Advanced Research Program for which grateful acknowledgement is made. The impetus and some continuing support has been supplied by Amoco Chemical Company and Texaco, U.S.A.

Literature Cited

1. Clearfield, A. In Surface Organometallic Chemistry: Molecular Approaches to Surface Catalysis, J. M. Basset, Eds.; Publishers: Kluwer Academic 1988; pp 198-271.
2. *Catalysis Today;* R. Burch, Ed.; Elsevier Amsterdam, 1988. Vol. 2.
3. *Pillared Layered Structures;* I. V. Mitchell, Ed.; Elsevier Appl. Sci., N.Y. 1990.
4. Ryland, L. B.; Tamele, M. W.; Wilson, J. N. In *Catalysis* Emmett, P. H. Ed.; Reinhold Publ. Co, New York 1960
5. Theng, B. K. G., The Chemistry of Clay-Organic Reactions; John Wiley & Sons; New York 1974.
6. Barrer, R. M. Zeolites and Clay Minerals as Sorbents and Molecular Sieves, Academic Press, New York 1978.
7. Vaughan, D. E. W. in Ref. 2, p 187.
8. Vaughan, D. E. W.; Lussier, R. J.; McGee, J. S. US Pat. 4 176 090, 1979; 4 248 739, 1981; 4 271 043, 1981 to W. R. Grace, Co.
9. Lahav, N.; Shani, V.; Shabtai, J. *Clays and Clay Minerals* **1978,** *26*, 107.
10. Bailey, S. W. In *Crystal Structures of Clay Minerals and Their X-ray Identification*; Brindley, G. W.; Brown, G. Mineral Society, London 1980.
11. Beson, G.; Mifsud, A.; Tchoubar, C.; Mering, J. *Clays and Clay Minerals* **1974,** *22*, 379.
12. Suquet, H.; Calle de la, C.; Pezerat, H. *Clays and Clay Minerals* **1975,** *23*, 1.
13. Clearfield,A.; Stynes, J.A. *J. Inorg. Nucl. Chem.* **1964,** *26*, 117.
14. Mortland, M. M. *Trans. 9th Int. Congr. Soil Sci.* **1968,** *1*, 691.
15. Van Olphen, H.*An Introduction to Clay Colloid Chemistry* ; John Wiley & Sons, New York 1977, 2nd Ed.
16. Pinnavaia, T. J. *Science* **1983,** *220*, 365.
17. Pinnavaia, T. J.; In *Heterogeneous Catalysis* ; Shapiro, B. Ed.; Texas A&M University Press; College Station, Texas 1984.
18. Brindley, G. W.; Semples, R. E. *Clays and Clay Minerals* **1978,** *26,* 229.
19. Brindley, G. W.; Yamanaka, S. *Clays and Clay Minerals* **1978,** *26,* 21.
20. Hem, J. D.; Robertson, C. E.; *U. S. Geol. Surv. Water-supply Pap.* **1967,** *1827-A*.
21. Smith, R. W.; Hem, J. D. *U. S. Geol. Surv. Water-Supply Pap.* **1972,** *1827-D.*.
22. Patterson, J. H.; Tyree Jr., S. Y. *J. Colloid Interface Sci.* **1973,** *43,* 389.
23. Johansson, G. *Acta chem. Scand.* **1960,** *14*, 769.
24. Pinnavaia, T. J.; Tzou, M.-S.; Landau, S. D.; Raythatha, R. H. *J. Mol. Catal.* **1984,** *27*, 195.
25. See Akitt, J. W.; Farthing, A. *J. Magn. Reson.* **1978,** *32*, 345; *J. Chem. Soc., Dalton Trans.* **1981,** 1617; **1981,** 1624 for further details on these assignments.
26. Vaughan, D. E. W.; Lussier, R. J. *5th Int. Conf. on Zeolites*, Naples, Italy, Heyden, London, 1980.
27. Lussier, R. J.; Magee, J. S.; Vaughan, D. E. W. *7th Canadian symp. Catal.*, Edmonton, Alberta, Oct. 19-22 1980.
28. Occelli, M. L.; Innes, R. A.; Hure, F. S. S.; Hightower, J. W. *J. Appl. Catal.*, **1985,** *14*, 69.

29. Occelli, M. L.; Parulekar, V.; Hightower, J. *Proc. 8th Int. Congr. Catal.,* Berlin, 1984.
30. Stul, M. S.; Mortier, W. J. *Clays and Clay Minerals* **1974,** *22,* 391.
31. Peogneur, P.; Maes, A.; Cremers, A. *Clays and Clay Minerals* **1975,** *23,* 71.
32. Fripiat, J. J. in Ref. 2, p 281 ff.
33. Plee, D.; Borg, F.; Gatineau, L.; Fripiat, J. J. *J. Am. Chem. Soc.* **1985,** *107,* 2362.
34. Pinnavaia, T. J.; Landau, S. D.; Tsou, M.-S.; Johnson, I. D.; Lipsicas, M. *J. Am. Chem. Soc.* **1985,** *107,* 7222.
35. Shabtai, J.; Lazar, R.; Oblad, A. G. *New Horizons Catal.* **1981,** *7,* 828.
36. Tennakoon, D. T. B.; Thomas, J. M.; Jones, W.; Carpenter, T. A.; Ramdas, S. *J. Chem. Soc., Faraday Trans. 1* **1986,** *82,* 545.
37. Tennakoon, D. T. B.; Jones, W.; Thomas, J. M.; *J. Chem. Soc., Faraday Trans. 1,* **1986,** *82,* 3081.
38. Burch, R.; Warburton, C. I. *J. Catal.***1986,** *97,* 503.
39. Bartley, G. J. J.; Burch, R. *J. Appl. Catal.* **1985,** *19,* 175.
40. Pinnavaia, T. J.; Tsou, M. S.; Landau, S. D. *J. Am. Chem. Soc.* **1985,** *107,* 4783.
41. Yamanaka, S.; Yamashita, G.; Hattori, M. *Clays and Clay Minerals* **1980,** *28,* 281.
42. Endo, T.; Mortland, M. M.; Pinnavaia, T. J. *Clays and Clay Minerals* **1980,** *28,* 105.
43. Christiano, S. P.; Wang, J.; Pinnavaia, T. J. *Inorg. Chem.* **1985,** *24,* 1222.
44. Clearfield, A.; Vaughan, P. A. *Acta Crystallogr.,* **1956,** *9,* 555.
45. Muha, G. M.; Vaughan, P. A. *J. Chem. Phys.* **1960,** *33,* 194.
46. Clearfield, A. *Inorg. Chem.* **1964,** *3,* 146.
47. Clearfield, A. *Rev. Pure Appl, Chem.* **1964,** *14,* 91.
48. Rijnten, H. In *Physical and Chemical Aspects of Adsorbents and Catalysts*; Lensen, B. G. , Ed.; Academic Press: New York, 1970.
49. Clearfield, A.; Nancollas, G. H.; Blessing, R. H. in *Ion Exchange and Solvent Extraction;* Marinsky, J. A.; Marcus, Y., Eds.; Marcel Dekker: New York, 1973, Vol. 5, Ch. 1.
50. Fryer, J. R.; Hutchinson, J. L.; Paterson, R. *J. Colloid Interface Sci.* **1970,** *34,* 238.
51. Johnson, J. S.; Kraus, K. A. *J. Am. Chem. Soc.* **1956,** *78,* 3937.
52. Jones, S. L. ref. 2, Ch. 3.
53. Sterte, J. ref. 2, Ch. 4.
54. Bartley, G. J. J. ref. 2 Ch. 5.
55. Farfan-Torres, E.M.; Grange, P.*J. Chem. Phys.* **1990,** *87,* 1547.
56. Tzou, M. D.; Pinnavaia, T. J. ref. 2, Ch. 6.
57. Reichle, W. T. *Chemtech,* Jan. 1986, p 58.
58. Reichle, W. T. *J. Catal.* **1985,** *94,* 547.
59. Pope, M. T. *Heteropoly and Isopoly Oxometallates;* Springer-Verlag: New York, 1983.
60. Drezdon, M. A. *Inorg. Chem.* **1988,** *27,* 4628.
61. Kwon, T.; Tsigdinos, G. A.; Pinnavaia, T. J. *J. Am. Chem. Soc.* **1988,** *110,* 3653.
62. Kwon, T.; Pinnavaia, T. J. *Chem. Mat.* **1989,** *1,* 381.
63. Dimotakis, E. D.; Pinnavaia, T. J. *Inorg. Chem.,* **1990,** *29,* 2393.
64. Jones, W.; Chibwe, M. in *Pillared Layered Structures;* Mitchell, I. V., Ed.; Elsevier: Amsterdam, 1990.
65. Clearfield, A. *Comments Inorg. Chem.* **1990,** *10,* 89.
66. Troup, J. M.; Clearfield, A. *Inorg. Chem.* **1977,** *16,* 3311.
67. Clearfield, A.; Duax, W. L.; Smith, G. D.; Thomas, J. R. *J. Phys. Chem.* **1969,** *73,* 3424.

68. Clearfield, A.; Tindwa, R. M. *Inorg. Nucl. Chem. Lett.* **1979,** *15,* 251.
69. Clearfield, A.; Roberts, B. D. *Inorg. Chem.* **1988,** *27,* 3237.
70. Maueles-Torres, P; Olivera-Pastor, P.; Rodriguez-Castellon, E.; Jimenez-Lopez, A.; Tomlinson, A. A. G. ref. 3, p 137.
71   MacLachlan, D. J.; Bibby, D. M. *J. Chem. Soc., Dalton Trans.* **1989,** 895.
72. Kuchenmeister, M.; Clearfield, A., work in progress.
73. Tomlinson, A. A. G., ref. 3, p 91.
74. Alberti, G.; Costantino, V.; Marmottini, F.; Vivani, R.; Zappelli, P. ref. 3, p 119.
75. Piffard, Y.; Oyetola, S.; Courant S,; Lachgar, A. *J. Solid State Chem.* **1985,** *60,* 209.
76. Piffard, Y.; Verbaere, A.; Lachgar, A.; Courant-Deniard, S.; Tournoux, M. *Rev. Chim. Mineral* **1986,** *23,* 766.
77. Wade, K.; Clearfield, A. work in progress.
78. Cheng, S.; Wang, T. C. *Inorg. Chem.* **1989,** *28,* 1283.
79. Dion, M.; Ganne, M.; Tournaux, M. *Mat. Res. Bull.* **1981,** *16,* 1429.
80. Jacobson, A. J.; Johnson, J. W.; Lewandowski, J. T. *Inorg. Chem.* **1985,** *24,* 3727.
81. Jacobson, A. J.; Johnson, J. W.; Lewandowski, J. T. *J. Less-Common metals* **1986,** *116,* 137.
82. Ram, R. A. Mohan; Clearfield, A. *J. Solid State Chem.* submitted.

RECEIVED January 16, 1992

Chapter 11

Organoclay Assemblies and Their Properties as Triphase Catalysts

Chi-Li Lin, Ton Lee, and Thomas J. Pinnavaia

Department of Chemistry and Center for Fundamental Materials
Research, Michigan State University, East Lansing, MI 48824

Organo hectorite clays interlayered with quaternary ammonium ions containing one long alkyl chain and three short chains exhibit a basal spacing (~18 Å) indicative of a bilayer assembly in which the axis of the long chain lies parallel to the silicate sheets. For hectorite derivatives in which the onium ion contains two or more long alkyl chains, more expanded "trimolecular" structures are adopted, as judged from basal spacings of ~21 Å. Both types of organo clay assemble as thin, membrane-like films at the interface of a water/toluene emulsion. The organo clay-stabilized emulsions are exceptionally active catalysts for the conversion of an alkyl bromide to the corresponding nitrile. In addition, a variety of other nucleophilic displacement processes can be catalyzed under triphasic reaction conditions in the presence of organo hectorites, including alcohol oxidations, C-alkylations of nitriles, and the conversion of alkyl bromides to thiocyanates, sulfides, and ethers.

Smectite clay minerals are members of a broad class of layered silicates with 2:1 mica-like layer lattice structures. As shown in Figure 1, the negative charge on the silicate layers is balanced by interlayer cations that occupy intercalated positions on the gallery surfaces. Alkali metal and alkaline earth cations in the pristine minerals can be replaced by almost any desired cation. Particularly novel wetting properties are obtained when the exchange ions are replaced by cationic surfactants and related long-chain alkyl ammonium ions. Such ion exchange forms are generally referred to as "organo" clays.

The work of Lagaly[1] has demonstrated that alkyl ammonium ions intercalated in smectite clay galleries form ordered assemblies in which the alkyl chains and the onium head groups adopt specific orientations with respect to the host layers. The structure of the ordered assemblies depends in part on the length of the alkyl chains and the charge density of the silicate host layers. Four different structural assemblies have been identified for smectite clays intercalated by quaternary ammonium ions that

0097–6156/92/0499–0145$06.00/0

contain one long-chain alkyl group and three-short chains groups. These include mono- and bi-layer structures in which the long alkyl chain axis lies parallel to the silicate layers, a tri-molecular structure in which the alkyl chains are kinked, and a paraffin-type structure in which the alkyl chain is inclined with respect to the silicate layers. These structures are schematically illustrated in Figure 2. The mono-layer structure is favored when the clay layer charge is low (~0.4 to 0.5 e$^-$ per $O_{20}(OH_4)$ unit) and the alkyl chain length is 18 to 10 carbon atoms long. As the steric congestion within the gallery increases with increasing layer charge, bi-layers assemblies are formed. As the charge increases still further, tri-molecular and paraffinic structures may be realized.

We have found that some of the organo clay assemblies shown in Figure 2 can exhibit novel properties as triphase catalysts for a wide range of nucleophilic displacement reactions.[2] In the present work we report the triphase catalytic properties of a series of alkyl ammonium ion exchange forms of hectorite. Among the members of the smectite families of clays, this mineral has a relatively low layer charge. Consequently, most quaternary ammonium ions containing one long and three short alkyl chains adopt bi-layer structures when intercalated in the clay galleries. However, tri-molecular assemblies can be obtained when two or more long-chain alkyl groups are present in the onium ion. These bi-layer and tri-molecular structures represent the two most promising organo clay assemblies for triphase catalytic applications.

EXPERIMENTAL

Materials. Naturally occurring sodium hectorite from Hector, California, with a particle ≤ 2 μm was obtained from the source clay mineral repository, University of Missouri, Columbia. The cation exchange capacity, as determined by the displacement of ammonia from an ammonium saturated sample using NaOH and an ammonia specific electrode, was 73 meq per 100 grams of air-dried clay. All reagents and solvents used in this work were obtained from Chemical Dynamics Corporation and Aldrich Chemical Company and were used without further purification.

Preparation of Organo Hectorites. Hectorite SHCa-1 was obtained from the Clay Mineral Society Repository at the University of Missouri, Columbia. The mineral was purified by sedimentation and subsequent treatment with $NaHSO_x$ solution to remove carbonates. A 1.0 wt. % aqueous suspension of sodium hectorite (100 mL, 0.73 meq) was mixed with a 0.073 M solution of the chloride or bromide salt of the desired onium ion (20 mL, 1.46 mmol) dissolved in either water or ethanol. The products were then washed free of excess onium salt with ethanol as determined by a bromide or chloride specific electrode. The clay was then resuspended in de-ionized water, centrifuged and air dried at room temperature.

Phase Transfer Catalysis. The reactions of pentyl bromide with potassium cyanide under triphase reaction conditions were carried out as follows. The air-dried organo clay was dispersed in 3.0 mL aqueous solution of potassium cyanide in a 15 x 150 millimeter Pyrex culture tube equipped with a magnetic stirring bar and fitted with a Teflon screw cap. After the mixture was stirred for a few hours, 2.0 mmol of pentyl bromide and 0.5 mL n-decane as a chromatographic standard in 6.0 mL of toluene

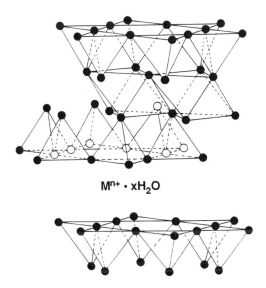

$M^{n+} \cdot xH_2O$

Figure 1. A schematic representation of the oxygen positions defining the mica-like layer lattice structure of a smectite clay. In hectorite, the tetrahedral positions are occupied by silicon; magnesium and lithium occupy octahedral positions. The gallery cations in the pristine mineral are alkali metal or alkaline earth cations.

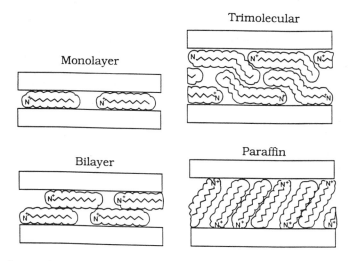

Figure 2. Structures adopted by assembled alkyl ammonium ions in smectite clay galleries. After Lagaly[1].

were added. The tubes were immersed in an oil bath maintained at 90 ± 0.5 °C and the reaction mixtures were stirred vigorously through the use of a magnetic stirrer. Reaction rates were determined by withdrawing 1 µL samples from the upper organic layer at various time intervals and analyzing the products by GLC. Analogous reaction procedures were used for the organo clay catalyzed conversion of pentyl bromide to pentyl phenol ether, the oxidation of benzyl alcohol, the C-alkylation of benzyl cyanide, the dehalogenation of <u>vis</u>-dibromides, alkyl halide exchange reactions and the synthesis of n-pentyl thiocyanate and dipentyl sulfide.

All product mixtures were analyzed by gas-liquid chromatography on a Hewlett Packard model 5880A chromatograph equipped with a flame ionization detector and a 25 m x 0.25 mm cross-linked dimethyl silicone column with GB-1 liquid phase and a 0.25 micrometer film thickness. Products were identified by comparing retention times with those for authentic samples. The percentage yields of products were determined by dual channel integration of the chromatographic peaks relative to an internal standard of n-decane.

X-ray basal spacings were determined for oriented film samples using a Phillips x-ray diffractometer and Ni-filtered Cu-K_α radiation. The samples were prepared by allowing an aqueous suspension of the clay to evaporate on a microscope glass slide. Diffraction was monitored over a 2Θ range from 2° to 45°.

RESULTS AND DISCUSSION

The hectorite clay used in this study has an idealized composition of $Na_{0.67}[Mg_{5.33}Li_{0.67}]$ $(Si_{8.0})O_{20}(OH, F)_4 \cdot x\ H_2O$. Replacing the gallery Na^+ ions in the pristine mineral by ion exchange reaction with an excess of a quaternary ammonium ion containing three short and one long alkyl chain results in a homoionic intercalate with a basal spacing of 18.0Å. Figure 3 illustrates the diffraction pattern for an oriented film sample of the $[C_{16}H_{33}NMe_3]^+$ exchange form of the mineral. Since the van der Waals thickness of the host layer is ~9.6Å, the gallery height is ~8.4Å. This value is in agreement with the height expected for a self-assembled bilayer of onium ions (<u>cf.</u>, Figure 2).

A series of hectorites interlayered with benzyl ammonium ions of the type [($\underline{n}$-$C_{14}H_{29})NMe_2(CH_2C_6H_5)]^+$, $[(\underline{n}-C_{16}H_{33}NMe_2(CH_2C_6H_5)]^+$, and $[(\underline{n}$-$C_{18}H_{37})NMe_2(CH_2C_2H_5)]^+$ also exhibited basal spacings of 18.2 ± 0.1 Å, indicative of an intercalated bilayer of onium ions. The absence of an appreciable increase in basal spacing upon increasing the alkyl chain length by four methylene units confirms the horizontal bilayer structure for the intercalated onium ions. However, quaternary ammonium ions containing two or more long alkyl chains exhibit basal spacings larger than the bilayer value. For instance, $[(\underline{n}-C_{12}H_{25})_2NMe_2]^+$ - and $[(\underline{n}$-$C_8H_{17})_3NMe^+$ - hectorites exhibit basal spacings of 20.1 and 21.0Å, respectively. These values approach the spacing of ~22 Å expected for pseudo trimolecular assembly of onium ions.

In a triphase catalytic system, reagents from two immiscible liquid phases are transferred to the interface of a solid where they react. The products formed at the interface are then returned to the liquid phases in which they are soluble. Typically, for a nucleophilic displacement reaction carried out under triphase conditions an

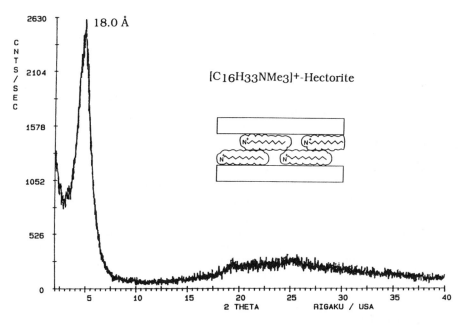

Figure 3.   X-ray diffraction pattern (Cu-Kα) for an oriented film sample of $[C_{16}H_{33}NMe_3]^+$ - hectorite.

aqueous phase contains the nucleophilic reagent and an organic phase accommodates the substrate. Functionalized organic polymers[3-9] and a variety of inorganic supports[10-13] containing immobilized organo groups have been utilized as catalysts for nucleophilic displacement reactions. Although triphase catalysts are easily recovered and convenient to use, their practical utility is often limited by poor diffusion properties, low mechanical strength or chemical and thermal instabilities[14-16].

Quaternary ammonium ion hectorites with bilayer and pseudo trimolecular structures exhibit two fundamental properties that make them attractive as potential triphase catalysts. Firstly, both structural forms are wetted by water, as well as by organic liquids such as toluene. Consequently, these organoclays stabilize oil/water-type emulsions by forming thin, membrane-like clay particle assemblies at the aqueous liquid/organic liquid interface. This feature is especially attractive for catalysis of nucleophilic displacement reactions under triphase conditions. Clay mediated emulsion formation provides a means of bringing ionic nucleophilic reagents that are soluble in water into contact with electrophilic reagents contained in the organic liquid phase.

A second important property of our organo hectorites is their inertness toward Na^+ and K^+ salts of nucleophilic reagents under reaction conditions. The lack of ion exchange reactivity restricts the clay-catalyzed nucleophilic reactions to the clay-liquid interface. Since little or no onium ion leaves the clay mineral surface, nucleophilic reactions by a biphasic reaction mechanism, wherein the onium ion is partitioned between the two liquid phases, is precluded. The emulsion forming ability of an organo hectorite most likely is related to the amphiphilic nature of onium ions on the <u>external</u> surfaces of the clay tactoids. By adopting parallel or perpendicular orientations at the clay-liquid interface, the onium ions may impart hydrophilic or organophilic wetting properties. Consequently, the clay exhibits the novel ability to assemble as a thin, membrane-like film at the aqueous liquid/organic liquid interface. This emulsion-forming property is associated with exceptional triphase catalytic activity for nucleophilic displacement reactions.

The conversion of pentylbromide to the corresponding nitrile according to equation 1 was used to probe the triphase catalytic activity of organo hectorites. Table I provides pseudo first order rate constants, k_{obs},

$$\underline{n}\text{-}C_5H_{11}Br_{(org)} + CN^-(aq) \rightarrow \underline{n}\text{-}C_5H_{11}CN_{(org)} + Br^-(aq) \qquad (1)$$

for pentyl bromide to pentyl cyanide conversion for several organo hectorite triphase catalysts. Included in the Table are the k_{obs} values for the same onium ion when utilized as a conventional biphase catalyst under equivalent reaction conditions in the absence of the clay support. For each organo clay reaction system, the reaction ceased when the clay was removed by filtration, verifying that catalysis occurs at the solid-liquid interface and not by desorption of the onium ion from the surface. An average value of $k_{obs} = 0.16 \pm 0.01h^{-1}$ was obtained for the seven related organo clay derivatives, suggesting that a similar interfacial interaction occurs for operates in each system. Both $[(\underline{n}\text{-}C_{14}H_{29})NMe_3]^+$ - and $[(\underline{n}\text{-}C_{16}H_{13})NMe_3]^+$ - hectorite exhibited a two-fold increase in reactivity relative to the corresponding biphase systems, but the five remaining clays were 2.3 to 6.1 times less reactive than the biphasic analog.

Table I.   Pseudo First Order Rate Constants (90°C) for the Cyanation of n-Pentyl Bromide in the Presence of a Hectorite-Supported Triphase Catalyst and Under Biphasic Reaction Conditions [a]

Cation	Hectorite-Supported Triphasic Catalyst, k_{obs}, hr^{-1}	Unsupported Biphasic Catalyst, k_{obs}, hr^{-1}
$[n - C_{14}H_{29}NMe_3]^+$	0.14	0.07
$[n - C_{16}H_{33}NMe_3]^+$	0.16	0.07
$[n - C_{14}H_{29}NMe_2(CH_2C_6H_5)]^+$	0.16	0.38
$[n - C_{16}H_{33}NMe_2(CH_2C_6H_5)]^+$	0.16	0.37
$[n - C_{18}H_{37}NMe_2(CH_2C_6H_5)]^+$	0.16	0.42
$[n - (C_{12}H_{25})_2NMe_2]^+$	0.16	0.54
$[n - (C_8H_{17})_3NMe]^+$	0.18	1.10

[a]All reactions were carried out at 90° C under the following conditions: 2.0 mmol of n- pentylbromide in 6.0 mL of toulene; 20 mmol of KCN in 3.0 mL of water; 0.073 mol of clay-bound onium ion.

A loss in reactivity upon onium ion immobilization is a general feature of triphase catalysis. The decrease in reactivity for onium ions immobilized on hectorite is comparable to the loss in activity observed for onium ions tethered to silica and alumina surfaces through the use of organosilane coupling agents[10].   However, organo clays are chemically much more robust than silane modified silica surfaces and can be utilized through numerous catalytic circles.   For instance, we have recycled $[(n-C_{16}H_{33})NMe_3]^+$- hectorite through more than 70 catalytic turnovers for the cyanation of n-pentyl bromide at 90°C before the reaction rate decreased to half of its initial value. Another practical advantage of organo clays is their facile synthesis by simple ion exchange reaction from readily available starting materials. Emulsion formation is not a serious disadvantage, because the emulsions can be easily broken by centrifugation or by filtration.

In addition to being catalysts for triphasic cyanation reactions, organo hectorite assemblies catalyze a variety of other nucleophilic substitution reactions. Equations 2-6 below illustrate the diverse range of reactions that can be catalyzed by $[(n-C_8H_{17})_3NMe]^+$- hectorite.   Reaction conditions and chemical yields for each conversion are provided in Table II.

Table II.   Nucleophilic Displacement Reactions Carried Out Under Triphase Reaction Conditions in the Presence of $[(\underline{n}\text{-}C_8H_{17})_3NMe]^+$- Hectorite

Substrate	Nucleophilic Reagent	Substrate Onium Ion Mole Ratio	Time/ Temp	Product	Yield
$\underline{n}\text{-}C_5H_{11}Br$	C_6H_5OH (3 mol)/ 2.5 $\underline{M}$ NaOH	100	1.5h/90°	$\underline{n}\text{-}C_5H_{11}OC_6H_5$	83%
$\underline{n}\text{-}C_5H_{11}Br$	NaSCN (10 mol)	200	1.5h/90°	$\underline{n}\text{-}C_5H_{11}SCN$	99%
$\underline{n}\text{-}C_5H_{11}Br$	NaS (6 mol)	100	0.5h/90°	$(\underline{n}\text{-}C_5H_{11})_2S$	91%
$\underline{n}\text{-}C_5H_{11}Br$	$C_6H_5CH_2CN$ (19 mol) 50% NaOH	100	65h/50°	$(\underline{n}\text{-}C_5H_{11})CH(CN)C_6H_5$	86%
$C_6H_5CH_2OH$	10%NaOCl (7 mol)	40	10h/50°	C_6H_5CHO	83%

$$\underline{n} - C_5H_{11}Br + C_6H_5OH \xrightarrow[-H_2O]{OH^-} \underline{n} - C_5H_{11}OC_6H_5 + Br^- \qquad (2)$$

$$\underline{n} - C_5H_{11}Br + SCN^- \rightarrow \underline{n} - C_5H_{11}SCN + Br^- \qquad (3)$$

$$\underline{n} - C_5H_{11}Br + S^{2-} \rightarrow (\underline{n} - C_5H_{11})_2S + Br^- \qquad (4)$$

$$\underline{n} - C_5H_{11}Br + C_6H_5CH_2CN \xrightarrow[-H_2O]{OH^-} (\underline{n} - C_5H_{11})CH(CN)C_6H_5 + Br^- \qquad (5)$$

$$C_5H_{11}CH_2OH \xrightarrow{OCl^-} C_6H_5CHO \qquad (6)$$

It is important to emphasize that the catalytic activity observed for the organo hectorites reported in this study is closely linked to the ability of these materials to stabilize organic/aqueous liquid emulsions. Emulsion formation seems to be correlated with the ability of the clay to adopt bilayer and trimolecular alkylammonium ion assemblies within the gallery region. Organo hectorites that form only monolayer onium ion assemblies do not form emulsions and are poor triphase catalysts. For instance, ($\underline{n}$ - Bu)$_4$N$^+$ - hectorite with a basal spacing of ~15 Å adopts a monolayer-type structure, does not stablize a toluene/water emulsion, and is a relatively poor triphase catalyst, even though the ion is a very good biphase catalyst for nucleophilic displacements reactions. We should note in this regard that Cornélius and Laszlo[11] first utilized an alkylammonium clay as a catalyst for the formation of formyl acetals from dihalomethanes and alcohols under strongly alkaline conditions. The organo clay utilized was a commercial organo montmorillonite, Thixogel VP®. All organo clays are likely to catalyze a nucleophilic displacement reaction under triphase reaction conditions. However, the results obtained here for organo hectorites indicate organo clays capable of stabilizing emulsion formation are exceptionally active triphasic catalysts. Other quaternary ammonium intercalation compounds, such as those formed by layered phosphates, titanates, vanadates and niobates, might also form interfacial assemblies and exhibit triphase catalytic properties.

Acknowledgment

The support of this research by the National Science Foundation through grant DMR-8903579 and by the Michigan State University Center for Fundamental Materials Research is gratefully acknowledged.

References

1. Lagaly, G. *Solid State Ionics* **1986**, *22*, 43.
2. Lin, C.-L.; Pinnavaia, T. J. Chem. Mater. **1991**, *3*, 213.
3. Regen, S. L. *Nov. J. Chem.* **1982**, 639.
4. Regen, S. L. *J. Am. Chem. Soc.* **1975**, *97*, 5956.
5. Tomoi, M.; Ford, W. T. *J. Am. Chem. Soc.* **1981**, *103*, 3821.

6.   Tomoi, M.; Ford, W. T. *J. Am. Chem. Soc.* **1981**, *103*, 3828.

7.   Montanari, F.; Tundo, P. *J. Org. Chem.* **1981**, *46*, 2125.

8.   Kimura, Y.; Regen, S. L. *J. Org. Chem.* **1983**, *48*, 195.

9.   Tomoi, M. *J. Poly. Sci. Polymer. Chem. Ed.* **1985**, *23*, 49.

10. Tundo, P.; Venturello, P.; Agelletti, E. *J. Am. Chem. Soc.* **1982**, *104*, 6547; ibid **1982**, *104*, 6551.

11. Cronelis, A.; Laszlo, P. *Synthesis,* **1982**, 162.

12. Kadkhodayan, A.; Pinnavaia, T. J. *J. Molec. Cata.* **1983**, *21*, 109.

13. Tundo, P.; Venturello, P.; Angeletti, E. *Is. J. Chem.***1985**, *26*, 283.

14. Ford, W. T. *Adv. Polym.Sci.* **1984**, *24*, 201.

15. Ford, W. T., *Polm. Sci. & Tech. (Plenum)* **1984**, *24*, 201.

16. Dehmlow, E. V.; Dehmlow, S. S. *Phase Transfer Catalysis*; Verlag Chemie, Basal, **1983**; p. 65.

RECEIVED February 18, 1992

Chapter 12

Thermal Analysis of Porphyrin–Clay Complexes

K. A. Carrado, K. B. Anderson[1], and P. S. Grutkoski

Chemistry Division 200, Argonne National Laboratory, 9700 South Cass Avenue, Argonne, IL 60439

Aluminosilicate smectite clays have been ion-exchanged with water-soluble, cationic porphyrins and metalloporphyrins. Characteristics of their thermal stability were measured by thermal gravimetric analysis in an inert atmosphere, which yielded approximately 60% weight loss of organics. Detailed structural information about the decomposition products was obtained by performing pyrolysis-gas chromatography-mass spectrometry on the clay-organic complexes.

The ability of clay structures to provide supramolecular organization in terms of catalysis, chiral reactions, colloid science, electron transfer and pillaring has been well recognized (1). The supramolecular architecture of clays is therefore of critical importance, and structural design can be finely tuned to suit specific applications. One approach is to modify natural clays (2) while another technique of potential is the preparation of tailor-made synthetic clays. Although synthetic zeolites can be routinely made in the presence of organic templating molecules (3), few references currently exist wherein the use of organics is employed during the synthesis of clays (4-6). Recently the magnesium silicate hectorite clay system was modified to incorporate water-soluble cationic porphyrins and metalloporphyrins by direct hydrothermal crystallization (6).

The thermal characteristics of synthetic organo-clay complexes are of interest for several reasons. First, since clay synthesis takes place at elevated temperatures (100-300°C), the stability of each constituent must be assured. The degree to which the thermal characteristics of a template are affected by a support is also of importance. In addition, the porphyrin-clay complexes are of interest as advanced materials in such areas as electrochemistry and catalysis (7), and as highly organized molecular assemblies (8). Since thermal stability

[1]Current address: Amoco Oil Co., Amoco Research Center, Mail Code H9, P. O. Box 3011, Naperville, IL 60566–7011

may be an important factor in these applications, thermal gravimetric analysis (TGA) was applied to characterize in detail the stability of the organic portion in porphyrin ion-exchanged montmorillonite clays. Further information about the degradation of porphyrin-clay complexes was obtained using the technique of pyrolysis-gas chromatography-mass spectrometry (Py-GC-MS).

Experimental

Synthesis. The cationic water-soluble porphyrins *tetrakis*(N-methyl-4-pyridyl) porphyrin (TMPyP), *tetrakis*(N,N,N-trimethyl-4-anilinium) porphyrin (TAP), and the metalloporphyrin Fe(III)TAP were purchased as chloride salts from Midcentury Chemicals, Posen, IL. Ion-exchanged clays were prepared by stirring 1-2 wt% clay in 1×10^{-3} M aqueous solutions of porphyrin overnight with subsequent isolation, washing and air-drying of products. Final porphyrin-exchanged clays contain about 10% organic by weight (see Table I). The natural montmorillonite used for these experiments, Bentolite L, is a Ca^{2+}-bentonite treated to remove excess iron, and was obtained from Southern Clay Products, Gonzales, TX. Detailed characterization of this clay has been published elsewhere (9). All starting materials were used without further purification.

TMPyPCl   TAPCl

Characterization. X-ray powder diffraction (XRD) was done on a Scintag PAD-V instrument using Cu K_α radiation. Scans were collected at either 0.5° or 1.0° 2θ/min. Oriented films were made by air-drying clay slurries from water on glass slides. Data was collected on a DG Desktop computer system.

Thermal gravimetric analysis was performed on a Cahn 121 electrobalance from 25-800°C at a rate of 10°C/min with data collection every two seconds, under a nitrogen atmosphere; Cahn software allowed processing of the data to yield first-derivative thermograms.

Pyrolysis-GC-MS analyses were carried out on an HP-5890 gas chromatograph coupled to an HP-5970 MSD (mass-selective detector), using a C.D.S. "pyroprobe" coil-type pryolyzer made from Pt wire. Samples were held in a specially-designed quartz boat, and the pyrolysis occured with a temperature rise time of 20 seconds. A 60m DB-1701 chromatographic column allowed excellent separation of components; the oven was ramped from 40-280°C at a rate of 8°C/min under a helium atmosphere.

Results and Discussion

Table I contains microanalysis and X-ray diffraction data for the porphyrin-exchanged complexes. The weight percentage of organics is typical for that expected considering the cation exchange capacity of this clay (80 meq/100gm). Comparison of the theoretically expected C/N ratios to the actual experimental values is excellent, considering experimental error, and the porphyrins are assumed to be incorporated fully intact. The results of UV-visible absorption spectroscopy which have been published elsewhere (7) confirm this assumption. The basal spacing or d_{001} value, which is the c-dimension of the unit cell, includes the clay layer (~9.6 Å for montmorillonites) and the height of the clay gallery. A complete description of the d_{001} values given in Table I, which are primarily based on orientation of the porphyrin macrocycles within the clay gallery, has already been provided (7).

Table I. Characterization of Porphyrin-Clays

Sample	Weight % %C	%N	C/N Ratio Theor.	Expt.	XRD d_{001}, Å
TMPyP-bentonite	7.5	1.5	5.5	5.8	14.5
TAP-bentonite	9.3	1.6	7.0	6.8	15.7
FeTAP-bentonite	8.5	1.4	7.0	7.1	18.0

Thermal Analysis. Clays, porphyrins, and porphyrin-clay complexes were analyzed by thermal gravimetric analysis under an inert nitrogen atmosphere. One previous study of the thermal stability of porphyrins reports that weight loss is greater and less complex in oxygen compared to nitrogen (10). Weiss and Roloff (11) observed that hemin was stable on the surface of montmorillonite at 300°C with air exclusion, with an implication that this was an enhanced thermal stability caused by the presence of clay. Results obtained under the more reactive, combustive oxygen environment will be reported later (12).

First, the characteristics of both the clay and porphyrin constituents were observed separately. Then, porphyrins loaded onto natural clays (e.g. bentonite) by ion-exchange were measured and compared to the separate components. The data will be used as a foundation for characterization of the synthetic porphyrin-hectorite systems in later studies (12). Weight loss and first-derivative curves from thermal gravimetric analysis in nitrogen of pure porphyrin chloride salts and ion-exchanged porphyrin-clays were collected. Figure 1a displays the TG curve for pure bentonite, which consists of three major transitions: a loss of surface water from 25-107°C (12.8%), interlayer water is lost from 110-167°C (1.8%), and dehydroxylation occurs from 610-760°C (2.5%). Figures 1b-c, 2a-b and 2c-d correspond to the TMPyP, TAP and Fe(III)TAP systems, respectively. The occurrence of two peaks in the 25-100°C temperature range is probably not real, but an artifact of the furnace heating at low temperatures, and is taken together as Transition A. Data is also tabulated in Table II.

The TG curve of TMPyPCl in Figure 1b has three main transitions indicated, with transition A being due to loss of waters of hydration. Except for determining the temperatures of maximum weight loss in the remaining transitions, in conjunction with the information in Table II, no other information can be gleaned. In other words, one cannot obtain with certainty any structural information about what is being lost upon decomposition from the technique of TGA alone. Consequently, as in Figure 1c for TMPyP loaded onto bentonite, the peak at 475°C cannot be assigned to either of the two main transitions shown from TMPyPCl in Figure 1b without further information.

For both the anilinium-porphyrins TAPCl and FeTAPCl, there is a major transition at 145°C and 215°C, respectively, which does not appear to occur when ion-exchanged into the corresponding clays. The total weight loss observed for the pure porphyrin chloride salts in this inert atmosphere is only about 60%. Note also that all the porphyrin-clay samples show a weight loss at roughly 650°C, which occurs about 60°C less than the pure clay dehydroxylation peak.

Py-GC-MS. The technique of pyrolysis-gas chromatography-mass spectrometry was used to definitively assign the weight loss peaks in TGA by identifying the decomposition products under an inert atmosphere. At least two different pyrolysis temperatures were used for each material in Table II (except for bentonite and TAP-bentonite). In this fashion, isolation and identification of the products evolved during at least two major TGA transitions was possible. Thus, TMPyPCl was pyrolyzed at 420°C and 600°C to determine transitions B and C (by subtraction), respectively; transition A is known to be due to water. The temperatures of pyrolysis for all samples are summarized in Table III, along with product assignments. The terms "pyridines" and "anilines" in

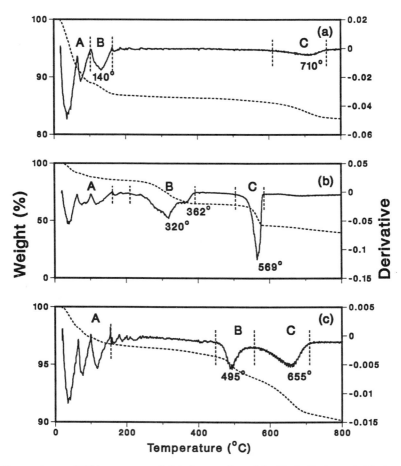

Figure 1.   TGA curves of (a) bentonite, (b) TMPyPCl and (c) ion-exchanged TMPyP-bentonite. Weight curve is dashed, derivative curve is solid; transitions A, B and C are indicated by vertical dashed lines.

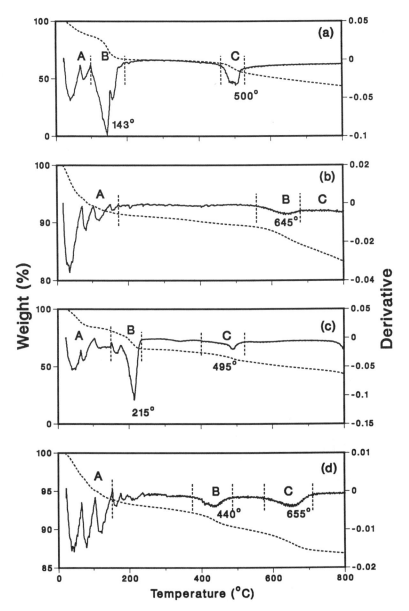

Figure 2.   TGA curves of (a) TAPCl, (b) ion-exchanged TAP-bentonite, (c) Fe(III)TAPCl and (d) ion-exchanged Fe(III)TAP-bentonite. Weight curves are dashed, derivative curves are solid; transitions A, B and C are indicated by vertical dashed lines.

Table III refer to the many derivatives and isomers observed, e.g. methyl-substituted pyridines or bipyridines. An example of these is indicated in Figure 3 for TMPyPCl pyrolyzed at 600°C. For this particular sample, two small peaks of m/e 249 are observed but are as yet not assigned, although they may be due to isomers of tripyridines. A small amount of pyrrole is also seen at this temperature (but not at 420°C, and therefore cannot be attributed to impurity), indicating a slight degree of decomposition of the porphyrin core.

Table II. TGA Data of Porphyrin Systems in Nitrogen*

Sample	Major Transitions Temp Range (°C) and Wt% Loss			Total Trans
	Trans A	Trans B	Trans C	
Bentonite	25-107 12.8%	110-167 1.8%	610-760 2.5%	25-800 18.8%
TMPyPCl	20-162 14.4%	211-392 19.7%	505-584 17.4%	20-800 60.2%
TMPyP-bentonite	19-148 3.3%	420-567 2.1%	570-800 4.8%	19-800 11.6%
TAPCl	20-104 13.8%	105-198 19.9%	468-520 5.3%	20-800 58.2%
TAP-bentonite	17-174 8.5%	558-680 3.1%	680-801 2.9%	17-801 16.8%
FeTAPCl	20-151 18.5%	151-236 15.5%	399-522 7.1%	20-810 55.7%
FeTAP-bentonite	20-154 6.0%	381-484 1.6%	584-710 2.2%	20-801 13.1%

*data represent an average of several runs.

Transition B in all of the pure porphyrins has been determined by Py-GC-MS to be due to release of chloromethane (see Table III). After ion-exchange of the porphyrin chloride salts into the clays, chloride ion is displaced into the aqueous solution; clay exchange sites now balance the charge on the porphyrin ring. As a result, the porphyrin-exchanged clays can no longer show chloromethane as a possible decomposition product. This information, which could be verified only by a supporting technique such as Py-GC-MS, greatly aids in the assignment of the remaining TGA transitions.

Table III. Py-GC-MS Data for Porphyrin Systems

Sample	Pyrolysis Temp (°C)	Products Observed
TMPyPCl	420	chloromethane
	600	chloromethane, pyridines
TMPyP-bentonite	540	pyridines
TAPCl	420	chloromethane
	600	chloromethane, anilines, some pyrrole
TAP-bentonite	---	TGA inconclusive
Fe(III)TAPCl	300	chloromethane
	600	chloromethane, anilines, some pyrrole
FeTAP-bentonite	540	anilines, some pyrrole

The %H_2O of each porphyrin as determined by TGA is listed in Table IV, and was used to calculate their molecular weights and formulas. From this data the expected %CH_3Cl was determined and then compared with the actual wt% loss seen by TGA. The results, also in Table IV, are in very good agreement. The same analysis can be applied to the pyridinium and anilinium substituents. For example, TMPyPCl loses 17.4 wt% from 505-584°C, which Py-GC-MS shows to be due primarily to a variety of pyridine analogs (see Figure 3). Since the calculated wt% of pyridine in TMPyPCl is roughly 32%, it is apparent that after partial decomposition and release of pyridine, the porphyrin rearranges in some unknown fashion because only about half of the pyridine is lost. Nearly 40% of the sample is not volatile under this nitrogen atmosphere at temperatures of 800°C.

For the porphyrins with anilinium substituents, even less wt% loss is seen in Transition C. TAPCl and FeTAPCl lose just 5.3% and 7.1%, respectively, in the region that Py-GC-MS reveals is primarily anilines (with some pyrroles). The reason for this difference between pyridinium and anilinium substituents is not quite clear. In addition, since porphyrins are loaded onto clays at only

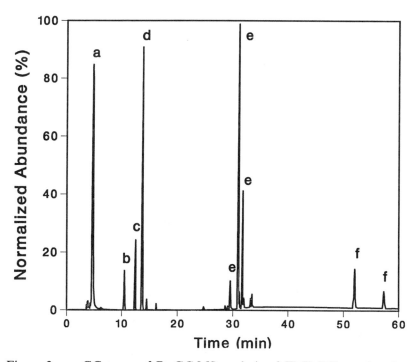

Figure 3.   GC trace of Py-GC-MS analysis of TMPyPCl pyrolyzed at 600°C: a = chloromethane, b = pyridine, c = pyrrole, d = methyl-pyridine, e = bipyridine isomers, and f = m/e 249 isomers (possible tripyridines).

about 10 wt%, this small weight loss is even further reduced. As a result, peaks due to porphyrin decomposition for TAP-bentonite could not be discerned by thermal gravimetric analysis.

Table IV. Calculation of Decomposition Products from Pure Porphyrins

Porphyrin	%H$_2$O[a]	Formula	MW (g/mol)	%CH$_3$Cl Expt.[a]	Calc.[b]
TMPyPCl	14.4	C$_{44}$H$_{38}$N$_8$Cl$_4$· 8H$_2$O	964	19.7	21.0
TAPCl	13.8	C$_{56}$H$_{62}$N$_8$Cl$_4$· 9H$_2$O	1150	19.9	17.6
FeTAPCl	15.7	FeC$_{56}$H$_{60}$N$_8$Cl$_5$·11H$_2$O	1275	15.5	15.8

[a]determined by TGA; [b]calculated from formulas.

Conclusions

The thermal stability of porphyrin-clay systems has been examined in detail. Results from TGA and Py-GC-MS indicate that the porphyrin nucleus is extremely stable in the presence of clay minerals, especially in an inert atmosphere like nitrogen. Substituents on the nucleus such as pyridinium or anilinium are, on the other hand, slightly destabilized, which can be attributed to the acidic nature of the clay surface. The use of pyrolysis-gas chromatography-mass spectrometry in conjunction with thermal gravimetric analysis greatly clarifies the assignment of weight loss peaks. Analysis of the resulting solids after thermolysis is currently underway.

Acknowledgments

Microanalyses were carried out by Mr. S. Newnam of the Analytical Chemistry Division of ANL. Helpful discussions with Dr. R. Hayatsu of ANL about MS data analysis are greatly appreciated. Mr. J. S. Gregar is acknowledged for fabrication of the quartz samples boats used for Py-GC-MS. The technical assistance of Mr. A. G. Stellpflug and his financial support from the Division of Educational Programs at ANL are also recognized. This work was performed under the auspices of the Office of Basic Energy Sciences, Division of Chemical Sciences, U.S. Department of Energy, under contract number W-31-109-ENG-38.

Literature Cited

1. Fripiat, J. J. *Clays Clay Miner.* **1986**, *34*, 501.
2. Pinnavaia, T. J. *Science* **1983**, *220*, 365.
3. Szostak, R. *Molecular Sieves: Principles of Synthesis and Identification* Van Nostrand Reinhold Catalysis Series; Van Nostrand Reinhold: NY, 1989.
4. Barrer, R. M.; Dicks, L. W. R. *J. Chem. Soc. (A)* **1967**, 1523.
5. Barrer, R. M.; Denny, P. J. *J. Chem. Soc.* **1961**, 971.
6. Carrado, K. A.; Thiyagarajan, P.; Winans, R. E.; Botto, R. E. *Inorg. Chem.* **1991**, *30*, 794.
7. Carrado, K. A.; Winans, R. E.; *Chem. Mater.* **1990**, *2*, 328.
8. Giannelis, E. P. *Chem. Mater.* **1990**, *2*, 627.
9. Carrado, K. A.; Kostapapas, A.; Suib, S. L.; Coughlin, R. W. *Solid State Ionics* **1986**, *22*, 117.
10. Said, E. Z.; Al-Sammerrai, D. *J. Analy. Appl. Pyrolysis* **1985**, *9*, 35.
11. Weiss, A.; Roloff, G. *Z. Naturforsch.* **1964**, *19b*, 533.
12. Carrado, K. A.; Winans, R. E.; Grutkoski, P. S.; Melnicoff, P. *to be published*.

RECEIVED January 16, 1992

Chapter 13

Synthesis and Study of Asymmetrically Layered Zirconium Phosphonates

David A. Burwell and Mark E. Thompson[1]

Department of Chemistry, Princeton University, Princeton, NJ 08544

$Zr(O_3PCH_2CH_2COOH)_{1.5}(O_3POH)_{0.5} \cdot 0.5H_2O$, **Zr-COOH-OH**, and $Zr(O_3PCH_2CH_2COOH)_{1.25}(O_3PH)_{0.75} \cdot 0.5H_2O$, **Zr-COOH-H**, were prepared by the reaction of $ZrOCl_2$ with a mixture of $(EtO)_2P(O)$- CH_2CH_2COOH and either phosphoric or phosphorous acid in the presence of HF. **Zr-COOH-OH** forms as a single phase material with an interlayer spacing of 13.8 Å. This phase has not been reported previously. Both **Zr-COOH-OH** and **Zr-COOH-H** react with thionyl chloride to give partial conversion of the carboxylic acid groups to acyl chlorides. These acyl chloride derivatives react readily with amines and alcohols to give amides and esters, respectively.

Intercalation of layered inorganic solids has been employed over the years to alter both the bulk physical properties and chemical reactivity of materials (1). Only recently, however, have investigators begun to take advantage of the 2-dimensionally-ordered framework of intercalation hosts to construct new materials with imposed solid-state structures. Some unique applications of this concept include the use of layered solids to assemble molecular multilayers at solid-liquid interfaces (2), study quantum-sized semiconductor particles (3), modify electrode surfaces (4), and prepare low-dimensional conducting polymers (5). In addition to these applications, layered solids continue to be investigated for uses in ion exchange (6) and catalysis (7).

In 1978, Alberti *et al.* reported the synthesis of a new class of layered metal phosphonates, $M^{IV}(O_3PR)_2$ (8). These compounds crystallize with layered structures similar to that of α-$Zr(O_3POH)_2 \cdot H_2O$ (α-ZrP). Each layer consists of a plane of metal atoms linked together by phosphonate groups. The metal atoms are octahedrally coordinated by oxygen atoms, with the three oxygens of each phosphonate tetrahedron

[1]Corresponding author

0097–6156/92/0499–0166$06.00/0

bound to three different metal atoms. This arrangement forces the organic groups to lie above and below the inorganic layer. The synthesis of these layered materials was extended by Dines and coworkers to include a wide range of organic functional groups such as $Zr(O_3PC_6H_5)_2$ and $Zr(O_3P(CH_2)_nX)_2$ (where n = 1 - 5; X = H, Cl, SH, SEt, OH, COOH, SO$_3$H, OC$_6$H$_5$) (*9*). Excellent reviews on the structure, properties, and applications of layered phosphonates have appeared recently (*10*).

With few exceptions, intercalation of organic guest molecules into inorganic hosts has been promoted by only three types of reaction: acid-base, ion-exchange and redox. However, with the variety of reactive groups available in layered group 4 metal phosphonates, it should be possible to promote intercalation using organic reactions of those groups. To this end, we have very recently reported the syntheses of amides and esters from an acyl chloride zirconium phosphonate (*11*). These new materials are well ordered, thermally stable and easily characterized.

Our first step in the preparation of new layered metal phosphonates from $Zr(O_3PCH_2CH_2COOH)_2$ is the synthesis of an acyl chloride derivative, $Zr(O_3PCH_2-CH_2COCl)_2$ (*12*). $Zr(O_3PCH_2CH_2COCl)_2$ is layered with an interlayer spacing *ca.* 0.6 Å larger than the starting acid. It reacts readily with amines and alcohols to give layered amides and esters, Eq. 1 (*11*). The structures of these new layered materials are similar to those obtained by amine intercalation of α-ZrP and $Zr(O_3PCH_2CH_2-COOH)_2$ except that the guest is held in the interlayer space by a covalent bond rather than an electrostatic one.

$$Zr(O_3PCH_2CH_2COCl)_2 + 2HXR \longrightarrow Zr(O_3PCH_2CH_2COXR)_2 + 2HCl \qquad (1)$$

$$X = NH, O; \; R = \textit{n}\text{-alkyl group}$$

Amide compounds with alkyl chains ranging in length from 1 to 18 carbons have been prepared. The interlayer spacing regularly increases, at increments consistent with a 59° angle between the alkyl chain and the surface of the host (Figure 1) (*11*). This angle is close to that seen for amine intercalation into both α-Zr and $Zr(O_3PCH_2CH_2COOH)_2$ (56° and 53°, respectively) (*13*). IR data suggest that the amides engage in *trans*-type hydrogen bonding, as shown in Figure 2.

The covalent bonding in amide and ester compounds leads to added thermal and chemical stability, as compared to materials prepared by Brønsted acid-base reactions. For example, $Zr(O_3PCH_2CH_2CONHC_6H_{13})_2$ shows the onset of its major weight loss *ca.* 200°C higher than $Zr(O_3PCH_2CH_2COO^-)_2(C_6H_{13}NH_3^+)_2$. The amide is also significantly less susceptible to hydrolysis. After five days at 60°C in 3M HCl only 10% of the amide groups are hydrolyzed, while the same conditions convert the ammonium carboxylate compounds completely to the acid in less than three hours.

In the preparation of $Zr(O_3PCH_2CH_2COCl)_2$ we found it necessary to first intercalate $Zr(O_3PCH_2CH_2COOH)_2$ with an amine, giving the carboxylate salt, and then react the salt with SOCl$_2$. The direct reaction of $Zr(O_3PCH_2CH_2COOH)_2$ with SOCl$_2$ does not lead to measurable conversion of the acid groups to acyl chlorides. A possible explanation for this is that the interlayer region is so tightly packed that SOCl$_2$

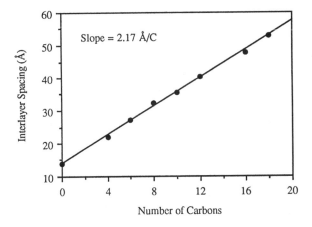

Figure 1. Plot of amide interlayer spacing versus number of carbons in alkyl chain.

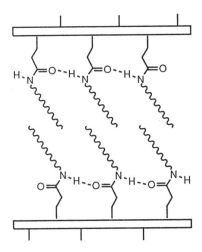

Figure 2. Schematic drawing of amide intralayer hydrogen bonding.

can not diffuse into the solid to react. If this is the case, making the material more porous should increase its rate of reaction with $SOCl_2$. With this idea in mind we have examined the chemistry of mixed component phosphonate-phosphate and phosphonate-phosphite compounds. The structures exhibited by these materials should be more porous than the pure phosphonates. We report herein the synthesis and characterization of mixed component carboxylic acids, acyl chlorides, amides and esters.

Results and Discussion

An alternative to amine pre-intercalation for converting pendant carboxylic acids to acyl halides involves the use of mixed component zirconium phosphonate-phosphate or phosphonate-phosphite compounds. While crystal structures have not been determined, it has been proposed that these materials are composed of asymmetrical α-type layers (*14*). That is, the inorganic portion of the layers, $Zr(O_3P-)_2$, is structurally similar to that of α-ZrP, but bound to the phosphorus are two different pendant groups (Figure 3a). Asymmetrically-layered zirconium phosphonates have previously been reported by Dines *et al.* (*15*), Alberti and co-workers (*16*), and Clearfield (*17*).

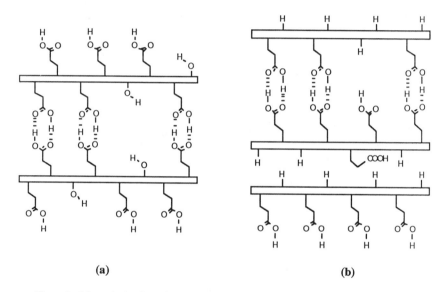

(a) **(b)**

Figure 3. Schematic drawings of (a) $Zr(O_3PCH_2CH_2COOH)_{1.5}(O_3POH)_{0.5} \cdot 0.5H_2O$ and (b) $Zr(O_3PCH_2CH_2COOH)_{1.25}(O_3PH)_{0.75} \cdot 0.5H_2O$.

The use of mixed component systems to prepare acyl chloride derivatives offers two potential advantages over analogous single component materials. First, substitution of the relatively large phosphonate groups with relatively small phosphate or phosphite groups may produce somewhat porous materials. This would allow a chlorinating agent easier access to carboxylic acid sites in the interior of the crystal-lites. Second, a marked decrease in total interlayer bonding is expected, as compared to the pure phosphonate phase, due to the lower concentration of hydrogen bonding carboxylic acid groups on the lamellar surfaces. Here we assume that the phosphate and phosphite groups in the mixed systems are similar to those found in α-ZrP and $Zr(O_3PH)_2$, respectively, where they do not participate in interlayer hydrogen bonding (*18*). The diminished interlayer bonding reduces the amount of energy required to spread the layers and, therefore, increases the likelihood of reaction. The combination of these effects could, in theory, alleviate the need for ammonia intercalation prior to reaction with $SOCl_2$ in our effort to prepare acyl chloride derivatives. Moreover, an acyl chloride prepared from a mixed material similar to that shown in Figure 3a, may lead to new porous phosphate compounds in which amide or ester groups act as pillars. A mixed component material can take on several distinct architectures. In one structure type, each surface of the host lamellae contains a mixture of two different pendant groups, as shown for example in Figure 3a. The observed interlayer spacing in this material is typically close to that for the pure phase of the largest pendant group (*e.g.*, $Zr(O_3PCH_2CH_2COOH)_2$ in Figure 3a).

In a second type of structure, one face of each lamella contains predominantly one pendant group, while the opposite face contains predominantly the other pendant group (Figure 3b). Here the repeat distance perpendicular to the layers is close to the sum of the interlayer spacings for the single-component phases (*e.g.*, $Zr(O_3PCH_2$-$CH_2COOH)_2$ and $Zr(O_3PH)_2$ in Figure 3b) (*10a*). This structure type is commonly observed when the two pendant groups are chemically very different (*e.g.*, hydrophobic and hydrophilic) (*16*).

Lastly, a mixed material may be composed of large regions which contain only one of the pendant groups, *i.e.*, the components are phase separated. In this instance, two phases will be observed, with interlayer spacings close to those for the single-component compounds.

To further investigate the reactivity of layered carboxylic acids, we prepared the mixed component systems $Zr(O_3PCH_2CH_2COOH)_{2-x}(O_3POH)_x \cdot yH_2O$ (**Zr-COOH-OH**) and $Zr(O_3PCH_2CH_2COOH)_{2-x}(O_3PH)_x \cdot yH_2O$ (**Zr-COOH-H**). The interlayer spacings of these compounds, along with others prepared in this study, are given in the following table.

Table. Interlayer spacings of compounds prepared in this study.

Compound	d (Å)
$Zr(O_3POH)_2 \cdot H_2O$	7.6
$Zr(O_3PH)_2$	5.6
$Zr(O_3PCH_2CH_2COOH)_2$	12.9
$Zr(O_3PCH_2CH_2COCl)_2$	13.5
$Zr(O_3PCH_2CH_2COOH)_{1.5}(O_3POH)_{0.5} \cdot 0.5H_2O$	13.8
$Zr(O_3PCH_2CH_2COO^-)_{1.5}(O_3PO^-)_{0.5}(C_6H_{13}NH_3^+)_2$	26.2
$Zr(O_3PCH_2CH_2C(O)NHC_6H_{13})_{1.5}(O_3POH)_{0.5}$	25.0
$Zr(O_3PCH_2CH_2COOH)_{1.25}(O_3PH)_{0.75} \cdot 0.5H_2O$	19.6†
$Zr(O_3PCH_2CH_2COO^-)_{1.25}(C_6H_{13}NH_3^+)_{1.25}(O_3PH)_{0.75}$	31.8†
$Zr(O_3PCH_2CH_2C(O)OC_6H_{13})_{1.25}(O_3PH)_{0.75}$	28.5†

† predominant phase

Zr-COOH-OH was prepared by precipitating Zr^{+4} with an aqueous solution equimolar in 2-carboxyethylphosphonic acid and phosphoric acid. The powder x-ray diffraction (XRD) pattern of **Zr-COOH-OH** exhibits three orders of sharp (00*l*) reflections corresponding to an average interlayer spacing of 13.8 Å. The single-component analogs of the individual components in **Zr-COOH-OH**, *i.e.*, $Zr(O_3PCH_2CH_2COOH)_2$ and α-ZrP, have interlayer spacings of 12.9 Å and 7.6 Å, respectively (*19*). Since no other reflections are seen in the powder XRD pattern at

Bragg angles corresponding to repeat distances larger than 4.4 Å, it appears that a single, regularly-spaced phase containing layers of a fixed percentage of each component is likely (Figure 3a).

Spectroscopic data suggest that **Zr-COOH-OH** is very similar to $Zr(O_3P-CH_2CH_2COOH)_2$. The C=O stretch in the IR spectrum of **Zr-COOH-OH** appears at 1701 cm^{-1}, as compared to 1700 cm^{-1} for $Zr(O_3PCH_2CH_2COOH)_2$. The ^{13}C cross-polarized (CP) magic angle spinning (MAS) NMR spectra of these two compounds are essentially identical. Both exhibit carbonyl isotropic chemical shifts at 180.0 ppm and broad resonances in the methylene region. The ^{31}P MAS NMR spectrum of **Zr-COOH-OH** clearly shows the presence of two different phosphorus groups. The phosphonate resonance is a broad peak centered at 9.2 ppm, while that of the phosphate is much sharper and falls at -22.1 ppm. The broadening of the phosphonate peak is probably due to a dispersion of local electronic environments (*i.e.*, static disorder), while the sharp phosphate peak suggests that all the phosphate groups have similar environments.

The thermogravimetric analysis (TGA) of **Zr-COOH-OH** closely resembles that of $Zr(O_3PCH_2CH_2COOH)_2$ and supports a stoichiometry of $Zr(O_3PCH_2CH_2CO-OH)_{1.5}(O_3POH)_{0.5} \cdot 0.5H_2O$. Both compounds convert to ZrP_2O_7 at *ca.* 1000°C in an oxygen atmosphere, as determined by powder XRD. For **Zr-COOH-OH**, the theoretical weight loss of this transformation is 29.5%, with 2.4% due to water; the observed weight losses are 29.5% and 2.5%, respectively.

C, H, N analysis by flash combustion is of limited use in determining the exact stoichiometry of **Zr-COOH-OH** and other compounds reported in this paper. The values obtained were 10.9% C, 2.3% H, and 0% N. The theoretical values for $Zr(O_3PCH_2CH_2COOH)_{1.5}(O_3POH)_{0.5} \cdot 0.5H_2O$ are 14.4% C and 2.4% H. For comparison, the theoretical values for $Zr(O_3PCH_2CH_2COOH)(O_3POH) \cdot H_2O$ are 10.1% C and 2.3% H. We believe the carbon value is low due to the trapping of organic pyrolysis products in the inorganic matrix; these products are only slowly lost at high temperature. Our TGA experiment shows that *ca.* 50% of the weight loss observed for **Zr-COOH-OH** is lost between 600° and 1100°C. In the short burning time used for flash combustion C, H, N analysis (*i.e.*, < 30 sec), a significant amount of the carbonaceous material presumably remains trapped in the inorganic matrix. Other investigators have also encountered this problem with elemental analyses of layered metal phosphonates (*16*). For this reason, we regard our thermal analysis data as a more accurate measure of composition when the TGA is run under an oxygen atmosphere.

Taken together, the above data suggest that **Zr-COOH-OH** is somewhat different than the mixed component systems reported by Alberti and co-workers. Alberti *et al.* reported a compound formulated as $Zr(O_3PCH_2CH_2COOH)(O_3P-OH) \cdot H_2O$ with an average interlayer spacing of 12.9 Å (*16*)— identical to that of $Zr(O_3PCH_2CH_2COOH)_2$. Subsequently, Alberti *et al.* reported a series of mixed compounds, $Zr(O_3PCH_2CH_2COOH)_{2-x}(O_3POH)_x$, with x ranging from 0.9 to 1.2 and interlayer spacings ranging from 9.8 Å to 10.2 Å (*20*).

Zr-COOH-H was prepared by precipitating Zr^{+4} with an aqueous solution of a 2:1 molar ratio of phosphorous acid and 2-carboxyethylphosphonic acid. The powder XRD pattern of this material contains reflections from three distinct phases. The interlayer spacings of these phases are: 19.6, 12.7 and 6.5 Å. The 12.7 Å phase likely contains mostly phosphonate pendant groups, while the 6.5 Å phase probably consists of mostly phosphite groups. The 19.6 Å phase may contain alternating mostly-phosphonate and mostly-phosphite layers since its interlayer spacing nearly equals the sum of the other two separations. This type of phase separation is expected since the carboxylic acid groups are hydrophilic, while the phosphite groups are relatively hydrophobic. For comparison, Alberti *et al.* reported a 12.2 Å interlayer spacing for $Zr(O_3PCH_2CH_2COOH)_{1.15}(O_3PH)_{0.85} \cdot 0.5H_2O$ (*16*).

The TGA of **Zr-COOH-H** is consistent with a stoichiometry of $Zr(O_3PCH_2-CH_2COOH)_{1.25}(O_3PH)_{0.75} \cdot 0.5H_2O$. However, since three distinct phases are present it is impossible to determine the exact stoichiometry of each phase. Again, the C, H, N analysis was of little help in determining the stoichiometry. Values of 7.04% C, 1.77% H, and 0% N were obtained. Theoretical values for $Zr(O_3PCH_2CH_2CO-OH)_{1.25}(O_3PH)_{0.75} \cdot 0.5H_2O$ are 12.8% C and 2.3% H; while for $Zr(O_3PCH_2CH_2-COOH)(O_3PH) \cdot H_2O$, they are 10.6% C and 2.4% H.

The IR spectrum of **Zr-COOH-H** looks like that of $Zr(O_3PCH_2CH_2COOH)_2$ with a superimposed P-H stretch at 2473 cm^{-1}. The 1703 cm^{-1} C=O stretch of **Zr-COOH-H** suggests a similarity between these carboxylic acid groups and those found both in **Zr-COOH-OH** and $Zr(O_3PCH_2CH_2COOH)_2$. The ^{13}C CP MAS NMR spectrum also closely resembles that of $Zr(O_3PCH_2CH_2COOH)_2$ with the carbonyl isotropic chemical shift at 179.8 ppm and an incompletely averaged methylene region. The ^{31}P MAS NMR spectrum of **Zr-COOH-H** shows three resonances: 11.8, 8.0, and –14.7 ppm of relative intensities 30:100:80, respectively. Owing to the similarities of these chemical shifts to those of the single component analogs, these peaks have tentatively been assigned to the mostly-phosphonate phase (12.7 Å), the phosphonate-phosphite phase (19.6 Å), and the mostly-phosphite phase (6.5 Å), respectively. The relative intensities of the ^{31}P peaks qualitatively match those observed in the powder XRD pattern. That is, the dominant phase, for which a sharp (00*l*) reflection is observed, is the 19.6 Å phase. A small, broad reflection of *ca.* 10% the intensity of the 19.6 Å reflection is seen at 12.7 Å, while the 6.5 Å reflection is of about 30% the intensity of the 19.6 Å reflection and is also rather broad.

The smaller percentage of phosphonate groups in **Zr-COOH-H** relative to **Zr-COOH-OH**, coupled with the smaller size of phosphite groups relative to phosphate groups, suggest that a guest molecule may have more access to the carboxylic acid sites of **Zr-COOH-H** than **Zr-COOH-OH**. In fact, we have found that under identical conditions **Zr-COOH-H** gave a significantly higher conversion to its acyl chloride derivative than did **Zr-COOH-OH**. Contacting the solids with refluxing $SOCl_2$ for two weeks did not fully convert either to the corresponding acyl chloride. However, while the IR spectrum of the product from **Zr-COOH-H** exhibits two C=O stretches at 1801 (–COCl) and 1703 cm^{-1} (–COOH) (Figure 4b), the IR spectrum of the product from **Zr-COOH-OH** shows a 1715 cm^{-1} absorption with only a high energy ($\approx$1800 cm^{-1}) shoulder. Absorptions at 1715 cm^{-1} have been observed during many reactions of $Zr(O_3PCH_2CH_2COOH)_2$, and have been attributed to a carboxylic acid phase in which the interlayer hydrogen bonding network has been disrupted (*21*).

Heating **Zr-COOH-OH** and **Zr-COOH-H** to 170°C under vacuum to drive off interstitial water prior to treatment with refluxing $SOCl_2$ increased the yield of both acyl chlorides, but still did not result in full conversion of either. The IR spectrum of the reaction product from *dehydrated* **Zr-COOH-OH** (Figure 4a) exhibited distinct

C=O stretches at 1794 and 1715 cm^{-1} in roughly the same ratio as those in the product obtained from *non-dehydrated* **Zr-COOH-H** (Figure 4b). Although neither mixed component system affords conversion of all carboxylic acid groups to acyl chlorides without pre-intercalation, both do yield partial conversion. We believe these results support the notion that strong interlayer hydrogen bonding is responsible for the lack of reactivity between $Zr(O_3PCH_2CH_2COOH)_2$ and $SOCl_2$.

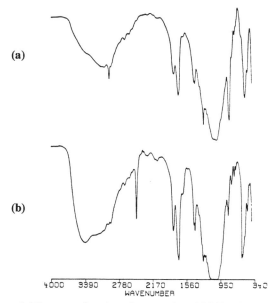

Figure 4. IR spectra of products from reaction of $SOCl_2$ with *dehydrated* **Zr-COOH-OH** (top) and *non-dehydrated* **Zr-COOH-H** (bottom).

When the partially converted acyl chloride derivative of **Zr-COOH-OH** was contacted with hexylamine, a material with a 25.0 Å interlayer spacing was formed. This represents an 11.2 Å increase in interlayer spacing over **Zr-COOH-OH**. For comparison, the analogous hexylamide derivative, $Zr(O_3PCH_2CH_2CONHC_6H_{13})_2$, has an interlayer spacing 14.1 Å greater than $Zr(O_3PCH_2CH_2COOH)_2$. Both IR and ^{13}C NMR spectroscopy showed the presence of amide and ammonium carboxylate salt, as expected.

When **Zr-COOH-OH** was treated with hexylamine directly, the product showed four orders of (00l) reflections in the powder XRD pattern from a 26.2 Å phase. This compares to a interlayer spacing of 28.7 Å for the hexylammonium intercalate of $Zr(O_3PCH_2CH_2COOH)_2$. Curiously, both the IR and ^{13}C NMR spectra of this product suggest that a small amount amide was formed in addition to the expected hexylammonium product.

When the partially converted acyl chloride derivative of **Zr-COOH-H** was contacted with hexyl alcohol, a phase of interlayer spacing 28.5 Å was formed— an increase of 8.9 Å over the 19.6 Å separation of the host. This can be compared to a 12.7 Å increase in interlayer spacing upon conversion of $Zr(O_3PCH_2CH_2COOH)_2$ to $Zr(O_3PCH_2CH_2COOC_6H_{13})_2$. The 12.7 Å phase of **Zr-COOH-H** appears to have intercalated some alcohol, as a broad reflection of *ca.* 5% the intensity of the 28.5 Å reflection is seen at 13.8 Å. A 13.8 Å phase has been noticed in alcohol intercalation compounds of $Zr(O_3PCH_2CH_2COOH)_2$ (*21*). A broad reflection at *ca.* 6.5 Å is also observed, indicating that the mostly-phosphite phase is unchanged. The IR spectrum shows two C=O absorptions at 1741 and 1702 cm^{-1}, corresponding to ester and acid, respectively. It should be noted that **Zr-COOH-H** as formed did not intercalate hexyl alcohol.

Contacting **Zr-COOH-H** with hexylamine resulted in intercalation of the amine to give the hexylammonium carboxylate compound. The 19.6 Å phase increased in interlayer spacing to 31.8 Å, while the 12.7 Å phase disappeared. The 6.5 Å phase also appears to have increased slightly to 6.7 Å.

Conclusion

The direct reactions of thionyl chloride with porous phosphonate-phosphate and phosphonate-phosphite compounds lead to partial conversion of the phosphonate carboxylic acids to acyl chlorides. This reaction does not occur in the single component phosphonate analog, $Zr(O_3PCH_2CH_2COOH)_2$. To form $Zr(O_3PCH_2CH_2COCl)_2$, $Zr(O_3PCH_2CH_2COOH)_2$ must be intercalated with an amine prior to treatment with thionyl chloride. While the porous materials did not give high yields of the desired acyl chloride, it is interesting to note that the partially-converted acyl chlorides behave in a similar fashion to $Zr(O_3PCH_2CH_2COCl)_2$ toward alcohols and amines. Thus, it may be possible to prepare pillared materials with amide or ester groups supporting the zirconium phosphate layers.

Experimental Section

General Methods. Room temperature ^{13}C and ^{31}P solid-state NMR spectra were recorded on a JEOL 270 MHz spectrometer (67.9 and 109 MHz for ^{13}C and ^{31}P, respectively) equipped with a 7-mm probe from Doty Scientific. High-power ^{1}H decoupling, cross polarization and magic angle spinning were employed for all ^{13}C spectra. A 50 kHz field strength was used for both ^{1}H decoupling and cross polarization; spinning speeds were 3.5-4.0 kHz. ^{13}C signals were referenced to TMS (downfield shifts positive) by using adamantane as a secondary external reference. High-power ^{1}H decoupling and MAS were employed for ^{31}P spectra. ^{31}P signals were externally referenced to 85 wt% H_3PO_4 (downfield shifts positive).

FTIR spectra were obtained on a Nicolet 730 FT-IR Spectrometer at a resolution of 4 cm^{-1}. A DuPont Instruments Thermogravimetric Analyzer 951 and Thermal Analyst 2000 workstation were used to obtain TGA thermograms. Thermograms were run in an air atmosphere from room temperature to 1150°C at a scan rate of 10°C/min.

Powder x-ray diffraction patterns were obtained using either a Scintag PAD-V diffractometer or a Philips Electronics Instruments diffractometer (both Cu Kα radiation). XRD analyses were typically conducted with powders which were preferentially aligned on a quartz single crystal or microscope slide. The calculated standard deviations for the interlayer spacings were no greater than 0.7 Å.

Materials. Except where noted, all starting materials were purchased from Aldrich Chemical Company and were used as received.

Synthesis of $Zr(O_3PCH_2CH_2COOH)_{1.5}(O_3POH)_{0.5} \cdot 0.5H_2O$.

$ZrOCl_2 \cdot 8H_2O$ (9.1 g) was dissolved in 50 ml distilled water and 4 ml HF solution (48 wt% in water) in a 250 ml plastic bottle. H_3PO_4 (3.1 g) and 2-carboxyethylphos-phonic acid (4.8 g) were dissolved in 75 ml distilled water and added to the $ZrOCl_2$-HF solution. Some precipitate formed on contact. The plastic bottle was filled with distilled water and sealed. The sealed bottle was placed in a 70°C oil bath for 4 days. The cooled mixture was centrifuged and the liquid decanted. The remaining white gel was dried at 100°C overnight. The resulting transparent, glass-like solid was ground with mortar and pestle to a coarse white powder.
IR (KBr): 3600-2500 (m, v br), 1701 (s), 1293 (m), 1266 (m), 1203 (m), 1040 (vs, v br), 815 (w), 551 (w), 519 (m), 470 (w) cm^{-1}. Powder XRD: d = 13.8 Å. ^{13}C CP MAS NMR: δ(COOH) = 180.0 ppm; δ(CH$_2$) = 27.8 ppm (v br). ^{31}P CP MAS NMR: δ(P-C) = 9.2 ppm, δ(P-OH) = -22.1 ppm. TGA: room temperature-200°C, –2.5%; 200-600°C, –14.2%; 600-1150°C, –12.7%.

Synthesis of $Zr(O_3PCH_2CH_2COCl)_{1.5}(O_3POH)_{0.5}$. Zr-COOH-

OH (0.5 g) was placed in a flask and dehydrated by heating to 170°C under vacuum for 3 hours. The flask was then cooled to 0°C under nitrogen and 15 ml SOCl$_2$ (previously cooled to 0°C) was added to the flask. The temperature of the mixture was raised to 90°C and held there for 16 days under an nitrogen atmosphere. The SOCl$_2$ was removed and the product dried under vacuum.
IR (KBr): 3600-2500 (m, br), 1794 (m), 1715 (m), 1434 (m), 1413 (m), 1263 (s), 1040 (vs, v br), 803 (s) 522 (s) cm^{-1}.

Synthesis of $Zr(O_3PCH_2CH_2COOH)_{1.25}(O_3PH)_{0.75} \cdot 0.5H_2O$.

$ZrOCl_2 \cdot 8H_2O$ (6.5 g) was dissolved in 20 ml distilled water and 3.3 g HF solution (48 wt% in water) in a round bottom flask. To this flask was added a solution of 4.6 g (HO)$_2$P(O)CH$_2$CH$_2$COOH and 5.0 g H$_3$PO$_3$ in 50 ml distilled water. Some precipitate formed on contact, and an additional 25 ml distilled water was added to fluidize the slurry. The mixture was refluxed for 24 hours under an argon atmosphere. The product was collected on a medium frit, washed with water, and dried to a constant weight at 100°C.
IR (KBr): 3600-2500 (m, br), 2473 (m), 1703 (s), 1266 (s), 1201 (s), 1075 (vs, br), 1035 (vs, br), 574 (m), 558 (m) cm^{-1}. Powder XRD: d = 19.6, 12.7 and 6.5 Å. ^{13}C CP MAS NMR: δ(COOH) = 179.8 ppm; δ(CH$_2$) = 28.4 ppm (v br). ^{31}P CP MAS NMR: δ(mixed P) = 8.0 ppm, δ(P-H) = -14.7 ppm, δ(P-C) = 11.8 ppm. TGA: room temperature-200°C, –2.8%; 200-600°C, –11.5%; 600-1150°C, –9.9%.

Synthesis of $Zr(O_3PCH_2CH_2COCl)_{1.25}(O_3PH)_{0.75}$. Zr-COOH-H

(1.2 g) was added to 25 ml SOCl$_2$ previously cooled to 0°C under a nitrogen atmosphere. The mixture was stirred 2 weeks at 90°C under a nitrogen atmosphere. The SOCl$_2$ was removed and the product dried under vacuum.
IR (KBr): 3600-2800 (s, v br), 2474 (m), 1801(m), 1703 (s), 1433 (m), 1413 (m), 1050 (vs, v br) cm^{-1}.

Synthesis of $Zr(O_3PCH_2CH_2C(O)OC_6H_{13})_{1.25}(O_3PH)_{0.75}$.

$Zr(O_3PCH_2CH_2COCl)_{1.25}(O_3PH)_{0.75}$ (1 g) was added to a flask along with 25 ml hexyl alcohol. The mixture was stirred for 2 weeks at 60°C under a nitrogen purge. The product was collected on a fritted glass filter, washed with acetone and water, and dried at 100°C to a constant weight.
IR (KBr): 2959 (s), 2931 (s), 2872 (s), 2860 (s), 2474 (s), 1741 (s), 1702 (m), 1468 (m), 1428 (m), 1365 (m), 1247 (s), 1176 (s), 1050 (vs, v br) cm^{-1}. Powder XRD: d = 28.5 Å.

Acknowledgement. Acknowledgement is also made to the Donors of the Petroleum Research Foundation, administered by the American Chemical Society (ACS-PRF#22051-G3), and to the Air Force Office of Scientific Research (AFOSR-90-0122) for the support of this project. The purchase of the JEOL 270 NMR spectrometer was made possible by NSF Grant CHE-89-09857.

Literature Cited

(1) (a) *Intercalation Chemistry*, Whittingham, M. S.; Jacobson, A. J.; Eds.; Academic Press Inc.: New York, 1982. (b) Schöllhorn, R. in *Inclusion Compounds*; Academic Press: London, 1984. (c) Dresselhaus, M. S. *Mater. Sci. Eng., B1* **1988**, 259. (d) Formstone, C. A.; Fitzgerald, E. T.; O'Hare, D.; Cox, P. A.; Kurmoo, M.; Hodby, J. W.; Lillicrap, D.; Goss-Custard, M. *J. Chem. Soc., Chem. Commun.* **1990**, 501.

(2) (a) Lee, H.; Kepley, L. J.; Hong, H.; Mallouk, T. E. *J. Am. Chem. Soc.* **1988**, *110*, 618. (b) Lee, H.; Kepley, L. J.; Hong, H.; Akhter, S.; Mallouk, T. E. *J. Phys. Chem.* **1988**, *92*, 2597. (c) Putvinski, T. M.; Schilling, M. L.; Katz, H. E.; Chidsey, C. E. D.; Mujsce, A. M.; Emerson, A. B. *Langmuir*, **1990**, *6*, 1567.

(3) Cao, G.; Rabenberg, L. K.; Nunn, C. M.; Mallouk, T. E. *Chem. Mater.*, **1991**, *3*, 149.

(4) (a) Li, Z.; Lai, C.; Mallouk, T. E. *Inorg. Chem.* **1989**, *28*, 178. (b) Rong, D.; Kim, Y. I.; Mallouk, T. E. *Inorg. Chem.* **1990**, *29*, 1531.

(5) (a) Kanatzidis, M. G.; Wu, C.; Marcy, H. O.; DeGroot, D. C.; Kannewurf, C. R. *Chem. Mater.* **1990**, *2*, 222. (b) Kanatzidis, M. G.; Hubbard, M.; Tonge, L. M.; Marks, T. J.; Marcy, H. O.; Kannewurf, C. R. *Synth. Metals* **1989**, *28*, C89. (c) Mehrotra, V.; Giannelis, E. P. *Solid State Comm.*, **1991**, *77*, 155.

(6) (a) Alberti, G.; Costantino, U. in ch. 5 of ref. 1a. (b) Alberti, G.; Costantino, U.; Marmottini, F. in *Recent Developments in Ion Exchange*, Williams, P. A.; Hudson, M. J.; Eds.; Elsevier Applied Science: New York, 1987. (c) Clearfield, A. *Chem. Rev.* **1988**, *88*, 125.

(7) (a) Alberti, G.; Costantino, U. *J. Mol. Catalysis* **1984**, *27*, 235. (b) Clearfield, A. *J. Mol. Catalysis* **1984**, *27*, 251. (c) Hattori, T.; Ishiguro, A.; Murakami, Y. *J. Inorg. Nucl. Chem.* **1978**, *49*, 1107. (d) Clearfield, A.; Thakur, D. *Appl. Catal.* **1986**, *26*, 1.

(8) Alberti, G.; Costantino, U.; Allulli, S.; Tomassini, N. *J. Inorg. Nucl. Chem.* **1978**, *40*, 1113.

(9) (a) Dines, M. B.; DiGiacomo, P. M. *Inorg. Chem.* **1981**, *20*, 92. (b) Dines, M. B.; Griffith, P. C. *Polyhedron* **1983**, *2*, 607. (c) Dines, M. B.; Griffith, P. C. *Inorg. Chem.* **1983**, *22*, 567.

(10) (a) Clearfield, A. *Comments Inorg. Chem.* **1990**, *10*, 89. (b) Alberti, G. in *Recent Developments in Ion Exchange*, Williams, P. A.; Hudson, M. J.; Eds.; Elsevier Applied Science: New York, 1987, 233.

(11) Burwell, D. A.; Thompson, M. E. *Chem. Mater.* **1991**, *3*, 730.

(12) Burwell, D. A.; Thompson, M. E. *Chem. Mater.* **1991**, *3*, 14.

(13)   Alberti, G.; Costantino, U. in ch. 5 of ref. 1a.

(14)   Alberti, G.; Costantino, U. *J. Mol. Catalysis* **1984**, *27*, 235.

(15)   Dines, M. B.; Cooksey, R. E.; Griffith, P. C.; Lane, R. H. *Inorg. Chem.* **1983**, *22*, 1003.

(16)   Alberti, G.; Costantino, U.; Környei, J.; Giovagnotti, L. *Reactive Polymers* **1985**, *4*, 1.

(17)   Clearfield, A. *J. Mol. Catalysis* **1984**, *27*, 251.

(18)   Troup, J. M.; Clearfield, A. *Inorg. Chem.* **1977**, *16*, 3311.

(19)   Dines, M. B.; DiGiacomo, P. M. *Inorg. Chem.* **1981**, *20*, 92.

(20)   Alberti, G.; Costantino, U.; Marmottini, F. in *Recent Developments in Ion Exchange*, Williams, P. A.; Hudson, M. J.; Eds.; Elsevier Applied Science: New York, 1987, 249.

(21)   Burwell, D. A.; Thompson, M. E. unpublished results.

RECEIVED January 16, 1992

Chapter 14

Polyether and Polyimine Derivatives of Layered Zirconium Phosphates as Supramolecules

Abraham Clearfield and C. Yolanda Ortiz-Avila

Department of Chemistry, Texas A&M University, College Station, TX 77843

Polyethylene oxides and polyethyleneimines are of great interest because of their ability to form a wide variety of metal and salt complexes. We have anchored polyethylene oxide oligomers (n=1-33) and polyimines (n=1-4) to zirconium phosphate type layers. The polymers are first converted to phosphates or phosphonates which in turn react with Zr(IV) solutions to form the layered derivatives. Cross-linking of the layers has also been accomplished. Preliminary structural and complexing behavior of these layered materials is presented. It is demonstrated that the NaSCN-polyethyleneoxide zirconium phosphate is an ionic conductor with $\sigma_{25} \cong 10^{-6}$ ohm^{-1} cm^{-1}.

Several years ago, Yamanaka [1] reported that the γ-form of zirconium phosphate (γ-ZrP) reacts with ethylene oxide to form ester adducts. γ-ZrP is a layered compound [2] with a basal spacing of 12.2 Å and has the composition $Zr(H_2PO_4)(PO_4) \cdot 2H_2O$. At the time of Yamanaka's report γ-ZrP was thought to be a monohydrogen phosphate but a subsequent NMR study [3] revealed the presence of two types of phosphorus and a subsequent X-ray study [4] on the isomorphous titanium compound confirmed the above formulation. The resultant ester thus can be represented as $ZrO_2P(OCH_2CH_2OH)_2(PO_4) \cdot H_2O$. The interlayer distance of this compound is 18.4 Å.

Polyethers are known to incorporate electrolytes, thereby becoming ionic conductors [5]. Thus, an original incentive for this investigation was to determine whether large polyether chains could be anchored to the layers producing, by electrolyte complexing, stable ionic conductors of unusual structure. The incorporation of the alkali metal salts arises from much the same forces which are operative in crown ethers [6-7]. Therefore, a further incentive was to attempt to cross-link the zirconium phosphate layers with polyether chains to create crown ether-like structures. In this paper we will discuss these polyether derivatives of zirconium phosphate and also their polyimine analogues.

0097–6156/92/0499–0178$06.00/0

Polyether Derivatives

Non-Cross-linked Polyether Derivatives of α-Zirconium Phosphate.

The α–phase of zirconium phosphate, like that of the γ-phase, is also layered, but of composition $Zr(HPO_4)_2 \cdot H_2O$ and with a smaller interlayer separation, 7.6 Å *(8)*. According to Yamanaka ethylene oxide did not react with α-ZrP. However, it was important that we be able to use the a-phase for our synthesis because it is more stable than the γ-phase and more likely to form cross-linked products since the phosphate groups are of one kind. Figure 1 shows that the OH groups are roughly perpendicular to the layers a position which is more likely to lend itself to layer cross-linking than the γ-type dihydrogen phosphate hydroxyls which are inclined at tetrahedral angles to the layers. The lack of reactivity of α-ZrP was found to arise because the ethylene oxide could not diffuse into the interlamellar space *(9)*. In point of fact reaction had occurred at the surface where no barrier to diffusion existed. As the surface area of the α-ZrP increased so did the carbon content of reacted product. In the case of a poorly crystalline sample, where some swelling of the layers occurs, almost all the phosphate groups were esterified. The interlayer spacing increased to 15.5 Å, a value expected for esterification of the phosphate groups but with no polymerization of the ethylene oxide groups. In more concentrated ethylene oxide solutions, some polymerization to form chains was observed at the surface but not in the interior.

We were interested in obtaining long chain polyether derivatives of α-ZrP and the ethylene oxide method did not appear to be the correct way to attain this end. Therefore, we decided to prepare polyethylene oxide (PEO) phosphates for direct reaction with Zr(IV). These phosphates were synthesized by the following sequence of reactions:

$$H(OCH_2CH_2)_nOH + POCl_3 \longrightarrow H(OCH_2CH_2)_nOPOCl_2$$

$$\begin{array}{c} \\ H_2O \mid H^+ \quad (1) \end{array}$$

$$H(OCH_2CH_2)_nOPO_3Ba \xleftarrow{\;Ba^{2+}\;} H(OCH_2CH_2)_nOPO_3H_2$$

An IR spectrum of the barium salt of the tetraethylene oxide phosphoric acid is shown in Figure 2 curve I. The wide band from 1480 to 1350 cm^{-1} due to C-O-H in plane bending appears very clear and almost without shift, but the broad band for $\nu(CH_2\text{-}O)_{ether}$, 1120 cm^{-1}, is shifted about 20 cm^{-1} relative to its original position in the pure glycol. The very strong $\nu(P=O)$ appears at 1000 cm^{-1}. Some other new bands, as compared with the glycol, appear below 800 cm^{-1}.

Addition of sulfuric acid precipitates the barium ion as sulfate, leaving behind the polyether phosphate. Elution of the filtrate through an ion-exchange column in the hydrogen form was found to remove any trace of Ba(II) not precipitated. The free acid was then reacted with Zr(IV) as follows:

$$2H_2O_3PO(CH_2CH_2O)_nH + Zr(IV) \rightarrow Zr[O_3PO(CH_2CH_2O)_nH]_2 \cdot xH_2O \quad (2)$$

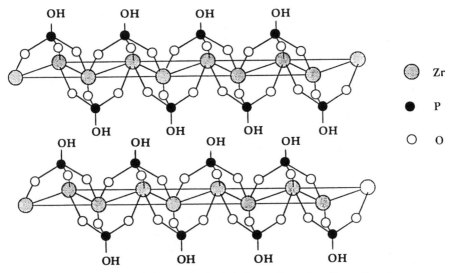

Figure 1. Schematic view of a portion of the layers in a-zirconium phosphate, $Zr(HPO_4)_2 \cdot H_2O$. The water molecule resides in the spaces created by three adjacent OH groups.

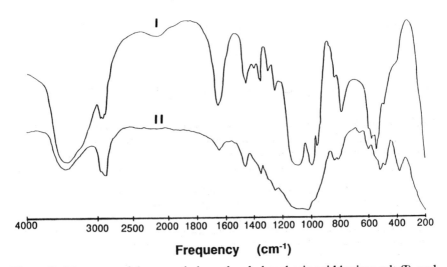

Figure 2. IR spectra of the tetraethylene glycol phosphoric acid barium salt (I), and the tetraethylene glycol zirconium phosphate derivative (II).

Polyether phosphates were prepared from commercially available polyethylene oxide glycols, $H(OCH_2CH_2)_nOH$. For n=1-4, pure products were available, but for n>5 only mixtures with an average molecular weight to $\pm 5\%$ were obtained. As the molecular weight of the polyether increased the zirconium phosphate ester became more difficult to isolate. This difficulty was a direct result of the greater hydrophilicity of the polyether chains. The technique developed to recover the product involved rotovaping the aqueous phase to a small volume and adding 2-3 times the volume of acetone.

Two types of products were obtained. If the reaction with zirconyl chloride was carried out at room temperature, the products were stoichiometric of composition $Zr(O_3PO(CH_2CH_2O)_nH)_2 \cdot xH_2O$. The water content depended upon how thoroughly the solids were dried. On standing they sorbed water. The second type of product was obtained when the mixtures were refluxed. Since the reaction between zirconyl chloride and the phosphate yielded HCl, acid cleavage took place to yield products of the type $Zr(O_3PO(CH_2CH_2O)_nH)_x(HPO_4)_{2-x} \cdot yH_2O$. The results are contained in Table I. Thermogravimetric analysis provided a quick and efficient way to determine the value of x. On heating to 1000 °C the end product is ZrP_2O_7. Thus, the amount of water and organic can be ascertained from the three weight losses observed as shown in Figure 3. The first is due to loss of sorbed water, the second to loss of organic and the final weight loss at ~400 °C represents condensation of HPO_4^{2-} to $P_2O_7^{4-}$. It is seen that there is excellent agreement between the results from elemental analysis and TGA. The infrared spectrum of the n=4, R=H compound has been published (9) and compared to that of the corresponding barium salt Figure 2 curve II. The interlayer distance of the compound with n=1, R=H is 11.9 Å, almost the same value is obtained when ethylene oxide esterifies α-ZrP. These results show that the α-form for the layers is obtained by this direct precipitation method.

If we focus on the interlayer spacing of the anhydrous compounds, there is a 2.9 Å increase in going from n=2, to n=3 and a 3.0 Å increase from n=3, to n=4. On the assumption that this trend continues the interlayer spacing for n=9 should be 35.8 Å. The larger value observed, 39.3 Å, is a reflection of our inability to dry the solid completely. This sample is extremely hygroscopic and picks up water the instant that it is removed from a dry box or oven. The sample with n=22 is a semi-gel. The derivatives with n ≥ 9 completely exfoliated when added to water.

The fully extended trans-trans conformation of the polyether chains would require an increase of 3.5 Å per monomer unit or 7 Å increase in interlayer spacing for every integral increase in the value of n because of the bilayer arrangement.. We observe less than half this value. Polyethylene oxide has a helical structure in the solid state *(10)*, the polymer chain having 7 monomer units in 2 turns of the helix with a repeat distance of 19.3 Å. Such a helix would not fit within the α-ZrP layers and still occupy 24 Å^2 of layer surface (the area of each phosphate group*(8)*). One possible arrangement would require that the chains be inclined at an angle to the layers and overlap. Another perhaps more satisfactory conformation is the cis-cis orientation of the chains. In this arrangement each monomer is 2.5 Å from oxygen to oxygen. The basal spacing data is plotted in Figure 4. The slope of the line in which n is plotted against the interlayer spacing of the dry zirconium polyether ester is 1.77 Å/per monomer unit. If we assume the all cis-conformation, then the chains would have to be inclined at 45° to the layers. Extrapolation of the line to n=o yields a value of 7.6 Å which is just the interlayer spacing of the layers in a-ZrP. Therefore, we assume that the chains do not overlap and the cis-conformation is the preferred one. This angle of tilt is considerably greater than that observed for n-alkylamine intercalates of α-ZrP, ~35° from the perpendicular to the layers *(11)*.

Table I. Polyethylene Glycol Esters of Zirconium Phosphate
$Zr[O_3PO(CH_2CH_2O)_nR]_x\,[HPO_4]_{2-x}$

n	R	Elemental analysis (EA)		(x)		Interlayer Spacing (Å)		TGA[a] T°C
		%C	%H	from EA	from TGA	anhy	wet	
1	CH3	17.91	3.48	1.99	2.00	14.9	15.3	200
2	H	20.87	3.47	2.00	2.01	14.9	17.6	197
3	H				2.00	17.8	21.3	190
3	CH2CH3	21.20	4.40	1.33	1.30	20.8	25.3	160
4	H	22.87	4.30	1.52	1.59	18.8		160
4	H	30.33	5.05	2.01	2.01	20.8	28.5	160
9	H				1.13	25.3	148	148
9	H	40.50	6.75	2.02	2.01	39.3	>44	147
12	CH3				1.00	39.3	>44	142
13	H	42.25	7.20	1.98	1.98	>44	>44	140
22	H				1.97	>44	>44	135

[a]Temperature at which decomposition begins.

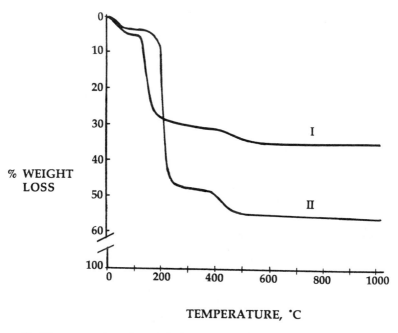

Figure 3. Thermogravimetric weight loss curves for (I) $Zr[O_3PO(CH_2CH_2O)_{1.22}H]_2 \cdot 1.18H_2O$ and (II) $Zr[O_3PO(CH_2CH_2O)_4H]_2 \cdot 1.37H_2O$ (from ref. 9, copyright 1985 American Chemical Society)

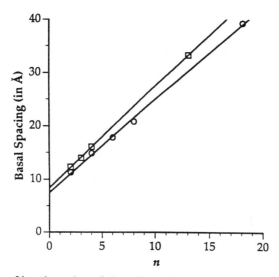

Figure 4. Plot of basal spacing of zirconium polyethylene oxide phosphates versus n, the degree of polymerization of the PEO: unbridged monophosphates, o; phosphate bridged structures, □.

In moist air the polyether phosphates swell, thereby relieving the crowding of the chains. The products that had chain lengths of $n=9$ or greater completely disperse in water. It is believed that these dispersions represent complete exfoliation of the layers. These dispersed solids can be recovered by centrifugation showing that they are not dissolved. It should also be noted that when x is less than 2, the interlayer spacing decreases. We interpret this to mean that the chains overlap or assume a new conformation which utilizes the available free space. Crown ethers, and presumably these polyether chains, exhibit a high degree of conformational flexibility in adapting to their surroundings. Thus, we note the great changes in interlayer spacing with moisture uptake may in part result from such conformation changes. Evidence for this hypothesis is being sought in studies of the IR spectra.

Diphosphate Esters. Diphosphates of the type $H_2O_3PO(CH_2CH_2O)_nPO_3H_2$ were prepared and reacted with Zr(IV) to produce pillared or cross-linked layers. The diesters were synthesized in essentially the same way as the monoesters but with the use of excess $POCl_3$. Unlike the monoposphate derivatives of Zr, the cross-linked products precipitate readily from the reaction solution in high yields and when dried yield free flowing powders. Table II summarizes the data for this class of compounds. Since the chains cannot overlap in this case, we consider only the cis conformation as reasonable. This information is summarized in the plot of interlayer spacing as a function of chain length in Figure 4. The slope of the line is 1.92 Å/monomer requiring an angle of tilt of 39.5 ° from the perpendicular to the planes. Extrapolation of the line to $n = 0$ yields a value of 8.2 Å. This value is considerably larger than the 7.6 Å basal spacing of α-ZrP. Furthermore, with these cross-linked layers there are no O-H end groups to consider so the layer thickness without the hydrogens in a-ZrP is 6.6 Å. To account for the 1.6 Å increase we assume that the layers are oriented such that a phosphate group in one layer is lined up so that it is above and below phosphate groups in adjacent layers. In α-ZrP the sequence in adjacent layers is P-Zr-P, etc. which allows for closer packing of the layers.

Although the polyether chains are anchored at both ends there is still some swelling of the layers with uptake of water. This could occur in a simple fashion by the layers shifting so that the chains do not tilt as much. For example, if we subtract 8.2 Å from 16.1 Å (n = 3, wet) and divide by 3 the value per monomer is 2.6 Å which could mean the chains are now perpendicular to the layers. For n = 4 this value is 2.25 per monomer Å. However, the expanded value for n = 2, 15.5 Å is too large to be explained this way and may indicate some breaking of the cross-links with further swelling.

Sorption of Electrolytes by Zirconium Polyether Compounds

Uptake of $CuCl_2$. A major purpose in preparing the polyether compounds was to immobilize the polyether chains so that they can be easily recovered after interactions in aqueous systems. In this regard several of the zirconium polyether compounds were contacted with O.1 M methanolic $CuCl_2$ solutions until saturated. The results are shown in Table III. Two types of reactions occurred, ion exchange at vacant HPO_4 sites and complexing by the polyether. During the course of these reactions hydrolysis of some of the chains occurred. The extent of hydrolysis is given as Δx in the Table and it is seen that Dx increases as n increases. The cross-linked diphosphates hydrolyse to a much lesser extent but with the larger n values it is still significant.

Consider the results with the monophosphate with n = 4, x = 1.36. In this sample 0.221 moles of Cu^{2+} were taken up while 0.28 moles of tetrameric polyether chains were lost. As a result the layers contained 0.92 moles of HPO_4 groups per

Table II. Polyethylene Glycol Esters of Zirconium Diphosphate
$Zr[O_3PO(CH_2CH_2O)_nPO_3]_x [HPO_4]_{2-x}$

n	Elemental analysis %C	%H	from (x) EA	from TGA	Interlayer Spacing (Å) anhy	wet	TGA[b] T°C
2	13.56	2.17	0.96	0.96	12.3	15.5	246
3	15.16	3.14	0.83	0.87	14.0	16.1	234
4	21.04	3.70	0.97	1.00	16.1	17.2	220
9	22.81	3.96	0.73	0.71	20.1[a]	21.3	195
13	37.97	6.32	1.02	1.00	33.3	>44	167
22	35.56	6.24	0.83	0.87	>44	>44	150

[a]This lower than expected interlayer spacing may be due to the lower organic content.
[b]Temperature of decomposition of organic groups as determined by thermogravimetric analysis.

Table III. Interaction of $CuCl_2$ with Polyethylene Oxide Zirconium Phosphates

n	x^a in $Zr(PEO)_x(HPO_4)_{2-x}$	%Cu	Cu(II) mmoles Cu / mole ZrP	Δx
monophosphates				
4	1.36	2.67	221	-0.28
9	2.00	3.99	660	-0.50
13	2.00	7.18	1004	-1.15
diphosphates				
4	1.0	2.64	200	-0.09
9	0.64	0.28	23	-0.06
13	0.90	0.43	53	-0.24
22	0.90	0.21	37	-0.27

[a]From C,H analysis

formula weight of zirconium compound. There are thus sufficient ion exchange sites to accommodate the Cu^{2+} and no complexing of $CuCl_2$ was required. In the second example with $n = 9$ and $x = 2$ a half mole of polyether chains were removed by hydrolysis to yield $Zr[O_3PO(CH_2CH_2O)_9H]_{1.5}[HPO_4]_{0.5} \cdot yH_2O$. The 0.66 moles of Cu^{2+} require an exchange capacity of 1.32 moles of HPO_4^{2-}. Thus, only 0.25 moles of Cu^{2+} were exchanged and 0.41 moles of $CuCl_2$ were complexed by the polyether chains. Similarly with the compound for which $n = 13$, $x = 2$ about a half of the copper was exchanged and half complexed.

In the diphosphate series only the compound with $n = 4$, $x = 1$ sorbed appreciable $CuCl_2$. Ion exchange can account for only 0.045 moles of Cu^{2+} of the 0.2 moles sorbed. The remainder must be held as $CuCl_2$ in a crown-ether like arrangement. In the case of the other diphosphates all of the Cu^{2+} could be accounted for by ion exchange. However, the low amounts sorbed may be due to an inability of the ions to diffuse into the interior.

Interaction of Zirconium Polyether Compounds with Alkali Thiocyanates. Polyethers are known to complex alkali thiocyanates to form ionically conducting films *(12)*. We determined the uptake of NaSCN and LiSCN from 0.1 M methanol solutions. The results are collected in Table IV. Again two types of reactions were observed, ion exchange and neutral salt complexing. In the case of NaSCN about half of the sodium ion is exchanged in the monophosphates and half complexed. These reactions are accompanied by extensive hydrolysis. With the diphosphates much less salt is taken up and very little is complexed, the predominant reaction being ion exchange. Apparently tying down the chains at both ends does not allow them to incorporate ions within the coiled chain as for free polyether chains. LiSCN yielded similar results to those obtained with NaSCN but a higher proportion of Li^+ was exchanged.

Measurement of Ionic Conductivity. The synthesis of solvent-free metal salt complexes of polyethylene oxides prompted detailed electrical measurements with the thought that these materials might prove to be useful electrolytes, in a hydrous environment, for high energy density batteries *(13-15)*. Many fundamental properties of these polymer electrolytes have been examined and a large literature on the subject is available *(16-17)*. We prepared a disk of one of our polyether complexes and measured its conductivity by impedance methods.

An actual plot of impedance in the complex plane for the polyethylene oxide NaSCN complex ($n = 22$, $x = 2.0$) is shown in Figure 5. This impedance plot represents an equivalent electrical circuit having a capacitor and a resistor in parallel with no resistor in series because the semicircle starts right at the origin. Here, the diameter of the semicircle gives the resistance of the circuit, the capacitance value is given by the formula:

$$\omega_{max} = \frac{1}{C_d R} \tag{3}$$

where ω_{max} is the frequency at the top of the semicircle. C_d is then obtained by solving this expression. For this case, $R = 6.5 \times 10^5 \, \Omega$ and $C_d = 2.5 \times 10^{-4} \, \mu F$.

An electrochemical system (solid-solution interface) can be represented by the same equivalent electrical circuit, displaying in most cases the same impedance plot as we have for polyethylene oxide zirconium phosphate. Here, the value of the capacitor and resistor is obtained in the same way as before. In addition, the

Table IV. Reaction of Polyethylene Glycol Esters of Zirconium
Phosphate with Electrolytes

| Sample | | NaSCN | | | LiSCN | | |
n	x	mmoleSCN mole ZrP	mmole Na mole ZrP	Δx	mmole SCN mole ZrP	mmole Li mole ZrP	Δx
monophosphates							
4	2.00	141	309	-0.29	75	114	-0.12
9	2.00	344	796	-0.53	158	306	-0.42
13	2.00	424	1005	-0.84	203	724	-0.78
22	2.00	654	1250	-0.86	331	938	-0.81
diphosphates							
4	1.00	36	52	0	28	46	0
9	0.64	57	264	-0.03	53	224	-0.04
13	0.90	93	566	-0.31	82	456	-0.23
22	0.90	194	887	-0.38	111	570	-0.30

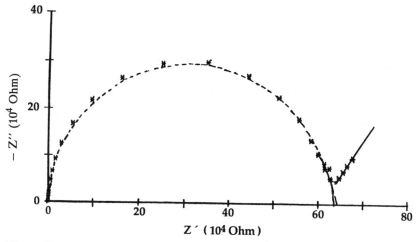

Figure 5. Representative impedance diagram for a polyethylene glycol zirconium phosphate-NaSCN complex.

semicircle region represents a kinetically control process taking place, and the straight line starting at low frequencies represents a mass transfer control process *(18)*. If the chemical system is kinetically sluggish, it will show a rather large R and may display only a very limited frequency region where mass transfer is a significant factor. This seems to be our case. On the other extreme, R may be inconsequentially small by comparison to the capacitor value over nearly the whole available range of frequency. Then the system is so kinetically facile that mass transfer always plays a role and the semicircle region is not well defined.

Conductivity measurements were performed on the polyethylene oxide zirconium phosphate-NaSCN complex (x = 22, n = 2.0) between room temperature and 100 °C. Conductivities of the order 10^{-6} ohm^{-1} cm^{-1} at room temperature were obtained, having an increment by an order of magnitude with an increase in temperature of about 100 °C. The data conformed to an Arrhenius regime, where σ vs $1/T$ is a straight line with Ea = 0.22 eV (Figure 6). The conductivity is associated with the motion of cations in the solid phase formed by chains of polyethylene oxide complexing the electrolyte in between the layers. In the polyethylene oxide-NaSCN helical complex, the formation of cation vacancies within the PEO helix is responsible for the conductivity at temperatures below 60 °C *(15)*. A higher conductivity is observed at temperatures greater than 60 °C, due to the melting of the PEO. At this temperature it becomes slightly soluble in the stoichiometric complex, creating more of the vacancies needed for conduction. In contrast it appears that no melting of PEO occurs in the zirconium PEO-phosphate. Even so it is surprising that the conductivity data conforms to an Arrhenius type dependence on temperature with an activation energy of 0.22 eV because there are two conductors present: Na$^+$ in the ion exchange sites and Na$^+$ in the PEO complex. The conductivity of sodium ion phases of α-zirconium phosphate are of the order of 10^{-6} to 10^{-8} ohm^{-1} cm^{-1} at 200 °C *(19)*. Activation energies for these phases are of the order of 0.7 eV. Thus, the conductivities in the range of 25-100 °C would be several orders of magnitude lower for Zr(NaPO$_4$)$_2$. Therefore, the Na$^+$ ions in the exchange sites can contribute very little to the overall conductivity. Given the changed environment in the present case one cannot be certain of this point.

Preparation and Behavior of Zirconium Polyimine Phosphonates

We have also prepared polyimine phosphonates, analogues of the polyethers. Table V provides data for some of these compounds. The phosphonates were prepared by refluxing a mixture of chloromethylphosphonic acid with ethyleneamines in aqueous sodium hydroxide. The products were then combined with zirconyl chloride solutions, generally at temperatures from 50 to 100°, to yield layered compounds in which a portion of the amino groups is protonated. The extent of protonation can be determined by analysis for Cl$^-$.

The interlayer distance or swelling of the polyimines depends upon the degree of protonation and is therefore very difficult to assess. We have undertaken a study of the degree of swelling as a function of protonation and will report these results in a subsequent communication. However, in both the cross-linked and uncross-linked forms these zirconium polyimine phosphonates are powerful complexing agents. In the protonated form they behave as anion exchangers. We have successfully exchanged Keggin ions, [PtCl$_4$]$^{2-}$, Fe(CN)$_6$$^{4-}$ and Fe(CN)$_6$$^{3-}$ into Zr(O$_3$PCH$_2$NHCH$_2$CH$_2$NH$_2$)$_2$. The procedure is to protonate the amino groups to the level desired for uptake of the anion. This causes the compound to swell and even to completely exfoliate. On addition of the anion salt, the anion is sequestered between the polyimine layers, displacing the Cl$^-$ counterion. Figure 7 shows the infrared spectrum of the Fe(CN)$_6$$^{4-}$ ion exchanged zirconium polyimine

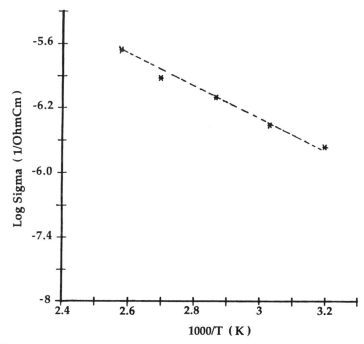

Figure 6. Variable temperature conductivity values for polyethylene glycol zirconium phosphate-NaSCN complex.

Table V. Interlayer Spacings of the Zrconium Polyimine Derivatives of the Type $Zr(RPO_3)_x(R'PO_3)_{2-x}$

Compound No.	R	R'	X	Interlayer Spacing (Å)
V-26	$NH_2CH_2NHCH_2$-	-	2	17.1
V-36	$NH_2CH_2NHCH_2$-	-OH	0.63	14.5
IV-96	$NH_2CH_2NHCH_2$-	-H	1.35	Amorphous
V-24	$NH_2CH_2NHCH_2$-	-	0.9	
IV-72	$O_3P-CH_2NHCH_2CH_2NHCH_2PO_3$	-	1[a]	13.8
IV-96	$O_3P-CH_2NHCH_2CH_2NHCH_2PO_3$	-OH	0.38	9.5
IV-97	$O_3P-CH_2NHCH_2CH_2NHCH_2PO_3$	H	0.94	Amorphous
V-38	$O_3P-CH_2NHCH_2CH_2NHCH_2PO_3$	-OH	0.5	

[a]The cross linked derivatives have the general formula $Zr(RPO_3)x(R'PO_3)_{2-2x}$

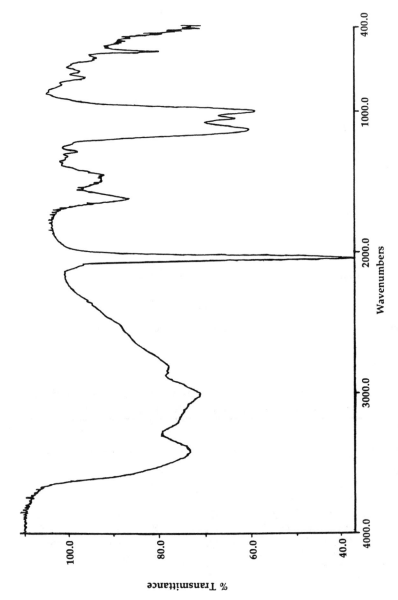

Figure 7. Infrared spectrum of a product containing Fe(CN)$_6^{4-}$ encapsulated between the layers of Zr(O$_3$PCH$_2$NHCH$_2$CH$_2$NH$_3^+$)$_2$.

phosphonate. This spectrum clearly reveals the presence of the ferrocyanide ion by the very strong $C \equiv N$ stretch band at 2036 cm^{-1} and the Fe^{II}-C stretch at 584 cm^{-1}. Also clearly shown are the H_2O and NH_2 stretch and bend bands. Elemental analysis for this complex showed that 0.517 moles of Fe(CN)$_6$$^{4-}$ had been incorporated between the layers and that only traces of Cl$^-$ remained. The interlayer spacing, originally at 15.3 Å increased to 16.4 Å, much lower than expected. Further studies with this interesting class of compounds are underway.

Experimental Section

Synthesis of Polyethyleneoxide Monophosphate.
To a solution of 0.3 moles of polyethylene oxide (Aldrich) in 100 ml of dry carbon tetrachloride (distilled over P_2O_5) contained in a three necked round bottom flask, was added dropwise 0.12 moles of phosphorus oxychloride (Mallinckodt) in 60 ml of carbon tetrachloride through a separatory funnel over a one hour period *(20)*. The HCl formed was eliminated by maintaining the system under suction while stirring was achieved magnetically. Solvent losses were prevented by means of a reflux condenser maintained near 0 °C. When the reaction was over, the solution was cooled down to 6 °C by means of a water-ice bath and 20 ml of water was added very slowly through the separatory funnel. The solvent was then evaporated on a rotary evaporator at 50 °C. After the solution had been cooled, it was neutralized by adding Ba(OH)$_2$ gradually with stirring until the solution gave a distinctive and permanent pink color with phenolphthalein *(21)*. The excess of barium was precipitated with dry ice. The barium phosphate and carbonate were filtered off and the clear filtrate was concentrated under reduced pressure to about 50 ml. To this solution was added 500 ml of ethanol and the resultant solution was left in the refrigerator overnight. The BaCl$_2$ which crystallized was removed by filtration and the filtrate was concentrated to about 100 ml on a rotary evaporator. To the clear, thick residue, 300 ml of acetone was added slowly until the mixture became cloudy. On stirring with a glass rod a soft solid started to form on the bottom. The acetone was decanted and 300 ml more added. The solid now began to harden. This was repeated until the solid could be broken into small pieces; it was then filtered off and washed with acetone and dried in the oven at 60 °C. A typical yield was 50%.

Synthesis of Polyethylene Oxide Diphosphate Ester.
To a solution of 0.06 moles of phosphorus oxychloride in 100 ml of dry carbon tetrachloride in a three necked flask, was added 0.05 moles of polyethylene glycol in 50 ml of dry carbon tetrachloride. After the addition was complete, the rest of the procedure was the same as that followed for the monophosphate.

Synthesis of Zirconium Polyetheroxide Phosphate.

For molecular weights less than 200: 40 mmoles of barium PEO phosphate was dissolved in 100 ml of water. If the solution was not completely clear, it was filtered through a fine filter. A calculated amount of H_2SO_4 was added to precipitate approximately 90% of the barium ion, and the barium sulfate separated by filtration. The clear filtrate was passed through a column packed with ten times the necessary amount of ion-exchange resin (Dowex 50W-X8) in the H$^+$ form to remove any trace of Ba^{2+}. To the resultant solution of the free acid and the combined washings, 10 mmoles of zirconyl chloride in 25 ml of water was added very slowly by means of a separatory funnel, while the solution was being magnetically stirred. Stirring was continued at room temperature for 10 hrs. The solid was separated by filtration and washed with distilled water until the pH of the filtrate was above 3, then air dried. The yield is essentially quantitative based on added zirconyl chloride.

For molecular weights higher than 200:   After the addition of zirconyl chloride no precipitate formed. Evaporation of the excess of water was accelerated by blowing dry N_2 through the solution. When the total volume had decreased to about 50 ml, acetone was added until cloudiness appeared. After leaving the mixture standing at 0 °C overnight, a white gel formed on the bottom of the flask. The acetone was decanted and the precipitate was washed with small portions of water and large portions of acetone until the pH of the liquid was above 3. The solid was then recovered by filtration and washed with acetone and air dried.

Reaction of Zirconium PEO Phosphate with Cu(II).   A stock solution of 0.1 M copper (II) chloride was prepared by dissolving a weighed amount of solid in distilled, deionized water and diluting to 250 ml. An accurately weighed sample of zirconium PEO phosphate was suspended in 10 ml of stock Cu(II) solution. The mixture was magnetically stirred at room temperature for 15 hrs. The solid was then filtered and washed with 10-ml portions of water until the filtrate was completely colorless. The combined filtrate and washings were collected and diluted to 100 ml for further analysis. The procedure was repeated two more times for each sample.

Reaction of Zirconium PEO Phosphate with NaSCN or LiSCN.   A stock solution of 0.1 M NaSCN or 0.1 M LiSCN (Fisher ACS reagent) was prepared in methanol. An accurately weighed sample of dry zirconium polyethylene glycol phosphate was suspended in 10 ml of the stock solution. The mixture was equilibrated for 15 hrs at room temperature. The solid was then filtered and washed with at least six 10-ml portions of methanol. The filtrate and washings were collected for analysis. The equilibration was repeated two more times.

Analysis and Instrumental. X-ray powder diffraction patterns were obtained using a Seifert Scintag Automated Powder Diffraction Unit (PAD-II) with a Ni filtered CuKα radiation. A Cahn electrobalance was used for TGA analysis. Thermograms were obtained in a nitrogen-oxygen atmosphere at a heating rate of 5 °C per minute. The pH measurements were carried out with a Fisher Accumet digital pH Meter model 144 fitted with a Fisher Microprobe combination electrode. The solids were analyzed for C, N, and H by the Center for Trace Characterization at Texas A&M University. IR spectra were obtained on a Perkin Elmer 580B Infrared Spectrophotometer; the samples were prepared as KBr pellets or run as neat liquids between NaCl windows.

Analytical.   Cu(II) was analyzed by the KI method *(22)*. Thiocyanate was determined by Volhard's method *(23)* using silver nitrate in the presence of an iron(III) salt as an indicator.

Conductivity Measurements.   Ion conductivity measurements were carried out by the complex impedance method. The sample was in the form of cold pressed pellets (8 mm dia. and 4-5 mm thick). The pellets were made at room temperature in an evacuable die at 10 tons/sq.in. and the blocking electrodes (Pt paint, Engelhard No. 6042) applied on both faces of the pellets. The pellets were sandwiched between two platinum plates which were spring loaded to ensure good electrode/electrolyte contact. The entire cell was controlled with a Barber-Coleman Controller (Model 520) and a Hewlett-Packard HP-559501B programmer. The temperature was measured using an HP-3421 data acquisition unit with internal calibration. Impedance measurements were performed in the frequency region 5Hz to 10 MHz using HP-4192A and 4800A impedance analyzers. The measuring instruments and the temperature controllers were interfaced to an HP 9816 desk-top computer. The impedance plots and the least squares fitting of the conductivity data were done with the aid of computer programs *(24)*. The electronic conductivity was checked by the low potential DC method and was found to be negligible in all samples investigated. Both cell and furnace were grounded to avoid stray A. C. Fields.

Acknowledgments. This work was supported by the National Science Foundation under grant No. CHE-8921859 and the Robert A. Welch Foundation Grant No. A-673 for which grateful acknowledgment is made. We wish to thank Dr. Chhaya Bhardwaj for carrying out the exchange of the ferrocyanide anion with zirconium polyimines.

Literature Cited

1. Yamanaka, S. *Inorg. Chem.* **1976**, *15,* 2811.
2. Clearfield, A.; Blessing, R. H.; Stynes, J. A. *J. Inorg. Nucl. Chem.* **1968**, *30,* 2249.
3. Clayden, N. J. *J. Chem. Soc., Dalton Trans.* **1987**, 1977.
4. Christensen, A. N.; Krogh-Andersen, E.; Krogh-Andersen, I. G.; Alberti, G.; Nielsen, M.; Lehmann, M. S. *Acta Chem. Scand.* **1990**, *44,* 865.
5. Wong, T.; Brodwin, M; Papke, B. L.; Shriver, D. F. *Solid State Ionics* **1981**, *5,* 689.
6. Lehn, J. M. *Structure and Bonding* **1973**, *16,* 1.
7. Poona, N. S.; Bajaj, A. V. *Chem. Rev.* **1979**, *79,* 389.
8. (a) Clearfield, A.; Smith, G.D. *Inorg. Chem.* **1985**, *24,* 1773.
   (b) Troup, J. M.; Clearfield, A. *Inorg. Chem.* **1977**, 16,3311
9. Ortiz-Avila, C. Y.; Clearfield, A. *Inorg. Chem.* **1985**, *24,* 1773.
10. Takahashi, Y.; Tadokoro, H. *Macromolecules* **1973**, *6,* 672.
11. Clearfield, A.; Tindwa, R. M. *J. Inorg. Nucl. Chem.* **1979**, *41,* 871.
12. Fenton, D. E.; Parker, J. M.; Wright, P. V. *Polymers* **1973**, *14,* 589.
13. Fouletier, M.; Degott, P. Armand, M. B. *Solid State Ionics* **1983**, *8,* 165.
14. Owen, J. R.; Lloyd-Williams, S. C.; Lagos G.; Spurdens, P. C.; Steele, B. C. H., Proc. 2nd Int. Workshop, Lithium Nonaqueous Battery Electrochem., The Electrochemical Soc. **1980**, 293.
15. Armand, M. B.; Chabagno, J. M.; Duclot, M. J. in "Fast Ion Transport in solids," Vashista, P.; Mundy, J. N.; Shenoy, G. K., Eds.; North Holland: New York, 1979, 131.
16. See for example the sections on polymer electrolytes in Solid State Ionics, Nazri, G.; Huggins, R. A.; Shriver, D. F., Eds; Mat. Res. Soc.: Pittsburg, PA 1989.
17. "Solid State Ionics," Boyce, J. B.; DeJonghe, L. C.; Huggins, R. A., Eds.; North Holland: New York, 1986, pp 253-325.
18. Bard, A. J.; Faulkener, L. R. *Electrochemical Methods, Fundamentals and Applications*; John Wiley: New York, 1980.
19. Jerus, P.; Clearfield, A. *Solid State Ionics*, **1982**, *6,* 79.
20. Slatin, L. A. *Synthesis* **1977**, 737.
21. Cherbuliez, E.; Probst, H.; Rabinowitz J. *Helv. Chim. Acta* **1960**, *43,* 464.
22. Vogel, A. *Textbook of Quantitative Inorganic Analysis*, 4th Ed.; Longman: London, 1978, p 379.
23. Ibid, p. 340.
24. Subramanian, M. A.; Rudolf, P. R.; Clearfield, A. *J. Solid State Chem.* **1985,** *60,* 172; Frase, K. Ph.D. Dissertation, University of Pennsylvania, 1983.

RECEIVED January 16, 1992

Chapter 15

Crystalline Inorganic Hosts as Media for the Synthesis of Conductive Polymers

M. G. Kanatzidis[1], C.-G. Wu[1], H. O. Marcy[2], D. C. DeGroot[2], J. L. Schindler[2], C. R. Kannewurf[2], M. Benz[1], and E. LeGoff[1]

[1]Department of Chemistry and the Center of Fundamental Materials Research, Michigan State University, East Lansing, MI 48824
[2]Department of Electrical Engineering and Computer Science, Northwestern University, Evanston, IL 60208

In this paper we present work on the insertion of selected conductive polymers into several crystalline inorganic layered hosts. Specifically, we describe the insertion of polyaniline and polyfuran in FeOCl. We also describe the reaction of the Hofmann-type inclusion compounds $Ni(CN)_2NH_3$(pyrrole) and $Ni(CN)_2NH_3$(aniline) with various oxidants which yield polypyrrole/$Ni(CN)_2NH_3$ and polyaniline/$Ni(CN)_2NH_3$ composites. Magnetic, infra-red spectroscopic, scanning and transmission electron microscopic (TEM) and X-ray diffraction (XRD) data on the state of the polymers inside FeOCl and $Ni(CN)_2NH_3$ are presented. Variable temperature electrical conductivity and thermoelectric power data are reported. Based on XRD and TEM data a structural model for the orientation of polyaniline in FeOCl is proposed.

The idea of having well ordered organic macromolecules inside inorganic host structures is intriguing and during the last five years notable progress towards producing such systems has been made. For example polypyrrole, polythiophene, polyaniline and polyacetylene, the four archetype conducting polymers, have been reported by several groups[1,2] to be synthesized into layered and three-dimensional hosts. Intercalated polymers in structurally defined hosts are of interest for several reasons, including (a) the possibility of direct structural characterization by crystallographic methods, (b) the opportunity for detailed spectroscopic studies on isolated polymer chains without interchain interactions (provided no interference is presented by the host) and (c) the prospect of obtaining conductive polymers with fewer defects and with greater orientation than otherwise possible. They also represent a new class of materials with anisotropic properties derived from the different structural and electronic nature of their individual components.[2]

A demonstrated method to insert polymer chains into host structures is by *in-situ* intercalative polymerization of a monomer using the host itself as the oxidant. FeOCl is one of the most convenient redox-intercalation hosts for a great variety of molecules[3] including conducting polymers.[4,5] The *in-situ* intercalative polymerization of pyrrole, 2,2'-bithiophene and aniline in the interlayer space of this material has been reported.[4,5] The resulting intercalation compounds are composed

0097–6156/92/0499–0194$07.50/0

of alternating monolayers of positively charged conductive polymer chains and negatively charged FeOCl layers. The mechanism of this intercalative process is not well understood.

Apart from the *in situ* intercalative polymerization into strongly oxidizing hosts, another synthetic method to obtain intercalated polymers is to oxidatively polymerize an appropriate monomer already intercalated into an inorganic host using an outside oxidant. This was applied in the case of anilinium ion which was included in the framework of a zeolite and then polymerized to polyaniline by reaction with $(NH_4)_2S_2O_8$.[1b] In this case an external oxidant acts as the electron acceptor instead of the host material. In yet another way, the host material is intercalated by an oxidant and then it is exposed to the appropriate monomer vapor. For example, the inclusion polymerization of polypyrrole, polythiophene and polyaniline in an non-oxidizing zeolite host was accomplished by first introducing an oxidizing ion (e.g. Fe^{3+}, Cu^{2+}) in the zeolite followed by exposure to monomer (pyrrole, thiophene, aniline) vapor.
  Here we report our recent progress in using crystalline hosts to insert conductive polymers by some of the methods mentioned above. Specifically, we describe the insertion of polyaniline and polyfuran in FeOCl. We also describe the reaction of the Hofmann-type inclusion compounds $Ni(CN)_2NH_3(C_4H_5N)$ and $Ni(CN)_2NH_3(C_6H_5NH_2)$ with various oxidants which yield electrically conductive polypyrrole/$Ni(CN)_2NH_3$ and polyaniline/$Ni(CN)_2NH_3$ type composites.

EXPERIMENTAL SECTION

Reagents.   FeOCl is prepared by heating at 380 °C in a sealed pyrex tube a mixture of $FeCl_3$ (anhydrous) and Fe_2O_3 in 1.2:1 ratio. The purple FeOCl crystals are isolated by washing away the excess $FeCl_3$ with acetone and ether. The synthesis of 2,2':5',2"-terfuran was reported earlier.[6] Pyrrole and aniline were distilled under reduced pressure. $Ni(CN)_2NH_3$(pyrrole) and $Ni(CN)_2NH_3$(aniline) were prepared according to the literature.[7]

Insertion of Polyaniline in Single Crystal FeOCl. 0.34 g (3.1 mmol) of single crystals were suspended in a solution of 2.00 g aniline in 50 ml acetonitrile. The mixture remained undisturbed for 60 days. The black crystals were collected by filtration and washed with acetone and ether. Elemental analysis gave (polyaniline)$_x$FeOCl **(I)**. The x can vary from 0.23-0.28.

Reaction of terfuran with FeOCl. 0.3g (2.8 mmole) of FeOCl and 0.2g (1 mmole) of terfuran were mixed with 1.5 ml MeOH, frozen in the bottom of a pyrex tube which was subsequently flame-sealed under reduced pressure. The tube was heated at 100 °C for 6 days. The black shiny solid was isolated by filtration washed with acetone and dried in vacuum. Elemental analysis gave $(C_4H_2O)_x(MeOH)_nFeOCl$ **(II)** with x ranging between 0.63-0.81. Due to the presence of methanol we do not know the exact ammount of polyfuran present. For simplicity we will refer to this material as polyfuran/FeOCl.

Reaction of $Ni(CN)_2NH_3$(pyrrole) (III), with Fe^{3+}. 0.20g (1 mmole) of $Ni(CN)_2NH_3$(pyrrole) were suspended in an aqueous solution of 0.8g (5 mmole) of $FeCl_3$. The mixture was stirred at room temperature for 1.5 hs and the black solid was isolated by filtration, washed with H_2O and acetone and dried in vacuum. Elemental analysis suggests the formula $[Ni(CN)_2NH_3]\{(C_4H_2NH)Cl_{0.22}\}_{0.9}$ **(V)**. The same reaction can also be run in $CHCl_3$. The corresponding reactions with single crystals were carried out in a similar manner as described above except that 3 days were required.

Reaction of $Ni(CN)_2NH_3$(aniline) (IV), with Fe^{3+}. The procedure was similar to the one employed for the pyrrole analog. Elemental analysis suggests the formula $[Ni(CN)_2NH_3]\{(C_6H_4NH)Cl_{0.52}\}_{0.15}$.**(VI)** The corresponding reactions with single crystals were carried out in a similar manner as described above except that 1 week was required.

Removal of the $Ni(CN)_2NH_3$ host. In a typical experiment $[Ni(CN)_2NH_3]$-$\{polymer\}_x$ is suspended in 50ml 1M aqueous HCl and stirred for 2 days until no CN vibration peak (2161 cm^{-1}) is detected in the Fourier transform infrared (FTIR) spectrum and no Ni by SEM-EDS. The black powder was washed with ethanol and ether and dried in vacuo. Ethylene diamine also can be used to dissolve the $Ni(CN)_2NH_3$ matrix if non-aqueous conditions are desired.

Physical Measurements. X-ray diffraction experiments were carried out with a Phillips XRG-3000 instrument using $CuK\alpha$ radiation. Infrared spectroscopy was performed with a Nicolet 740 FTIR spectrometer. A Nicolet P3F four-circle diffractometer was used for single crystal studies. Direct current electrical conductivity and thermopower measurements were obtained from 5 to 320 K using a data acquisition and analysis system described elsewhere.[8a] Samples were measured as pressed pellets using conditions and protocols described earlier.[8b] Scanning electron microscopic (SEM) and transmission electron microscopic (TEM) studies were performed on JEOL-JSM35CF and JEOL-100CX-(II) instruments respectively.

RESULTS AND DISCUSSION

FeOCl as an Intercalation Host: FeOCl possesses a two-dimensional polymeric structure in which FeOCl layers are separated by van der Waals gaps between chlorine atoms.[9] The coordination number of iron in this compound is six and the geometry is distorted octahedral. There are two axial *trans* chloride ligands and four equatorial oxygen ligands. The oxygen atoms are shared by four iron atoms while the chlorine atoms are shared by two iron atoms. The crystallographic b-axis is 7.92 Å and runs perpendicular to the layers. It is this axis which expands upon intercalation, while the a- and c- axes at 3.30 Å and 3.78 Å respectively do not change significantly. The structure of FeOCl and its unit cell (a- and c- axes) are shown in Figure 1. The average Fe-O and Fe-Cl distances are 2.032 Å and 2.368 Å respectively. The Fe-Fe distance is 3.30 Å. The ability of FeOCl to intercalate guest ions and molecules derives by its redox properties and specifically its ability to be reduced ($Fe^{3+} \longrightarrow Fe^{2+}$) to a certain extent, without undergoing significant structural transformations. When intercalated with species such as pyridinium, ferrocenium, K^+, Rb^+, Cs^+ etc, the FeOCl layers become semiconducting due to the thermally activated carrier hopping between Fe^{3+} and Fe^{2+} sites.[10] Over-reduction of the framework (>~30%) can result in destruction of its crystallinity.

Insertion of Polyaniline in FeOCl Single Crystals. The intercalation of single crystals of many materials at room temperature is known to be very slow, it is seldom complete, and usually it destroys the crystals. Nevertheless we still attempted the reaction of FeOCl with aniline to see if similar difficulties existed in this system as

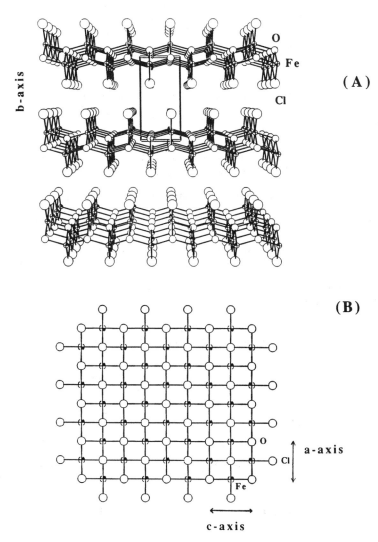

Figure 1. (A) The layered nature of FeOCl. View perpendicular to the b-axis.
(B) Projection of a single FeOCl layer on the ac-plane. The unit-cell in this plane
is shown.

well. After several unsuccessful tries, we have been able to insert polyaniline into single crystals of FeOCl by careful slow reaction using a CH_3CN solution of aniline in air. Surprisingly, the resulting compound, (polyaniline)$_{0.28}$FeOCl , retains significant single crystal character, despite the fact that the overall crystallinity has decreased and most crystals appear to have been exfoliated. The crystals cleave perpendicular to the [010] direction, consistent with their lamellar structure. This can be observed clearly in the SEM photographs of (I) which are shown in Figure 2. The insertion of polyaniline causes an interlayer expansion of 6.54 Å.[5] The formation of polyaniline (emeraldine salt) in the interlayer space can be easily confirmed by FTIR spectroscopy.

Examination of crystals of (I) by X-ray diffraction revealed that the crystal quality, though inferior to that of starting FeOCl, was sufficiently good for preliminary X-ray diffraction experiments. Oscillation photographs of such crystals show broad but intense diffraction peaks. The broadest peaks are those of the *00l* class of reflections. The peak width by class varies as follows: *00l>0kl,h0l>hkl>hk0*. Axial X-ray photographs from single crystals are shown in Figure 3. This is to be expected considering that the FeOCl crystals have undergone a topotactic intercalation reaction.

Interestingly, the diffraction pattern of the *hk0* zone reveals a set of strong reflections associated with the parent 3.30 X 3.78 Å cell, and a set of several weak reflections half way between the strong ones, suggesting that the periodicity along the a- and c-axes have doubled relative to the original FeOCl unit cell. The new unit cell is orthorhombic with a=6.60 Å, b=28.86 Å, c=7.56 Å and V=1459 Å^3. The origin of this superlattice is due to substantial long-range order of polyaniline in FeOCl. Long range order of polyaniline in the intralamellar space of FeOCl could be achieved by orientation of the polymer chains along certain crystallographic directions, such that a doubling of the periodicity of a- and c- axes is caused (vide infra).

Although the structure of polyaniline is not yet known in detail, it has been proposed, based on X-ray diffraction studies on powders, that it is similar overall to that of poly-phenyleneoxide.[11,12] This is shown in scheme (A).

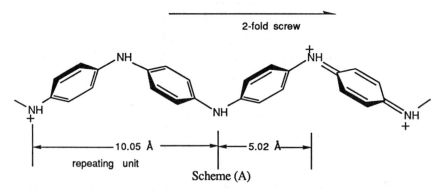

Scheme (A)

Based on single crystal crystallographic studies of oligoanilines, by Baughman *et al*[12], the repeating unit of polyaniline is estimated to be 10.05 Å (ignoring the details in individual phenyl groups). The highest symmetry possible for the polymer chain corresponds to a 2-fold screw axis in the chain direction. The step of the 2-fold screw axis is 10.05/2=5.02 Å. If we now attempt to orient such a chain along either the a or the c-axes of the FeOCl layers in Figure 4 we see that the repeating unit of the polymer

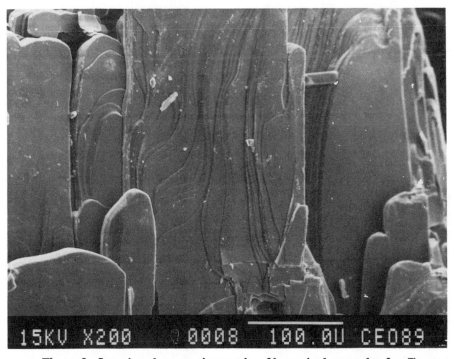

Figure 2. Scanning electron micrographs of large single crystals of α-(I)

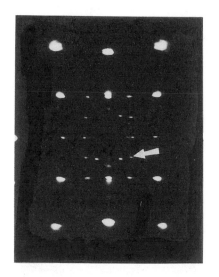
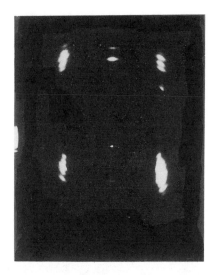

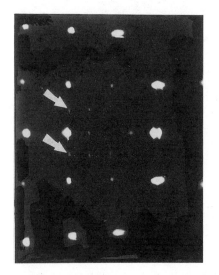
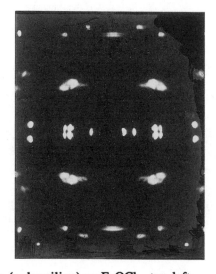

Figure 3. Oscillation photographs of α-(polyaniline)$_{0.28}$FeOCl. <u>top left</u>: random orientation;, <u>top right</u>: along the a-axis. The arrows indicate weak superlattice reflections that correspond to a doubling of the original a-axis; <u>bottom left</u>: along the b-axis. This is the axis perpendicular to the layers; <u>bottom right</u>: along the c-axis. The arrows indicate weak superlattice reflections that correspond to a doubling of the original b-axis   The new orthorhombic cell is a=6.60 Å, b=26.45 Å and c=7.56 Å.

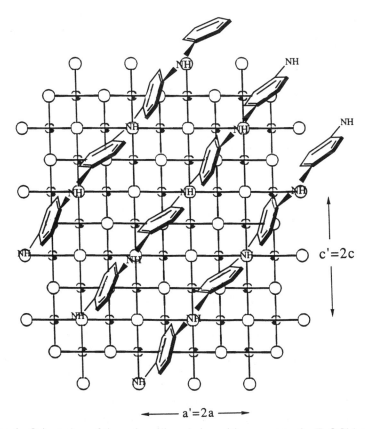

Figure 4. Orientation of the polyaniline chains with respect to the FeOCl lattice. This particular arrangement results in doubling of the periodicity of the a- and c-axes. The polymer chains are arranged side by side at ~5 Å intervals.

and the inorganic layers are not commensurate. A doubling of the a- or c- axes cannot be achieved. A close examination of the structural details of FeOCl reveals that the minimum Cl---Cl' distance along the [101] direction (diagonal to the a-,c- axes) is 5.2 Å, almost one-half of the repeating unit of polyaniline. By orienting parallel to the [101] direction, the polymers can provide each NH unit with a Cl "partner" thus establishing H---Cl hydrogen bonding which can act as an additional stabilizing force, the main stabilizing force of course being the coulombic attraction of the positively charged polymer to the negatively charged FeOCl layers. This is shown in scheme (B). We see that a diagonal orientation of polyaniline produces a system in which the positions of the NH groups are commensurate with those of the Cl atoms in the layers. This produces a new orthorhombic unit cell with crystallographic axes a'=2a and c'=2c. In this fashion, the polymer chains can stack side by side lying above every other row of Cl atoms with an interchain distance of ~5 Å. Because the doubling of the unit cell is caused by the light carbon and hydrogen atoms, the superlattice reflections would be weak, consistent with the experimental data. What the experimental data cannot tell us, is how the polyaniline chains are oriented (i.e. along the [101] or [-101]) in the interlayer space below and above the one shown in Figure 4.

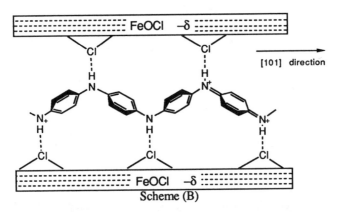

Scheme (B)

Continuing Polymerization. Interestingly, fresh samples of (I) transform in air at ambient temperature to another more oxidized phase. This change appears to be continuous rather than abrupt and does not occur when the material is kept under inert atmosphere or vacuum. Spectroscopically this change is visible by IR spectroscopy by the disappearance of the strong peak at 480 cm^{-1} and by the appearance of a new peak at ~680 cm^{-1} [5]. For the sake of convenience we will call this oxidized material the "β"-phase. By comparison the freshly prepared samples of (I) will be called the α- phase. The "β"-phase can also be distinguished by Mössbauer[13] spectroscopy, EXAFS measurements[13] and charge transport measurements. Energy dispersive microprobe analysis (EDS) of the oxidized sample reveals significantly decreased Cl content which varies from sample to sample. This could be attributed to hydrolysis and formation of $FeOCl_{1-x}(OH)_x$. which, if left to proceed, forms β-FeOOH.

The change from α– to "β"- though unexpected, can be rationalized if reaction with oxygen is taken into account. Oxidation of aniline oligomers by O_2 in the intralamellar space of the FeOCl could result in further coupling of these molecules and an increase in the average polymer molecular weight, as shown in Eq. 1. Recent experiments involving the measurement of oxygen consumption during the reaction are positive and are consistent with the above interpretation. This is a new development and is currently under investigation.

Eq. 1

Polyfuran in FeOCl. Following the successful insertion of polypyrrole, polythiophene and polyaniline we attempted to extend this chemistry to polyfuran using furan and 2,2'-bifuran. We found that these two molecules do not intercalate probably because they are hard to oxidize. Subsequently, we used terfuran and tetrafuran for intercalation in iron oxychloride. The redox potentials of terfuran and tetrafuran are ~1.3 and ~1.0 volts vs. SCE respectively as determined by cyclic voltammetry.[14] These species should oxidatively polymerize in FeOCl to yield polyfuran, according to Eq. 2. In fact, the preparation of polyfuran from such oligomers has not been reported previously and work is in progress in our laboratory to properly characterize polyfuran prepared from these monomers. Literature reports on polyfuran indicate that when prepared from furan or 2,2'-bifuran the polymer is ill-defined, inhomogeneous, chemically unstable and insulating.[15]

Eq. 2

Terfuran reacts slower with FeOCl than pyrrole or 2,2'-bithiophene, presumably because of its larger size and higher oxidation potential. In the beginning stages of this work we used acetonitrile as solvent and reflux in air but we failed to obtain reproducible and complete intercalation, even after several weeks. After a systematic search for suitable condition, we found that methanothermal conditions (sealed tube, 100 °C) give the best results in terms of crystallinity and phase homogeneity. Sonication of the FeOCl in methanol for ~2-5 min was found to accelerate the reaction considerably. In the intercalated product, the FeOCl layers are expanded by ~7.7 Å to accommodate the intercalated polyfuran species. Interestingly this expansion is considerably larger than those observed for other polymers inserted in FeOCl, see Table I. The reason for the larger expansion is not known but there is spectroscopic infrared evidence that methanol co-intercalates in the host material. The intercalation was deemed complete when the *002* reflection of FeOCl disappeared completely from the X-ray powder diffraction pattern of the product.

Tetrafuran intercalated partially when it was heated with FeOCl in methanol in a sealed tube at 100 °C. It did not intercalate when refluxed in acetonitrile. This was attributed to the large size of this monomer and the resulting exceedingly slow kinetics of the reaction.

The Fourier transform infrared spectrum (FTIR) of (II) shows absorptions both from the organic and FeOCl portions of the material. The presence of polyfuran in (II) is apparent when comparing its FTIR spectrum with that of chemically prepared bulk polyfuran and that of polyfuran extracted from (II). This is shown in Figure 5.

TABLE I   Observed Expansion in the Intralamellar Space of FeOCl Upon Insertion of Various Conducting Polymers and Organic Molecules

System	d-spacing (Å)	Expansion (Å)	Ref.
(Ppy)/FeOCl	13.15	5.23	4
(Pth)/FeOCl	13.25	5.33	4
(PANI)/FeOCl	14.46	6.54	5
(Pfu)/FeOCl	15.62	7.70	14
(Py)(PyH$^+$)/FeOCl	13.45	5.53	24
(An)(AnH$^+$)/FeOCl	14.32	6.40	16
(Fc)/FeOCl	13.03	5.11	22a
(TTF)/FeOCl	12.8-13.6	4.9-5.7	23

Ppy=polypyrrole; Pth=polythiophene; PANI=polyaniline; Pfu=polyfuran; Py=pyridine; An=aniline; Fc=ferrocenium; TTF=tetrathiafulvalene

The crystalline nature of polyfuran/FeOCl is also evident in the SEM images of the samples which also show the absence of impurity amorphous phases. Selected area diffraction (SAED) patterns of polyfuran/FeOCl obtained from TEM studies show clearly the structural integrity of the material is maintained after intercalation. Figure 6 shows a SAD pattern of the *(h0l)* zone with diffraction spots which can be indexed according to the original a- and c- axes of the FeOCl lattice. This is consistent with the expected invariance of the a- and c- axes of the crystal upon intercalation. The orientation of polyfuran in the intralamellar space remains unknown as no diffraction spots corresponding to a superlattice are evident.

Polymerization in Hofmann type inclusion compounds

The Hofmann-type inclusion compounds $Ni(CN)_2NH_3(X)$ (X=benzene, pyrrole, aniline, thiophene etc.) have been known for almost 100 years.[7] Extensive structural and spectroscopic studies on this class of compounds unequivocally established their layered structure and the presence of the small aromatic molecules at well defined crystallographic positions.[17] Our interest in developing a general methodology to obtain conductive polymers in various ordered host media and the ready availability of the Hofmann-type compounds with oxidatively polymerizable guest molecules, prompted us to examine the reactivity of these compounds with external oxidants, in the hope of obtaining electrically conductive polymer inclusion $Ni(CN)_2NH_3$ complexes.

The structure of a Hofmann type inclusion compounds is shown in scheme (C). The reaction of (III) and (IV) with oxidants such as Fe^{3+} in water results in a color change from pale violet to black. When single crystals are used as starting materials, the resulting products, retain the original crystal shape (pseudomorphs), but they lose their single crystal nature and under SEM microscopic examination show small cracks and other defects. The proper formulation of the oxidation products would be $[Ni(CN)_2NH_3]\{(polymer)(A)_x\}_y$ where polymer=polypyrrole, polyaniline;

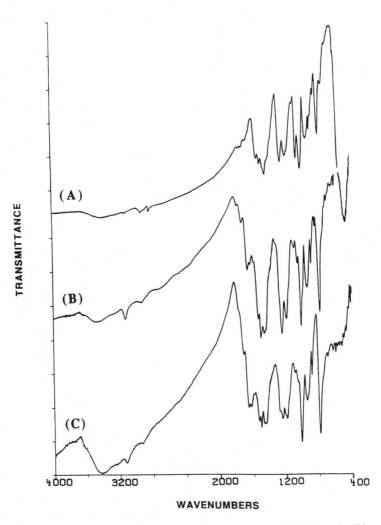

Figure 5. FTIR spectra (KBr pellets) of (A) polyfuran/FeOCl (II) (B) polyfuran isolated from (II) and (C) polyfuran prepared by the oxidation of terfuran with FeCl₃.

Figure 6. Selected area diffraction pattern of crystallites of polyfuran/FeOCl (II) with the beam perpendicular to the (101) plane showing the crystalline nature of the material.

A=dopant, e.g. Cl⁻ or NO₃⁻. There is still some uncertainty as to the exact amount of dopant in these materials.

Scheme (C)

X-ray powder diffraction data from $[Ni(CN)_2NH_3]\{(polymer)(A)_x\}_y$ show that the stacking order in the $Ni(CN)_2NH_3$ layers has been disrupted, resulting in turbostratic and interstratified systems. The presence of polypyrrole and polyaniline and absence of the corresponding monomers was established by FTIR spectroscopy.[18] The FTIR spectra of (III), (V), bulk polypyrrole and the polypyrrole itself after isolated from the host $Ni(CN)_2NH_3$ matrix are shown in Figure 7. The $Ni(CN)_2NH_3$ framework vibrations at 2161, 1610 and 440 cm⁻¹ are present in the FTIR spectra of (V) indicating that it remains intact after the polymerization process. In addition, the strong narrow peaks due to the pyrrole monomer at 1168, 745, 612 and 528 cm⁻¹ are completely absent from the spectra of (V). Instead the characteristic polypyrrole vibration peaks[18] are prominent (Figure 7B). Similar spectroscopic data (not shown) are obtained from the polyaniline analog.

Selected area electron diffraction (SAED) patterns taken in a TEM with the electron beam perpendicular to the (001) crystallographic plane from crystals of (V) and (VI), show a tetragonal lattice as expected from the $Ni(CN)_2NH_3$ sheets. This is shown in Figure 8. The estimated a- axis length as measured from the SAED patterns agrees very well with a- axis of $[Ni(CN)_2NH_3]\{monomer\}$.[20] This supports the IR and XRD conclusions that the structural integrity of the inorganic framework is preserved after oxidation.

Generally, the oxidation of aniline included in the $Ni(CN)_2NH_3$ network is much slower than that of pyrrole. We consistently found significantly lower amounts of polyaniline than expected, had all the aniline converted to polyaniline. Only about 15% of the original aniline was found polymerized in the $Ni(CN)_2NH_3$ network. In water we observe, by NMR spectroscopy, that significant amounts of aniline diffuse out of the structure of $[Ni(CN)_2NH_3]\{aniline\}$ into the aqueous phase. Although pyrrole also diffuses out of the $Ni(CN)_2NH_3$ network when suspended in water, its rapid oxidation to polypyrrole results in high polymer (but still less than theoretical) content in the final product.

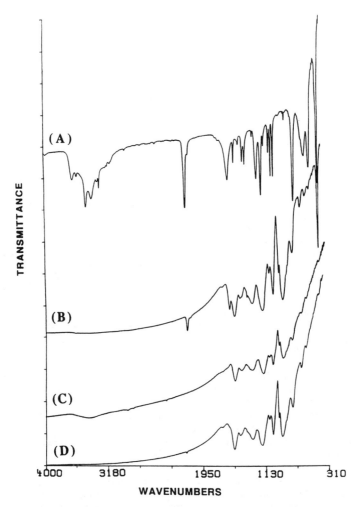

Figure 7. FTIR spectra (KBr pellets) of: (A) $Ni(CN)_2NH_3$(pyrrole), (B) $[Ni(CN)_2NH_3]\{(C_4H_2NH)Cl_{0.22}\}_{0.9}$ (V), (C) Bulk polypyrrole prepared by oxidation with aqueous $FeCl_3$, (D) Polypyrrole extracted from (V) by dissolving the inorganic framework with ethylene diamine.

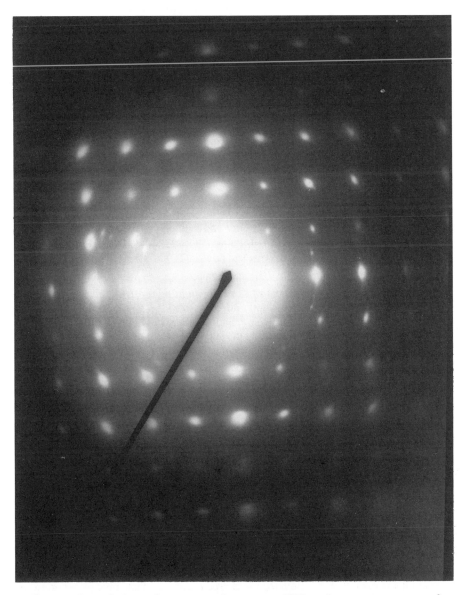

Figure 8. Selected area electron diffraction patterns of $[Ni(CN)_2NH_3]\{(polymer)(A)_x\}_y$ and $[Ni(CN)_2NH_3]\{monomer\}$ for polypyrrole showing the crystalline nature of the material. The e-beam is parallel to the [001] direction.

We used EPR spectroscopy to address the question of the location of the polymer: *inside versus outside*. The EPR spectra of $[Ni(CN)_2NH_3]\{(polymer)(A)_x\}_y$ show broad and weak resonances. For example, the typical narrow and strong EPR resonance of bulk polyaniline is not observed here (see Figure 9), suggesting that the unpaired electron density in the polymer is very close to, and magnetically interacts with the paramagnetic $Ni(CN)_2NH_3$ matrix.[21] This exchange broadened EPR signal suggests that polyaniline is indeed intercalated into, and not intimately mixed (separate phases) with, $Ni(CN)_2NH_3$. If the $Ni(CN)_2NH_3$ matrix is removed by dissolving in aqueous HCl we observe the expected EPR signal of bulk polyaniline. This effect has also been observed in conductive polymers encapsulated in zeolite[1], $V_2O_5 \cdot nH_2O$ [2] and FeOCl [4]. Control experiments show that intimate mixtures of bulk polymer and $Ni(CN)_2NH_3(H_2O)_{0.25}$, show the same EPR spectra as bulk polyaniline. Analogous effects are seen in the corresponding polypyrrole products.

Charge Transport Properties

Polyaniline/FeOCl. Electrical conductivity data from single crystals of α–(I) show a room temperature value of $3 \times 10^{-2} \, \Omega^{-1} cm^{-1}$ and thermally activated behavior, Figure 10. Interestingly the conductivity is not significantly larger than that of pressed pellets of the same material.[5] The thermoelectric power (TP) of single crystals of α-(I) is positive at room temperature (~7 $\mu V/K$) and increases linearly with falling temperature reaching the value of 100 $\mu V/K$ at 140 K, as shown in Figure 10. This is typical of a p-type semiconductor and it is distinctly different from the p-type metallic character found previously in $(Ppy)_{0.34}FeOCl$ [4] and $(Pth)_{0.24}FeOCl$.[4] Similar TP vs. T dependence was found in single crystals of the p-type semiconductor $(Fc)_{0.50}FeOCl$.[22b] Since bulk protonated emeraldine shows a metal-like TP behavior[25], it is likely that in α–(I) the FeOCl section and not polyaniline is responsible for the charge transport properties. The reason for this, may be the short chain length of the intercalated species which would inhibit facile chain to chain charge transport to the point that it becomes more favorable through the FeOCl layers.

The transformation to the "β-" form is accompanied by a *significant increase in conductivity* (Figure 11) and a dramatic change in the TP behavior (Figure 11). The single crystal conductivity data of "β"-(I) show that with falling temperature behavior of the material is in fact metallic down to 250 K, suggesting oriented polyaniline chains. At this temperature a metal to semiconductor transition is observed down to 220 K below which the decrease in conductivity changes slope suggesting a second transition. The thermoelectric power of the material is small and positive (~2 $\mu V/K$ at 300 K) and decreases with temperature down to 260 K below which it remains constant at ~0.6 $\mu V/K$. This suggests a metallic system with holes being the charge carriers. Bulk polyaniline is known to exhibit positive small thermopower at less than maximum protonation levels. The TP data suggest that in "β"-(I) the polyaniline is mainly responsible for the charge transport. This must originate from the increase in MW by O_2 oxidation (Eq. 1) which enhances the effective conjugation length in individual polymer chains and reduces the frequency of carrier hopping from chain to chain, thus facilitating charge transport through the intercalated polymer. Work is continuing to further characterize "β"-(I).[13]

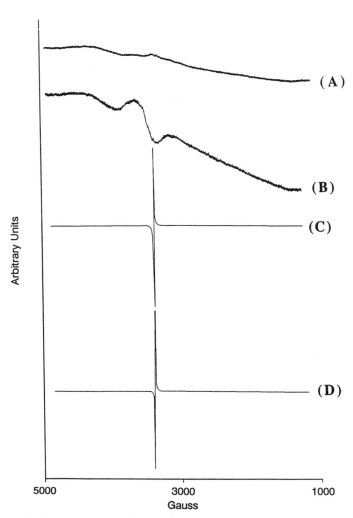

Figure 9. EPR spectra (at 25 °C, frequency, 9.13 GHz, power 20mW) of: (A) Ni(CN)$_2$NH$_3$(aniline) (II), gain=10^3; (B) [Ni(CN)$_2$NH$_3$]{(C$_6$H$_4$NH)Cl$_{0.52}$}$_{0.15}$ (VI), gain=10^3; (C) bulk polyaniline, gain=120; (D) polyaniline extracted from (VI) by dissolving the inorganic framework with ethylene diamine, gain=45.

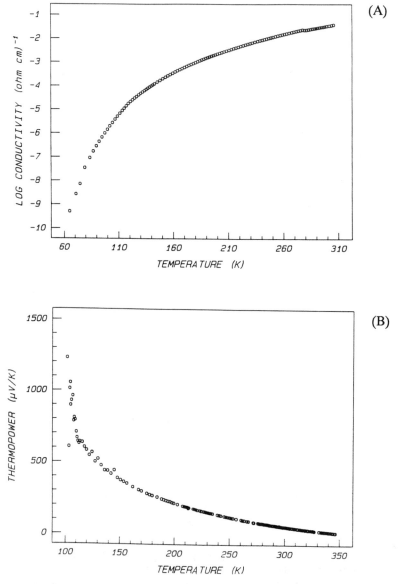

Figure 10. (A) Single crystal four probe electrical conductivity data of α-(I) as a function of temperature. (B) Single crystal four probe thermoelectric power data of α-(I) as a function of temperature.

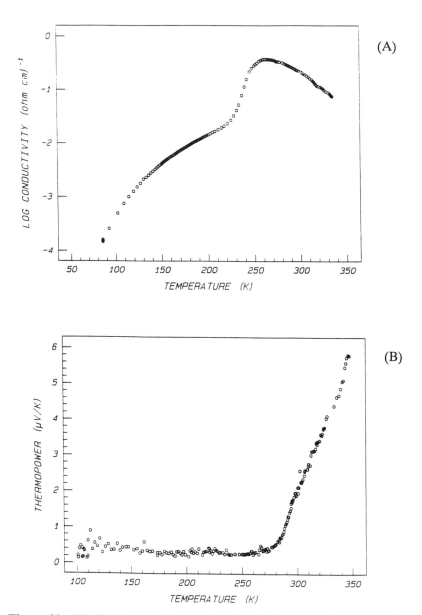

Figure 11. (A) Single crystal four probe electrical conductivity data of "β"-(I) as a function of temperature. (B) Single crystal four probe thermoelectric power data of "β"-(I) as a function of temperature.

Polyfuran/FeOCl. Electrical conductivity data on pressed pellets of (II) show a room temperature value in the range of 10^{-3}-10^{-4} Ω^{-1}cm^{-1} and thermally activated behavior. The general behavior is thermally activated charge transport, similar to that observed in other intercalated conductive polymers.[2,3] However, (II) is considerably less conductive than its polypyrrole and polythiophene analogs.[4] A possible reason for this may be the anticipated shorter chain-lengths of polyfuran in (II). Short chain lengths are expected considering that terfuran is a relatively large molecule and its reactivity will slow down significantly in the constrained intralamellar space of the host thus avoiding extensive polymerization.

The thermoelectric power of (II) is positive at room temperature (~7 μV/K) and increases linearly with falling temperature reaching the value of 45 μV/K at 230 K (see figure 12). This behavior suggests a p-type semiconductor and it is distinctly different from the p-type metallic character found previously in polypyrrole/FeOCl [2] and polythiophene/FeOCl.[2] Similar temperature dependence of the thermopower was found in α-(I). Based on the thermopower results, it is likely that in (II) the FeOCl and not polyfuran, is responsible for the charge transport properties. The reason for this, may be the short chain length of the intercalated polyfuran which would inhibit facile chain to chain charge transport.

Polypyrrole and Polyaniline in Ni(CN)$_2$NH$_3$. Variable temperature electrical conductivity measurements on single crystals and pressed pellets of (V) show a relatively high room temperature value of 0.6 and 0.2 Ω^{-1}cm^{-1} and thermally activated behavior, as shown in Figure 13A. It should be noted however that "single" crystals of (V) remain conductive down to fairly low temperatures, while the corresponding powder samples become insulating below 50 K (see Figure 13A). Since the host Ni(CN)$_2$NH$_3$ matrix is an insulator, the electrical properties of (V) must be due entirely to the included polypyrrole. This is further supported by the TP measurements which show a small positive Seebeck coefficient (~10 μV/K at RT) which decreases linearly with decreasing temperature (2 μV/K at 90 K), typical of metallic behavior (see Figure 13B). Consistent with the higher conductivity of "single" crystals, variable temperature TP data from "single" crystals of (V) show a smaller Seebeck coefficient than corresponding powder samples (see Figure 13B). The TP data indicate a p-type metal-like conductor similar to that of bulk polypyrrole.[26] The higher electrical conductivity observed in the "single" crystal samples compared to the powder samples suggests a greater degree of orientation of the polypyrrole chains in the Ni(CN)$_2$NH$_3$ lattice. It should be noted that, although the room temperature conductivity of the single crystals is only slightly higher than that of the powder, at low temperatures the difference in conductivity approaches five orders of magnitude. The thermally activated behavior found in the conductivity of single crystals (instead of say metal-like) probably reflects a) the finite polymer chain length in the host Ni(CN)$_2$NH$_3$ matrix which is associated with the inevitable carrier hopping from chain end to chain end, and/or b) resistance effects found at microscopic crystal domain boundaries (microcracks).

Despite the fact that polyaniline is present as the conductive emeraldine salt in Ni(CN)$_2$NH$_3$, the material was found to be essentially an insulator. This is consistent with the small amount of polymer present. The polymer is inaccessible for electrical contact with the probe electrodes. When the Ni(CN)$_2$NH$_3$ matrix is removed

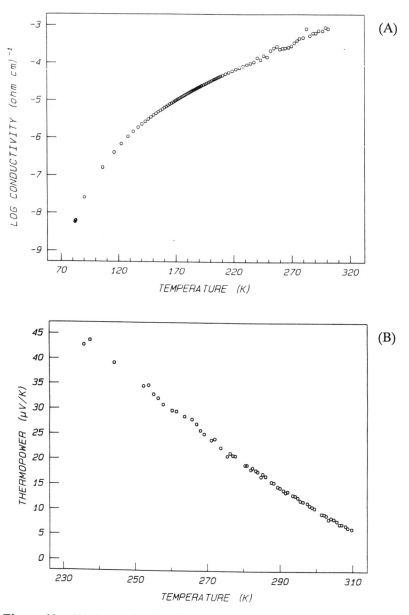

Figure 12. (A) Pressed pellet four-probe electrical conductivity data as a function of temperature of (II) (B) Thermoelectric power (Seebeck coefficient) data for a pressed pellet of (II) as a function of temperature.

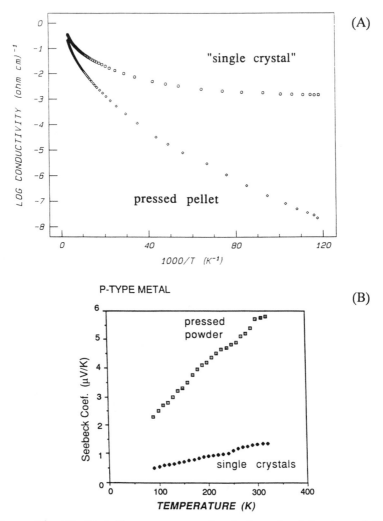

Figure 13. (A): Variable temperature electrical conductivity data of (V) as pressed polycrystalline pellet and in "single" crystal form. (B): Variable temperature thermolectric power data (Seebeck coefficient) of (V) as pressed polycrystalline pellet and in "single" crystal form.

by dissolution in ethylene diamine (or aq. HCl) the black polymer which is isolated is a good electrical conductor with room temperature conductivity of 1 S/cm.

Location of the polymer chains in (V) and (VI). The inside vs. outside question.

We note here that we could not obtain unequivocal evidence that the conductive polymers in $[Ni(CN)_2NH_3]\{(polymer)(A)_x\}_y$ are indeed completely intercalated between the $Ni(CN)_2NH_3$ sheets. The X-ray powder diffraction (XRD) patterns of (V) and (VI) are very similar. The original 002 peak present in $[Ni(CN)_2NH_3]\{monomer\}$ is absent in the corresponding patterns of (V) and (VI) and no reflection can be identified which can be assigned to the interlayer spacing in $[Ni(CN)_2NH_3]\{(polymer)(A)_x\}_y$. The XRD patterns show primarily $hk0$ type reflections. There are several explanations for these observations. First, the absence of any order along the stacking c-direction may be due to random stacking of the $Ni(CN)_2NH_3$ sheets. This could happen if there was random staging (not every interlayer gap was occupied by polymer). A second possibility is that the polymers are not intercalated at all but intimately mixed with a crystalline phase such as $Ni(CN)_2NH_3(H_2O)_{0.25}$.[27] This however is in contrast with the EPR data discussed above. The XRD patterns cannot rule out such a phase and cannot confirm that the oxidation reaction is really topotactic. It is likely that in the case of (V) both of the above possibilities occur simultaneously. This is suggested by the following. When single crystals of (V) are cleaved and examined under the scanning electron microscope a separate polypyrrole phase (small spheres of amorphous material ~1-2 μm in diameter) is visible at the edges and surface of the crystals. The spheres however are well separated from each other and do not seem to make a continuous path from one end of the sample to the other. Nevertheless, the fact that "single" crystals of (V) are more conductive than powder samples, argues that at least a significant portion of polypyrrole is aligned with, and intercalated in the $Ni(CN)_2NH_3$ framework. The above data suggest that the oxidative polymerization reaction is not truly topotactic. Auger and X-ray photoelectron spectroscopic (XPS) studies are needed to further probe this point. In contrast to (V), we see no evidence, under SEM conditions, of separate polyaniline phase in crystals of (VI). This, in combination with our EPR spectroscopic data (vide supra) suggests that polyaniline is intercalated between the $Ni(CN)_2NH_3$ layers.

Acknowledgement. Financial support from the National Science Foundation (grant DMR-8917805 to MGK) is gratefully acknowledged. At Northwestern University this work made use of Central Facilities supported by NSF through the Materials Research Center. This work made use of the SEM facilities of the Center for Electron Optics at Michigan State University.

REFERENCES

1)   (a) Bein, T.; Enzel, P. *Synth. Met.* **1989**, *29*, E163-E168 (b) Enzel, P.; Bein, T. *J. Phys. Chem.* **1989**, *93*, 6270-6272   (c) Bein, T.; Enzel, P. *Angew. Chem. Int. Ed. Engl.* **1989**, *28*, 1692-1694 (d) Brandt, P.; Fisher, R. D.; Martinez, S. E.; Calleja, R. D. *Angew. Chem. Int. Ed. Engl.* **1989**, *28*, 1265-1266 (e) Caspar, J. V.; Ramamurthy, V.; Corbin, D. R. *J. Am. Chem. Soc.* **1991**, *113*, 600-610 (f) Pereira, C.; Kokoteilo, G. T.; Gorte, R. J. *J. Phys. Chem.* **1991**, *95*, 705-709 (h) Cox, S. D.; Stucky, G. D. *J. Phys. Chem.* **1991**, *95*, 710-720 (i) Soma, Y.; Soma, M.; Furukawa, Y.; Harada, I. *Clays and Clay Minerals* **1987**, *35*, 53-59

2)   (a) Kanatzidis, M.G.; Wu, C.-G.; Marcy, H.O.; DeGroot, D.C.; Kannewurf, C.R. *J. Am. Chem. Soc.* **1989**, *111*, 4139-4141   (b) Wu, C.-G.; Kanatzidis, M.G.; Marcy, H.O.; DeGroot, D.C.; Kannewurf, C.R. *Polym.*

Mat. Sci. Eng. **1989**, 61, 969-973. (c) Wu, C.-G.; Kanatzidis, M.G.; Marcy, H.O.; DeGroot, D.C.; Kannewurf, C.R. NATO Adavced Study Institute *"Lower Dimensional Systems and Molecular Devices"* R. M. Metzger, Ed. Plenum Press, Inc. 1990 in press (d) Kanatzidis, M.G.; Wu, C.-G.; Marcy, H.O.; DeGroot, D.C.; Kannewurf, C.R. *Chem. Mater.* **1990**, 2, 222-224 (e) Wu, C.-G.; Marcy, H.O.; DeGroot, D.C.; Kannewurf, C.R.; Kanatzidis, M.G. *Mat. Res. Soc. Symp. Proc.* **1990**, 173, 317-322

3)   (a) Herber, R. H.; Maeda, Y. *Inorg. Chem.* **1981**, 20, 1409-1415 (b) Weis, A.; Choy, J.-H. *Z. Naturforsch.* **1984**, 39b, 1193-1198 (c) Kauzlarich, S. M.; Stanton, J. L.; Faber, J. Jr.; Averill, B. A. *J. Am. Chem. Soc.* **1986**, 108, 7946-7951

4)   (a) Kanatzidis, M. G.; Tonge, L. M.; Marks, T. J.; Marcy, H. O.; Kannewurf, C. R. *J. Am. Chem. Soc.* **1987**, 109, 3797-3799 (b) Kanatzidis, M. G.; Marcy, H. O.; McCarthy, W. J.; Kannewurf, C. R.; Marks, T. J. *Solid State Ionics* **1989**, 32/33, 594-608

5)   Kanatzidis, M.G.; Wu, C.-G.; Marcy, H.O.; Kannewurf, C.R.; Kostikas, A.; Papaefthymiou V. *Adv. Mater.* **1990**, 2, 364-366

6)   (a) Leung, W.-Y.; LeGoff, E. *Synthetic Communications*, **1989**, 19, 789-791 (b) Leung, W-Y. Ph.D. Thesis, Michigan State University, Sept. 1988.

7)   (a) Hofmann, K. A.; Küspert, F. A. *Z. Anorg. Allg. Chem.* **1897**, 15, 204 (b) Hofmann, K. A.; Höchtlen, F. *Chem. Ber.* **1903**, 36, 1149-1151 (c) Hofmann, K. A.; Arnoldi, H. *Chem. Ber.* **1906**, 39, 339-344

8)   (a) Lyding, J. W.; Marcy, H. O.; Marks, T. J.; Kannewurf, C. R. *IEEE Trans. Instrum. Meas.* **1988**, 37, 76-80 (b) Almeida, M.; Gaudiello, J. G.; Kellogg, G. L.; Tetrick, S. M.; Marcy, H.O.; McCarthy, W. J.; Butler, J. C.; Kannewurf, C. R.; Marks, T. J. *J. Am. Chem. Soc.* **1989**, 111, 5271-5284

9)   Lind, M. D. *Acta Cryst.* **1970**, B26, 1058-1062

10)   (a) Bannwart, R. S.; Phillips, J. E.; Herber, R. H. *J. Solid State Chem.* 1987, 71, 540-542 (b) Halbert, T. R.; Johnston, J. C.; McCandlish, L. E.; Thompson, A. H.; Scanlon, J. C.; Dumesic, J. A. *Physica* **1980**, 99B, 128

11)   Jozefowicz, M. E.; Laversanne, R.; Javadi, H. H. S.; Epstein, A. J.; Pouget, J. P.; Tang, X.; MacDiarmid, A. G. *Phys. Rev. B* **1989-I**, 39, 12958-12961

12)   Baughman, R. H.; Wolf, J. F.; Eckhart, H.; Shacklette, L. W. *Synth. Met.* **1988**, 25, 121-137

13)   Wu, C.-G.; Prassidis, K.; Papaefthymiou, V.; Kanatzidis, M. G. manuscript in preparation.

14)   Wu, C.-G.; Marcy, H.O.; DeGroot, D. C.; Schindler, J. L.; Kannewurf, C. R.; Leung, W.-Y.; Benz, M.; LeGoff, E.; Kanatzidis, M. G. *Synthetic Metals*, **1991**, 41/43, 797-803

15)   (a) Zotti, G.; Schiavon, G.; Comisso, N.; Berlin, A.; Pagani, G. *Synth. Met.* **1990**, 36, 337-351 (b) Tourillon, G.; Garnier, F. *J. Electroanal. Chem.* **1982**, 135, 173-178 (c) Oshawa, T.; Kaneto, K.; Yoshino, K. *Jpn. J. Appl. Phys.* **1984**, 23, L663-L665

16)   Bissessur, R.; Wu, C.-G.; Kanatzidis, M. G. unpublished results

17)   Iwamoto, T. in *"Inclusion Compounds"* ; Atwood, J.L.; Davies, J. E. D.; MacNickol, D. D.M. Eds.;Academic Press, Landon, 1984. References therein.

18)   Wu, C.-G.; Marcy, H. O.; DeGroot, D. C.; Schindler, J. L.; Kannewurf, C. R.; Kanatzidis, M. G. *Synthetic Metals*, **1991**, 41/43, 693-698

19)   Wynne, K. J.; Street, G. B. *Macromolecules* **1985**, 18, 2361-2368

20)   Rayner, J. H.; Powell, H. M. *J. Chem. Soc.* **1958**, 3412-3418

21) (a) Kondo, M.; Kubo, M. *J. Phys. Chem.* **1957**, 61, 1648-1651 (b) Drago, R. S.; Kwon, J. R.; Archer, R. D. *J. Am. Chem. Soc.* **1958**, 80, 2667-2670

22) (a) Schäfer-Stahl, H.; Abele, R. *Angew. Chem. Int. Ed. Engl.* **1980**, 19, 477-478 (b) Villeneuve, G.; Dordor, P.; Palavadeau, P.; Venien, J. P. *Mat. Res. Bull.* **1982**, 17, 1407-1412

23) (a) Averill, B. A.; Kauzlarich, S. M. *Mol. Cryst. Liq. Cryst.* **1984**, 107, 55-64 (b) Kauzlarich, S.M.; Ellena, J.; Stupik, P.D.; Reiff, W. M. and Averill, B. A. *J. Am. Chem. Soc.* **1987**, 109, 4561-4570

24) (a) Kanamaru, F.; Yamanaka, S.; Koizumi, M.; Nagai, S. *Chem. Lett.* **1974**, 373-376 (b) Salmon, A.; Eckert, H.; Herber, R. H. J. Chem. Phys. 1984, 81, 5206-5209

25) (a) Javadi, H. H. S.; Cromack, K. R.; MacDiarmid, A. G.; Epstein, A. J. *Phys. Rev. B* **1989-II**, 39, 3579-3584 (b) Roth, S.; Bleier, H. *Adv. Phys.* **1987**, 36, 385-462 (c) Epstein A. J.; MacDiarmid, A. G. *Mol. Cryst. Liq. Cryst.*, **1988**, 160, 165-173 (d) Zuo, F.; Angelopoulos, M.; MacDiarmid, A. G.; Epstein, A. J. *Phys. Rev. B* **1987**, 36, 3475-3478

26) (a) Maddison, D.S.; Roberts, R. B.; Unsworth, J. *Synth. Met.* **1989**, 33, 281-287 (b) Maddison, D.S.; Unsworth, J. ; Roberts, R. B. *Synth. Met.* **1988**, 26, 99-108

27) (a) Bhatnagar, V. M.; Clouter, J. A. R. *Can. J. Chem.* **1962**, 40, 1708-1710 (b) Uemasu, I.; Iwamoto, T. *J. Inclusion Phenomena* **1983**, 1, 129-134

RECEIVED January 28, 1992

Chapter 16

Intercalation and Polymerization of Aniline in Layered Protonic Conductors

Deborah J. Jones, Raja El Mejjad, and Jacques Rozière

Laboratoire des Agrégats Moléculaires et Matériaux Inorganiques, URA
Centre National de la Recherche Scientifique 79, Université Montpellier II,
34095 Montpellier Cedex 5, France

Aniline has been intercalated into the acidic host matrices of
$HFe(SO_4)_2.4H_2O$, and of α-$Zr(HPO_4)_2.H_2O$ and γ-$Zr(HPO_4)_2.2H_2O$,
and their partially copper exchanged phases. The nature of the
guest species: neutral, protonated or polymerised, depends on the
inorganic template. Polymerised aniline has been identified in both
α and γ modifications of zirconium copper hydrogen phosphates.

Amongst the strategies used for the optimisation of physicochemical properties
of materials, one approach is the molecular level modification of the constitutive
chemical units. A pathway to attaining this objective is through the intercalation
by ion exchange, or by the insertion of new ionic or molecular species into open
host structures having either a pronounced two-dimensionality, as in clays,
certain mineral classes or their synthetic analogues, or a three-dimensional
framework structure, as in zeolites. The use of inorganic lamellar solids in
electron transfer reactions between intercalated species and the layers has
recently been demonstrated (1-11), and the elaboration of occluded low dimensional conductors has been realised (12-15).

It may be desirable to associate two potential properties of the inorganic host
namely, electron transfer and proton transfer characteristics. This may be achieved by following the intercalation of the organic monomer into a Brønsted acid
matrix by an oxidation step, or by the use of an acidic electron-active host matrix.

This paper reports studies of the intercalation of aniline in various layered
proton conducting host matrices: ferric hydrogen sulphate, $HFe(SO_4)_2.4H_2O$ and
two crystalline modifications of zirconium hydrogen phosphate α-$Zr(HPO_4)_2.H_2O$
and γ-$Zr(HPO_4)_2.2H_2O$, as well as in the copper ion-exchanged derivatives of the
zirconium phosphates. The first of these, and the zirconium-copper phosphates,
combine the availability of transferable protons and potential electron accepting
metallic sites, whilst the zirconium hydrogen phosphates only provide an acidic
host framework for aniline molecules.

0097–6156/92/0499–0220$06.00/0

The proton conduction properties of layered matrices $HM(III)(SO_4)_2 \cdot nH_2O$, M=Fe, In and n = 1 and 4, have previously been reported *(16)*, and the redox transfer characteristics of the ferric tetrahydrate member have been demonstrated. Facile insertion of monovalent or divalent ions occurs, with retention of the layered structure, and reduction of Fe(III) centres *(17)*. The negatively charged layers of $HFe(SO_4)_2 \cdot 4H_2O$ contain octahedral Fe(III), the coordination sphere of which is formed by 4 oxygen atoms from sulphate groups and two from water molecules *(18)* (Figure 1).

The interlayer zone is occupied solely by diaquahydrogen ions $H_5O_2^+$, which are responsible for the proton transfer properties, and which are characterised by their hydrogen bond length of 2.44 Å. The basal spacing is 9.2 Å.

The zirconium hydrogen phosphates have interlayer distances of 7.6 Å (α-form) and 12.2 Å (γ-form). The structure of the former is made of sheets containing octahedral zirconium atoms with HPO_4 groups lying alternately above and below the plane *(19)*. No proton transfer to interlayer water molecules occurs *(20)*.

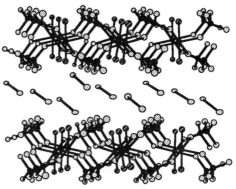

Figure 1. Structural arrangement of $HFe(SO_4)_2 \cdot 4H_2O$ (filled circles, Fe; hatched circles, H_2O; grey circles, O). Only the oxygen atoms of interlayer $H_5O_2^+$ ions are represented *(18)*.

The structure of $\gamma\text{-}Zr(HPO_4)_2 \cdot 2H_2O$ is undetermined as yet, but partial resolution of the titanium analogue by Rietveld profile refinement of a powdered sample *(21)* agreed with previous conclusions from ^{31}P MAS nmr *(22)*, in the existence of two chemically distinct types of phosphate groups, leading to a formulation $\gamma\text{-}Ti(H_2PO_4)(PO_4) \cdot 2H_2O$.

Experimental

Preparation of host structures $HFe(SO_4)_2 \cdot 4H_2O$ was synthesised as described previously *(16)*, and α and γ zirconium hydrogen phosphates were prepared according to published methods *(23,24)*. Copper exchange on these latter matrices has been fully reported *(25,26)*. For $\alpha\text{-}Zr(HPO_4)_2 \cdot H_2O$, a 60% copper loading was used, which corresponds to a biphasic system *(25)*. 70% copper loadings were used for the γ modification. Despite partial exchange, $\gamma\text{-}ZrCu_{0.7}H_{0.6}(PO_4)_2 \cdot 4H_2O$ is monophasic, having an interlayer distance of 12.4 Å. The retention of a certain degree of acidity in both cases was intentional for the subsequent intercalation reaction.

Reaction with aniline $HFe(SO_4)_2 \cdot 4H_2O$ was placed in contact with solutions of aniline in acetone with molar ratios $HFe(SO_4)_2 \cdot 4H_2O$: aniline of 1 : 4, and reacted either at room temperature or under reflux. The progress of the reaction was monitored by powder X-ray diffraction (XRD, Philips instrument, Cu Kα radiation). In the case of zirconium hydrogen phosphates and their copper derivatives, intercalation reactions were performed either by contacting the solids with the pure liquid or with aniline diluted in acetone, and the progress of the insertion reactions were followed as a function of the reaction temperature and time using XRD. All reactions were performed in an air atmosphere. The suspensions were then filtered and the solids washed with acetone until excess amine was removed. The products were air dried and stored. Elemental and thermogravimetric analyses were performed on the resulting materials for the determination of compound stœchiometry, aniline and water content.

Spectroscopic investigation Diffuse reflectance spectra of powdered samples in the region 200 - 2000 nm (50,000 - 5000 cm^{-1}) were recorded on a Cary instrument model 2300. Infrared spectra were recorded at room temperature as KBr discs using a Bomem FTIR DA8 spectrometer between 4000 and 400 cm^{-1}. Incoherent inelastic neutron scattering (INS) spectra were recorded on powdered samples at the spallation neutron source ISIS (U.K.) on the TFXA spectrometer.

Results and Discussion

Intercalation in HFe(SO$_4$)$_2$.4H$_2$O, α-Zr(HPO$_4$).H$_2$O and γ-Zr(HPO$_4$)$_2$.2H$_2$O The intercalation of aniline in acetone solution into ferric hydrogen sulphate occurs readily at room temperature. The appearance of new X-ray diffraction lines indicates the progressive formation of a new phase of interlayer spacing 14.7 Å (phase I), with total consumption of the host matrix after 48h reaction time for a host : guest mole ratio of 1/4 (Figure 2 a, b). This is accompanied by partial amorphisation indicative of a rearrangement of the macrostructure. Reflux conditions, or the use of higher host : aniline ratios, leads to the appearance of X-ray diffractions lines characteristic of anilinium sulphate, resulting from progressive hydrolysis of the phase I intercalate. If the reaction is continued even further, a second phase is observed, which has a basal spacing of 13.7 Å (phase II) (Figure 2 c).

The X-ray diffractograms of the intercalates indicate that a laminar structure is retained in both cases. The increase in basal spacings amounts to 5.5 and 4.5 Å respectively, suggesting that the orientation of intercalated aniline molecules in the interlayer space is such that the C$_2$ axis is almost perpendicular to the inorganic layers. The chemical compositions of the two aniline intercalates are different. Phase I contains the same Fe : SO$_4$ stœchiometry as the starting host compound (1 : 2), which encloses *ca.* 2 moles of aniline per mole of iron. On the other hand, phase II contains Fe : SO$_4$: C$_6$H$_5$NH$_2$ in the ratio 1 : 1 : 1. When heated in air from room temperature, the phase I intercalate eliminates a small and variable amount of water up to 90°C, without any appreciable change in the interlayer distance, indicating that the amines form organic pillars, and that any water is either superficial, or occupies interpillar sites.

The number and positions of the maxima in the electronic spectra are characteristic of Fe(III) in phase I (1680, 830 and 340 nm) and of Fe(II) in phase II (1450, 1020 nm), and indicate that the iron atoms occupy octahedral sites in both phases. Similar conclusions may be drawn from the Mössbauer spectra, which show an evolution from resonance from Fe(III) ions (phase I) to a single final Fe(II) site in phase II.

The intercalation of aniline into the solid acid matrices of α and γ zirconium hydrogen phosphates occurs at room temperature to give, in the case of α-$Zr(HPO_4)_2.H_2O$, a new product of interlayer distance 18.4 Å, of which brief mention has been made previously *(27)*, after 7 days contact with the pure liquid. Reducing the reaction time, or the use of dilute aniline solutions leads to an intermediate compound of basal spacing 13.7 Å. Elemental and thermal analyses of the intercalate of 18.4 Å interlayer separation are compatible with its formulation as $Zr(HPO_4)_2.2C_6H_5NH_2.0.4H_2O$, and the aniline molecules probably form a bilayer in the interlaminar zone, anchoring through N-H---O hydrogen bonding to the mineral layers, while the packing of organic rings is stabilised through Van der Waals interactions.

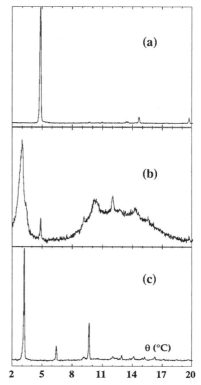

Figure 2. Reaction of aniline with $HFe(SO_4)_2.4H_2O$: X-ray diffractograms recorded after reaction times of : (a) time = 0, $HFe(SO_4)_2.4H_2O$ (b) 24h, 25°C: host + phase I intercalate (c) 8 days reflux: basal spacing 13.7 Å (phase II)

Reaction of aniline with γ-$Zr(HPO_4)_2.2H_2O$ leads to intercalates of composition $Zr(HPO_4)_2.C_6H_5NH_2.0.7H_2O$ however, variation of the experimental conditions gives rise to two different XRD patterns, giving interlayer distances of 17.05 Å (using pure aniline), and 18.15 Å (dilute acetone solutions) probably corresponding to two different phases. The average increase in gallery height, 5.4 Å, is exactly half that observed for α-$Zr(HPO_4)_2.H_2O$ (18.4 - 7.6 = 10.8 Å) and geometrical considerations are thus in agreement with the insertion of only one mole of aniline per mole of zirconium ions in γ-$Zr(HPO_4)_2.2H_2O$. An inter-digitating monolayer arrangement of the aniline molecules can be envisaged whereby hydrogen bonding alternately to the layer above and the layer below is favoured.

Vibrational spectroscopy   The question thus arises as to the nature of occluded aniline in the ferric hydrogen sulphate and zirconium hydrogen phosphate hosts: neutral, protonated, polymerised, or a mixture of these forms.
This problem can be usefully addressed using vibrational spectrosopic methods: infrared, Raman, and incoherent inelastic neutron scattering. The last is sensitive only to vibrational modes involving large amplitude displacement of hydrogen, and is essentially "transparent" to other vibrations of a molecule. The particular advantage in the case of intercalation chemistry is the absence of any signal in the INS spectrum resulting from an *aprotic* host lattice, which confers almost "guest-specific" characteristics to the method. Furthermore, any vibrational regions masked in the infrared or Raman spectrum (strong SO_4 and PO_4 stretching modes around 1000 - 1100 cm^{-1} in the present case) become accessible.
The infrared spectra of $HFe(SO_4)_2.2C_6H_5NH_2$ (phase I), α-$Zr(HPO_4)_2.2C_6H_5NH_2.0.4H_2O$ and γ-$Zr(HPO_4)_2.C_6H_5NH_2.0.7H_2O$ indicate that proton transfer to intercalated aniline has occured. Characteristic absorption bands are observed in infrared in the N-H stretching region: ~ 2930 cm^{-1} [v_s,v_{as} (NH_3^+)] and a combination band at ~ 2650 cm^{-1}, and in the region of deformation vibrations: $\delta_{as}(NH_3^+)$ at ~1550 cm^{-1} and $\delta_s(NH_3^+)$ at ~1500 cm^{-1}. The C-N stretching vibration is observed at ~1200 - 1220 cm^{-1}. The exact positions vary slightly according to the nature of the host matrix. Other vibrations of the phenyl group can be identified, and differ little from those of of neutral aniline, for example. Maxima in the INS spectra should arise from librational modes of the NH_3^+ group, in addition to phenyl ring vibrations. The torsion, $\tau(NH_3^+)$, dominates the INS spectrum of α-$Zr(HPO_4)_2.2C_6H_5NH_2.0.4H_2O$; it is observed at 506 cm^{-1}, close to its position in the spectra of γ-$Zr(HPO_4)_2.C_6H_5NH_2.0.7H_2O$ and $HFe(SO_4)_2.2C_6H_5NH_2$: 483 and 502 cm^{-1} respectively. The frequency range characteristic of the torsional mode is superimposed on those of certain ring vibrations, in particular mode 16a at 405 cm^{-1}, 16b at 502 cm^{-1}, 6a at 527 cm^{-1} (where the numbers are those of Wilson's notation (28)), nevertheless $\tau(NH_3^+)$ may be identified by its high intensity.
The presence of neutral aniline is also detected to a variable extent in the different phases. This is expected for $HFe(SO_4)_2.2C_6H_5NH_2$, in view of charge balancing considerations on the basis of elemental analyses, and indicates that proton transfer is incomplete in α and γ zirconium hydrogen phosphate intercalates. Usually intense bands due to the NH_2 group are observed in α-$Zr(HPO_4)_2.2C_6H_5NH_2.0.4H_2O$ at 3299 and 3357 cm^{-1} ($v_s(NH_2)$ and $v_{as}(NH_2)$ respectively), and the scissoring vibration $\delta(NH_2)$ is seen at 1610 cm^{-1}. In all cases, the infrared spectra show two bands for the C-N stretching vibration at 1200 - 1220 and 1290 cm^{-1}, corresponding to protonated and "free" aniline respectively.
The strong and sharp absorptions at 3329 and 3264 cm^{-1} in the infrared spectrum of $FeSO_4.C_6H_5NH_2$ (phase II) attest to the presence of neutral aniline only. However, these wavenumbers are lower than those in free aniline (vapour

or solution), where typical values are 3481 and 3360 cm^{-1}, which suggests the coordination of aniline to metal centres. The different infrared spectral characteristics of aniline in γ-Zr(HPO$_4$)$_2$.C$_6$H$_5$NH$_2$ .0.7H$_2$O and FeSO$_4$.C$_6$H$_5$NH$_2$ are shown in Figure 3.

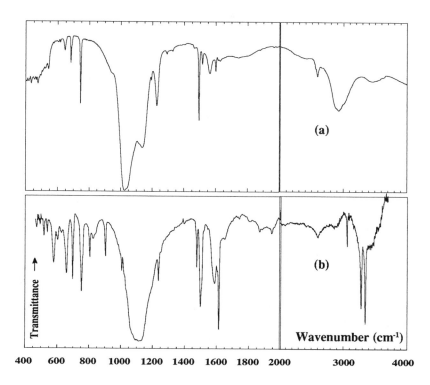

Figure 3. Infrared spectrum of (a) anilinium ion in γ-Zr(HPO$_4$)$_2$.C$_6$H$_5$NH$_2$.0.7H$_2$O and (b) coordinated aniline in Fe(SO$_4$).C$_6$H$_5$NH$_2$

In none of the above intercalation compounds would it appear that electron transfer from the organic guest has occured. This is not unexpected for the zirconium hydrogen phosphates, although the possibility has been envisaged in the case of the insertion of cobaltocene in α-Zr(HPO$_4$)$_2$.H$_2$O (29). However, in the potentially electron-active matrix of ferric hydrogen sulphate, neither phase I nor phase II has a colouration suggesting that polymerisation of aniline monomers has occured: phase I is brick red and phase II beige in colour. Reduction to Fe(II) has nevertheless taken place in the latter phase.

Understanding the retention of a layer structure for $FeSO_4C_6H_5NH_2$ is not straightforward, but a structural model can be proposed in which sulphate groups and an aniline molecule participate in the coordination sphere, analagous to that of divalent metals in monohydrated M(II) phosphonates *(30)*.

Intercalation in copper-exchanged α and γ zirconium hydrogen phosphates

Despite the biphasic nature of $\alpha\text{-}ZrCu_{0.6}H_{0.8}(PO_4)_2.4H_2O$, its reaction with aniline leads to a single phase product of slightly reduced crystallinity and dark green colour, having an interlayer distance of 18.5 Å. Similarly, the reaction of $\gamma\text{-}ZrCu_{0.7}H_{0.6}(PO_4)_2.4H_2O$ with aniline leads to a colour change from white to dark blue, whereas intercalation into the fully protonated matrix gives a white product. The dark blue-black material shows some amorphous character, and has an interlayer distance of 19.1 Å, slightly variable with the duration of the reaction with aniline. Figure 4 compares the X-ray diffractograms of the intercalates obtained with and without prior exchange of copper into the $\gamma\text{-}Zr(HPO_4)_2.2H_2O$ host matrix.

Thermal and elemental analyses show that less aniline is inserted into pre-copper exchanged matrices than into their fully protonated analogues, and less copper is retained than in the starting material, such that the product after intercalation into $\gamma\text{-}ZrH_{0.6}Cu_{0.7}(PO_4)_2.4H_2O$ may be formulated as $Zr(PO_4)_2Cu_{0.3}(C_6H_4NH)_{0.7}$. This stœchiometry does not exclude the participation of an additional oxidant besides Cu(II) during the reaction, as reported for the FeOCl-polyaniline system, where only ~ 25% of the electrons removed were transferred to Fe(III) *(1)*.

Spectroscopic analyses As described above, the nature of the guest can be readily identified from its infrared spectrum. Figure 5 shows the infrared spectra of aniline intercalated γ-zirconium copper phosphate over the region 1200 - 1900 cm⁻¹. In this spectrum, the absorption maxima at 1309, 1496 and 1576 cm⁻¹ are characteristic of the emeraldine salt form of polyaniline, (typical frequencies for bulk polyaniline formed by oxidation using $H^+/(NH_4)_2S_2O_8$ are 1302, 1496, and 1578 cm⁻¹ *(12,31,32)*). In α-zirconium copper hydrogen phosphate intercalate, the band at 1309 cm⁻¹ is weak, but more intense

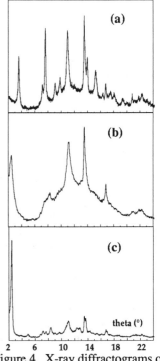

Figure 4. X-ray diffractograms of
(a) $\gamma\text{-}ZrH_{0.6}Cu_{0.7}(PO_4)_2 4H_2O$
(b) $Zr(PO_4)_2Cu_{0.3}(C_6H_4NH)_{0.7}$
(c) $Zr(PO_4)_2H.C_6H_5NH_3.0.7H_2O$

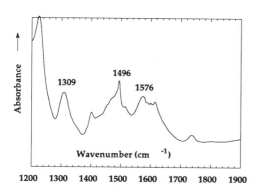

Figure 5. FTIR spectrum of γ-Zr(PO$_4$)$_2$Cu$_{0.3}$·(C$_6$H$_4$NH)$_{0.7}$

absorption is observed at 1396 cm^{-1}. According to vibrational studies on polyanilines and model oligomeric species, the wavenumber of this mode is sensitive to electron delocalisation around the C-N-C part. It is lower for a more delocalised structure, converging to a value of 1310 cm^{-1}. For short oligomers, where electron delocalisation is limited, characteristic positions are 1401 - 1383 (32). The observation of two bands with different relative intensities indicates the presence of various parts of different electron delocalised structures. It may be concluded that polymerised aniline is formed in both host lattices, and that the nature of the polymer is different, particularly in respect of its length and the extent of oxidation.

Similar conclusions may be drawn from the electronic spectra, reproduced in Figure 6. Diffuse reflectance measurements are similar to those on bulk polyaniline, with absorptions indicating the presence of phenyl groups (~ 300-320 nm), radical cations (~ 390 - 410 nm) and quinone diimine units (~ 610 nm) (33,34). The relative intensity of the maxima at 400 and 600 - 650 nm, and the tail of absorption at lower energies are consistent with a higher degree of oxidation of polyaniline in the host derived from the γ-zirconium copper hydrogen phosphate than the α-analogue, and less relative protonation in the latter. Pretreatment conditions affect the extent of protonation of the hosts which may influence the extent of polymerisation and oxidation. Pressed discs of both samples show no measurable conductivity, for reasons probably related to the inhibition of chain to chain charge transport, because of the short chain length, and the presence of an insulating inorganic matrix.

Figure 6. Electronic absorption spectra of zirconium copper hydrogen phosphate/aniline samples (a) from α-modification (b) from γ-modification

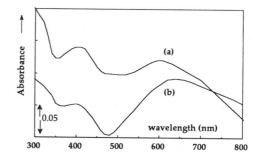

As described above, optical vibrational spectra undergo considerable modification in the mid-frequency range on polymerisation of aniline. Between 740 - 1600 cm^{-1} the modes most affected have been summarised as: 8a, 19b, 13, 9a and 11, of Wilson's notation (35). However other, low frequency, vibrations are also modified when aniline is polymerised; these correspond to modes which are known to be affected by the nature of the substituent on the phenyl ring (36)and which are also modified, not unexpectedly, by oligomerisation. Some are skeletal vibrations modified by coupling with C-N modes: 16b, (out-of plane skeletal vibration), 1 and 6a (radial skeletal vibrations), in addition to the C-N in-plane (15) and out-of-plane (10b) deformations and C-N stretching mode (13) (36), which might be expected to show enhancement or reduction in intensity, or displacement. Spectral evolution of these modes can be detected using INS spectroscopy. The INS spectra of various forms of intercalated aniline within the various hosts described here are compared in Figure 7.

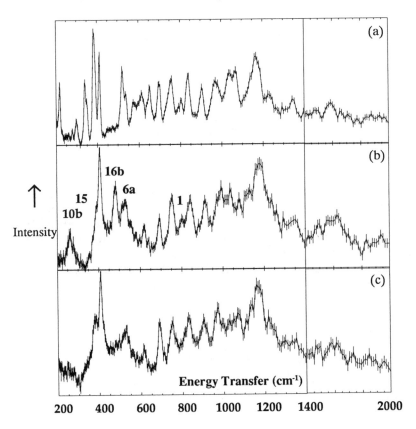

Figure 7. Incoherent inelastic neutron scattering spectra of (a) Fe(SO$_4$).C$_6$H$_5$NH$_2$ (b) γ-ZrH(PO$_4$)$_2$.C$_6$H$_5$NH$_3$.0.7H$_2$O (c) γ-Zr(PO$_4$)$_2$.Cu$_{0.3}$.(C$_6$H$_4$NH)$_{0.7}$ containing coordinated, protonated, and polymerised aniline respectively.

Modes 10b and 16b respectively, are observed to radically decrease in intensity from maxima at 262 cm^{-1} and 485 cm^{-1} of medium intensity in the INS spectrum of the aniline intercalate in γ-Zr(HPO$_4$)$_2$.2H$_2$O, where anilinium ion is the sole organic species, to virtual absence in that containing polyaniline. In contrast, mode 15 increases in intensity, whereas the radial skeletal vibrations, expected to be coupled with the C-N stretch, undergo little modification.

It is interesting to note that the same modes experience very similar intensity transformations by *coordination* of aniline to the metal centre, as in FeSO$_4$.-C$_6$H$_5$NH$_2$.

Conclusion

The above results demonstrate the existence of three well-differentiated groups of intercalates. In the poorly oxidising α and γ zirconium hydrogen phosphates, intercalation of aniline readily proceeds with proton transfer from the host matrix. As in zeolites Y and mordenite *(13)*, pre-inserted Cu(II) ions in the host structure act as oxidants in the reaction with organic monomers, such that a certain degree of charge transfer from aniline and polymerisation seem to have taken place, as evidenced by the spectroscopic results. Considerable modification of the X-ray diffractograms suggests polymer formation within the interlayer region. It is surprising that under conditions seemingly appropriate which are united in the layered acidic redox ion exchanger HFe(SO$_4$)$_2$.4H$_2$O, intercalation with proton transfer is favoured relative to electron transfer, and even prolonged reflux leads instead to modification of the inorganic layers by a grafting reaction involving coordination of aniline. The question of the inactivity of Fe(III) towards aniline in HFe(SO$_4$)$_2$.4H$_2$O, in view of the reported polyaniline formation in FeOCl *(1)*, needs to be further investigated.

Acknowledgments

We thank the Science and Engineering Research Council (U.K.) for access to the spallation neutron source, ISIS. This research is financially supported in France by the Centre National de la Recherche Scientifique.

Literature Cited

1. Kanatzidis, M. G.; Wu, C-G.; Marcy, H. O.; DeGroot, D. C.; Kannewurf, C. R.; Kostikas, A.; Papaefthymion, V. *Adv. Mater.* **1990** 2 364
2. Kanatzidis, M. G.; Wu, C-G. *J. Am. Chem. Soc.* **1989** 111 4139
3. Kanatzidis, M. G.; Hubbard, M.; Tonge, L. M.; Marks, T. J.; Marcy, H. O.; Kannewurf, C. R. *Synth. Metals* **1989** 28 C89
4. Kanatzidis, M. G.; Marcy, H. O.; McCarthy, W. J.; Kannewurf, C. R.; Marks, T. J. *Solid State Ionics* **1989** 32/33 594
5. Bringley, J. F.; Averill, B. A. *Chem. Mat.* **1990** 2 180
6. Bringley, J. F.; Fabre, J. M.; Averill, B. A. *Mol. Cryst, Liq. Cryst.* **1988** 170 215

7. Kauzlarich, S. M.; Ellena, J.; Stupik, P. D.; Reiff, W. M.; Averill, B. A. *J. Am. Chem. Soc.* **1987** *107* 4561

8. Kauzlarich, S. M.; Stanton, J. L.; Faber Jr.; J., Averill, B. A. *J. Am. Chem. Soc.* **1986** *108* 7946

9. Kauzlarich,S. M.; Teo, B. K.; Averill, B. A. *Inorg. Chem.* **1986** *25* 28

10. Averill, B. A.; Kauzlarich, S. M.; Teo, B. K.; Faber Jr., J. *Mol. Cryst. Liq. Cryst.* **1985** *120* 259

11. Averill, B. A.; Kauzlarich, S. M.; Antonio, M. R. *J. Phys., (Paris)* **1983** *44* C3 1373

12. Enzel, P.; Bein, T. *J. Phys. Chem.* **1989** *93* 6270

13. Enzel, P.; Bein, T. *J. Chem. Soc. Chem. Commun.* **1989** 1326

14. Brandt, P.; Fischer, R. D.; Sanchez-Martinez, E.; Diaz-Calleja, R. *Angew. Chem. Int. Ed. Engl.* **1989** *28* 1265

15. Wu, C.-G.; Marcy, H. O.; DeGroot, D. C.; Schindler, J. L.; Kannewurf, C. R.; Kanatzidis, M. G. *Int. Conf. Synth. Metals,* Tübingen **1990**

16. Brach, I.; Jones, D. J.; Rozière, J. *Solid State Ionics* **1989** *34* 181

17. Jones, D. J.; Rozière, J. *Solid State Ionics* **1989** *35* 115

18. Mereiter, K. *Tschermaks Min. Petr. Mitt.* **1974** *21* 216

19. Clearfield, A.; Troup, J. M. *Inorg. Chem.* **1977** *16* 3311

20. Jones, D. J.; Rozière, J.; Penfold, J.; Tomkinson, J. *J. Mol. Struct.* **1989** *197* 113

21. Christensen, A. N.; Krogh Andersen, E.; Krogh Andersen, I. G.; Alberti, G.; Nielsen, M.; Lehmann, M. S. *Acta Scandinavia* **1990** *44*

22. Clayden, N. J. *J. Chem. Soc. Dalton Trans.* **1987** 1877

23. Alberti, G.; Torraca, E. *J. Inorg. Nucl. Chem.* **1968** *30* 77

24. Clearfield, A.; Blessing, R. H.; Stynes, J. A. *J. Inorg. Nucl. Chem.* **1968** *30* 2249

25. Clearfield, A.; Kalnins, J. M. *J. Inorg. Nucl. Chem.* **1976** *38* 849

26. Clearfield, A.; Kalnins, J. M. *J. Inorg. Nucl. Chem.* **1978** *40* 1933

27. Costantino, U. In *Inorganic Ion Exchange Materials*; Clearfield, A., Ed.; CRC Press, Boca Raton, FA, 1982; pp 112-132

28. Wilson, E. B. *Phys. Rev.* **1934** *45* 706

29. Johnson, J. W. *J. C. S. Chem. Comm.* **1980** 263

30. Cao. G.; Lee, H.; Lynch, V. M.; Mallouk, T. E. *Inorg. Chem.* **1988** *27* 2781

31. Furukawa, J.; Ueda, F.; Hyodo, Y.; Haroda, I.; Nakajima, T.; Kawagoe, T. *Macromolecules* **1988** *21* 1297

32. Ueda, F.; Mukai, K.; Harada, I.; Nakajima, T.; Kawagoe, T. *Macromolecules* **1990** *23* 4925

33. Glarum, S. H.; MacDiarmid, A. G. *J. Phys. Chem.* **1988** *92* 4210

34. McManus, P. M.; Cushman, R. J.; Yang, S. C. *J. Phys. Chem.* **1987** *91* 744

35. Raupach, M.; Janik, L. J. *J. Colloid Interface Sci.* **1988** *121* 449

36. Whiffen, D. H. *J. Chem. Soc.* **1956** 1350

RECEIVED January 16, 1992

Chapter 17

Electrocrystallization of Low-Dimensional Molecular Solids

Michael D. Ward

Department of Chemical Engineering and Materials Science, University of Minnesota, 421 Washington Avenue S.E., Minneapolis, MN 55455

The preparation of crystalline low-dimensional molecular solids, commonly is performed by electrocrystallization techniques wherein redox active molecules are reduced or oxidized at a working electrode in the presence of appropriate counterions. Very little is known, however, about the effect of electrochemical parameters and interfacial structure on the self assembly processes that lead to crystallization on the electrode surface. This work will describe the electrocrystallization of various crystalline molecular solids, focusing on the control of nucleation, growth, morphology and stoichiometry of these materials through manipulation of the electrochemical growth conditions and interfacial properties of the electrode.

An increasingly popular method for the preparation of low-dimensional molecular solids involves crystallization of these materials directly at conventional electrochemical electrodes(1). Electrocrystallization is conveniently used when only one of the molecular components is electrochemically active at potentials either applied or incurred under potentiostatic or galvanostatic conditions, respectively. Oxidation of a donor molecule in the presence of an anion or, conversely, reduction of an acceptor in the presence of a cation results in crystal growth at the electrode (equations 1,2). The process is generally performed at platinum electrodes in conventional H-cells, in which the working and counter electrode compartments are separated by a glass frit in order to minimize contamination by counterelectrode processes (Figure 1). The crystals are harvested from the working electrode upon completion. This approach is tantamount to electrochemically controlled metathesis, wherein the rate of introduction of the redox-active component can be adjusted by either the current at the working electrode. This offers unparalleled control over the solution crystallization process. For example, we have demonstrated that crystal size can be readily adjusted by the applied current density during electrochemical crystal growth(2). Electrocrystallization has been successfully employed for numerous molecular solids with interesting electronic properties, including the Krogmann's salts $K_2Pt(CN)_4X_{0.3} \cdot 3H_2O$(3), conducting and superconducting phases of $(TMTSF)_2X$ (TMTSF = tetramethyltetraselenafulvalene; X = PF_6^-, AsF_6^-, SbF_6^-,

0097–6156/92/0499–0231$06.00/0

BF$_4^-$, NO$_3^-$)(4), and more recently, (BEDT-TTF)$_2$X (BEDT-TTF = bis(ethylenedithio)tetrathiafulvalene; X = ReO$_4^-$, BrO$_4^-$, I$_3^-$, AuI$_2^-$)(5).

$$\overset{A^-}{D - e^- ----------> D^+A^-} \qquad (1)$$

$$\overset{D^+}{A + e^- ----------> D^+A^-} \qquad (2)$$

Almost without exception, electrocrystallization has been performed using poorly characterized electrodes under galvanostatic conditions at somewhat arbitrarily chosen current densities. In fact, actual electrochemical growth conditions are rarely reported in detail and detailed mechanistic investigations are sorely lacking, unlike the numerous investigations describing electrocrystallization of metals(6). It has been our goal to investigate the electrocrystallization process in greater detail. There are numerous reasons for this interest. Firstly, the simultaneous growth of different phases is not uncommon; for example numerous polymorphs are frequently observed in the growth of BEDT-TTF salts. If preparation of a single phase is desirable, it is important to understand the mechanistic aspects that affect the growth of the different phases, and how to alter the electrochemical conditions to control growth in polymorphic or multi-stoichiometric systems. Secondly, little is known of the effect of electrochemical conditions or solution environment on crystal morphologies of materials grown by this method (the conducting low-dimensional compounds generally grow as highly anisotropic needles). Thirdly, the role of solution composition (i.e. electrolyte concentration) and counter electrode orientation has not yet been addressed. The last two aspects are particularly important for fabricating macroscopic crystals for convenient measurement of electronic properties as well as for actual applications. We also have an interest in determining the feasibility of exploiting these highly anisotropic, conducting crystalline materials as interconnects in *in situ* electrochemical micro-, and possibly, nano-fabrication processes. The purpose of this paper is to outline the fundamental aspects of electrocrystallization that have an impact on these issues, and to describe some recent results that address them.

Elementary Steps of Electrocrystallization

Electrocrystallization of low-dimensional molecular solids can be described as consisting of several elementary steps that need to be considered when synthesizing materials in this manner. One possible scheme is depicted in Figure 2, in which an acceptor molecule is reduced to its anion at the electrode. In the kinetically limited regime the rate of A$^-$ generation will be limited by k_{et}, whereas at high overpotentials the generation of A$^-$ will become diffusion limited. The generation of A$^-$ may be followed by formation of ion aggregates of A$^-$, or A and A$^-$ (to give mixed valent aggregates), which eventually complex with D$^+$ present in solution to form ion pairs. Alternatively, A$^-$ and D$^+$ may form ion pairs directly. The ion pairs subsequently form aggregates; for low-dimensional molecular solids it is likely that these aggregates result from favorable charge-transfer interactions between molecules within stacks and the weaker interactions between molecules of separate stacks. Nucleation of a given crystalline phase occurs when the dimension of these aggregates exceed the critical radius of nucleation; that is, when the volume free energy becomes sufficiently large to overcome the increase in energy associated with the surface boundary of the nucleus. The nucleus evolves into the macroscopic

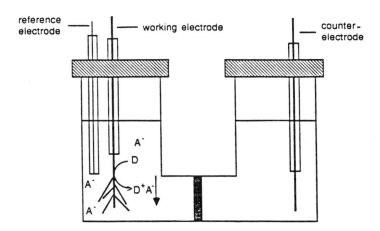

Figure 1. Schematic representation of a conventional electrocrystallization cell. In this example, donor molecules are oxidized at the anode in the presence of A⁻, resulting in crystal growth of the electrode.

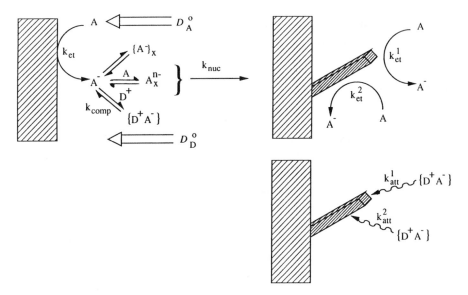

Figure 2. Possible elementary steps involved in electrocrystallization of a low-dimensional molecular conductor.

crystal that is evident on the electrode surface. In general, under conditions of slow growth crystallization is observed on relatively few sites on the electrode, which probably indicates nucleation on high energy surface sites or defects where the current density is higher. In fact, whereas (pyrene)$_2$ClO$_4$ crystals are readily grown on untreated platinum electrodes, crystal growth could not be induced on fire-polished platinum electrodes, indicating the need for nucleation sites (edges, kinks, etc.)(7). The role of these elementary steps is evident from general observations reported for electrocrystallization of molecular solids. There generally is a critical current density below which crystal growth will not occur; below this value the concentration of A$^-$ or subsequently formed aggregates does not exceed the solubility product at the electrode surface because diffusion of these species away from the electrode competes with nucleation. On the other hand, if the current density is too high, the nucleation rates will be rapid and microcrystalline product will result. High current densities can also result in deleterious high electrode potentials because the flux of A is not sufficient to overcome concentration polarization, resulting in overoxidation and decomposition.

The following steps depend upon the electronic nature of the crystalline phase. The growth of non-conductors must necessarily proceed by reduction of A$^-$ at the electrode with migration of A$^-$ or the subsequently formed {D$^+$A$^-$} ion pair to a growing face of the crystal on the electrode. Conducting materials, however, can behave as extensions of the electrode as electron transfer can, in principle, take place on the conductive faces of these deposits. This is indicated in the scheme for a monoclinic crystal, with k_{et}^1 and k_{et}^2 the electron transfer rate constants at the fastest and slowest growing faces, respectively. Crystal growth occurs via attachment of ion pairs or aggregates to the respective crystal faces. The rates of attachment on the different faces, k_{att}^1 and k_{att}^2, will be largely controlled by the binding strength of the solution species on the different faces, E_{att}^1 and E_{att}^2, respectively(8). For low-dimensional conductors it is expected that $k_{et}^1 > k_{et}^2$ due to the higher conductivity along the needle axis of these materials ($\sigma_{\parallel} / \sigma_{\perp} \approx 10^3$), and $k_{att}^1 > k_{att}^2$ due to the favorable intermolecular $\pi-\pi$ interactions between stacking molecules. Both conditions favor the observed growth morphology of these materials, but the relative contributions of these factors is not yet understood.

The above discussion indicates a relatively poor understanding of the mechanistic aspects of electrocrystallization, clearly suggesting opportunities in both experimental and theoretical (modeling) areas. This will require careful studies of the role of electrochemical parameters and solvent composition in crystal growth, as well as methods that can probe the influence of these factors, preferably in a dynamic fashion.

Role of Electrochemical Parameters in Electrocrystallization

The mechanism depicted in Figure 2 suggests that the relative concentrations of A and A$^-$ near the electrode surface can be adjusted by changing the electrode potential. At very negative potentials the concentration of A at the electrode surface will be severely depleted and the generation of A$^-$ controlled by diffusion. At less negative potentials, however, the relative concentrations of A and A$^-$ near the electrode will be dictated by the Nernst equation, which predicts that [A] and [A$^-$] will be equivalent at the formal reduction potential E^0, provided the removal of either of these species by nucleation and crystallization processes is not greater than k_{et}. This suggests that adjustment of electrochemical potential during potentiostatic crystal growth can be exploited to control the stoichiometric composition of crystalline deposits. We have demonstrated potential control of stoichiometry in the synthesis of [(C$_6$Me$_6$)M^{2+}][TCNQ]$_x$$^{n-}$ (M =

(n = 2; x = 2,4)(9). At negative potentials, fully reduced $\rho = 1$ phases (ρ = degree of charge per molecule) grew at the electrode, whereas at $E_{app} \geq E^{o}$ the mixed valent $\rho = 0.5$ phases were observed (Scheme 1). In the former the TCNQ acceptors are fully reduced and singly charged, but the mixed valent phases formally contain equal concentrations of TCNQ and TCNQ⁻. Therefore, the stoichiometry of the crystallized phases parallels the composition of the solution near the electrode surface.

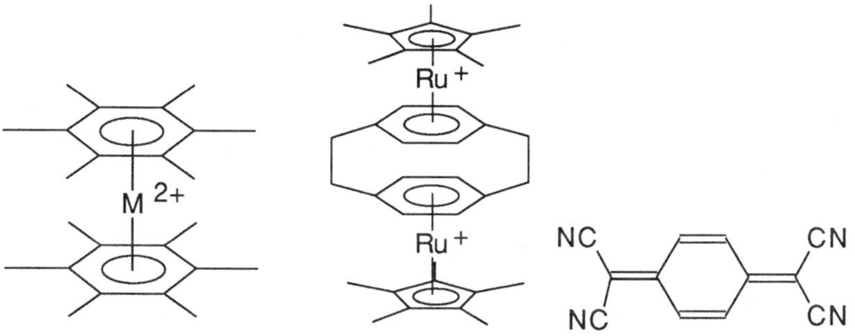

$(C_6Me_6)M^{2+}$ $\qquad$ $(C_5Me_5)Ru(2_2\text{-}1,4\text{-}$ cyclophane$)Ru(C_5Me_5)^{2+}$ $\qquad$ TCNQ

Inspection of the crystal structures of these compounds indicates that it is likely that aggregates of $[TCNQ]_2{}^{2-}$ formed in solution (although probably in low concentrations) are responsible for the formation of the $\rho = 1$ phases, whereas mixed valent aggregrates $[TCNQ]_{2n}{}^{n-}$ are responsible for the $\rho = 0.5$ phases. The observations of both phases suggests that the products $k_{\rho=1}[TCNQ]_2{}^{2-}$ and $k_{\rho=0.5}[TCNQ]_4{}^{2-}$ are not appreciably different, but potential dependent concentration of the aggregates influences the crystal growth process. These effects have been observed recently for other organic solids(10,11). The potential dependent selectivity demonstrates that careful attention to electrochemical parameters can result in greater control of the crystallization process.

Scheme 1

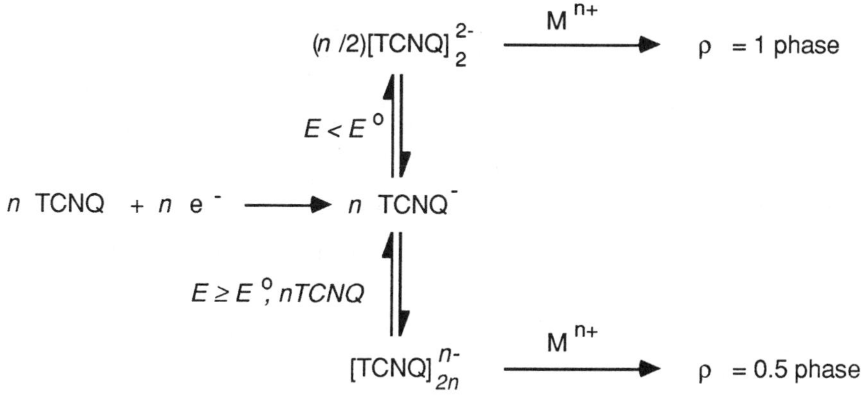

Morphology of Electrocrystallized Low-Dimensional Molecular Solids.
The morphology of molecular crystals during electrochemical crystal growth has received scant attention, other than the importance of morphology in identifying different phases when multiple phases grow simultaneously. One report has commented on the effect of solvent on the quality and morphology of TTF, TMTSF and TMTTF phases(12). To our knowledge there have not been any reports describing the effect of electrochemical conditions on morphology, other than the aforementioned effects of current density on crystal size. We have begun more detailed investigations of the role of current densities and applied potential on crystal morphology, which we detail below.

Determination of Morphological Index and Crystal Growth Rates. We previously reported(13) that the electrocrystallization of $TTFBr_{0.7}$, a low-dimensional conductor, could be monitored readily with the mass-sensing electrochemical quartz crystal microbalance (EQCM)(14). The EQCM, fabricated from a thin wafer of piezoelectric AT-cut quartz sandwiched between two gold electrodes that serve to excite the crystal at its resonant frequency, enables the determination of mass changes at one of the gold electrodes from shifts in the resonant frequency of the crystal. This electrode can also be employed as a electrochemical working electrode, which allows simultaneous, and dynamic, measurements of mass, and electrochemical current and charge. Under most conditions, the mass change taking place at the electrode can be determined from the Sauerbrey equation (equation 3), where

$$\Delta f = - \frac{2 f_0^2 \Delta m}{A \sqrt{\rho_q \mu_q}} \tag{3}$$

Δf is the measured frequency change, f_0 the parent frequency of the quartz crystal, Δm the mass change, A the piezoelectrically active area, ρ_q the density of quartz and μ_q the shear modulus. The key experiments in this study were performed by applying a double potential step waveform (+0.13 <-> -0.2 V vs SCE; 5 sec intervals) in which $TTFBr_{0.7}$ crystals were alternately formed (by generation of TTF^+ in the presence of Br^-) and redissolved. The crystallization step was performed at low overpotential so that the amount of crystallized material was small. In a typical experiment the measured Δf, which corresponds to the mass change associated with crystal growth on the electrode, increased with each successive cycle (Figure 3). In addition, the cathodic baseline return frequency gradually shifted downward, indicating a retention of mass during the cathodic step. The observed frequency responses during this potential waveform were attributed to formation of $TTFBr_{0.7}$ (Δf decrease) followed by slower removal of $TTFBr_{0.7}$ (Δf increase). This results in persistent $TTFBr_{0.7}$ growth centers that serve as crystal growth sites in the following cycle; the presence of these growth centers results in large frequency changes in the following cycle owing to the larger surface area for crystal growth coupled with the lower Gibbs energy for crystal growth on these sites compared to nucleation.

$$\text{TTF } - \text{ 0.7e}^- + \text{ 0.7Br}^- \xrightarrow{\quad +0.13 \text{ V} \quad} \text{TTFBr}_{0.7} \tag{4}$$

$$\text{TTFBr}_{0.7} + \text{ 0.7e}^- \xrightarrow{\quad -0.2 \text{ V} \quad} \text{TTF } + \text{ 0.7Br}^- \tag{5}$$

The relative rates of growth along the different crystallographic directions could be estimated by modeling this frequency response. $TTFBr_{0.7}$ crystallizes in the monoclinic C2/m space group and crystallizes as needle shaped crystals, with stacks of TTF molecules along the c axis(15). A model based on cylindrically shaped crystals was therefore employed to describe the growth parallel (short face) and perpendicular to the stacking axis (long face) (Figure 3b). A simulated frequency response matching the observed response was consistent with a growth rate along the stacking needle axis that was 15 times greater than that perpendicular to this direction. This value represents this best fit to the observed data. Numerous trials indicated that the increases in Δf was extremely sensitive to this ratio; small excursions from the value exhibited dramatic changes in the simulations. These results demonstrated that the morphological indices of crystalline materials could be determined *during* crystal growth, instead of relying solely on visual inspection of the crystals after the growth process.

The morphology of $TTFBr_{0.7}$ deposits was also affected by the applied potential during single potential step experiments. Figure 4 depicts current transients for $TTFBr_{0.7}$ electrocrystallization at gold electrodes held at different potentials. Upon application of the potential step, the current increased sharply, declined and then increased again. With increasing overpotential the value of the maximum current at early times increased and the time of the maximum current shifted to shorter times. This behavior can be attributed to an increase in electroactive area at early times due to the formation of conductive nuclei. When these nuclei cover the surface sufficiently or enough time elapses so that their diffusion spheres merge, the current falls due to the onset of planar diffusion. The second current increase (at long times) can be explained by an increase in the effective electrode area due to a second nucleation process and the onset of dendritic growth. The steeper rise in current at larger overpotentials is a manifestation of greater excess free energy (i.e. larger $[TTF^+]$) and the resulting larger driving force for nucleation, which increases the rate of formation of dendrites. Under these conditions, the driving force exceeds that required for "normal" crystal growth along the preferred directions and nucleation and growth can occur on slip planes of the crystalline deposits.

The role of electrochemical potential on dendritic growth in low-dimensional conductors is especially evident when electrocrystallization is performed on a microelectrode with a 10 μm diameter. These electrodes were used in order to minimize the number of active crystal growth sites, in the hope that this would simplify analysis and interpretation considerably. When TTF was oxidized under galvanostatic conditions at a microelectrode in a solution of n-$Bu_4N^+Br^-$, crystal growth of $TTFBr_{0.7}$ could be observed readily with a microscope. We have employed video microscopy to follow the crystallization process dynamically, and analysis of individual frames allows interpretation of the various stages of the growth process. Figure 5 illustrates one such example of $TTFBr_{0.7}$ crystallization. When the applied current is 1 μA, several crystal growth sites are evident (< 5), and the crystals grow apparently as single crystals with the typical rod-like morphology. Increasing the current to 3 μA results in a second nucleation event and the emergence a new crystal growth site, and the resulting crystal has a much greater anisotropy than the initial crystals. The growth of the initially formed crystals appears to have ceased along the stacking direction, but continues perpendicular to the long axis, resulting in "thickening;" that is, the anisotropy becomes less pronounced due to growth on the faces parallel to the more conducting stacking axis. At a higher current of 10 μA the morphology of the crystals changes more dramatically. The crystals thicken to a greater extent, and the onset of some dendritic instability, although slight, can also be observed. The reduction in anisotropy at 3 and 10 μA may indicate that in this range the oxidation of TTF at the crystal tip becomes diffusion limited, and electron transfer at the less conductive faces parallel to the stacking axis becomes more significant.

(a)

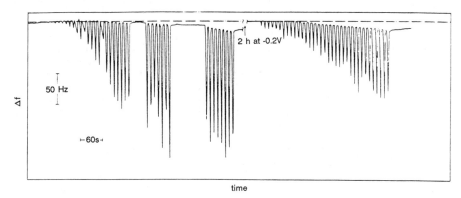

(b)

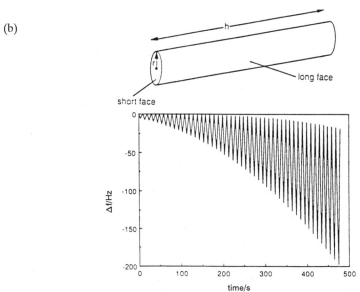

Figure 3. (a) Frequency response of the EQCM during repetitive double potential steps between -0.2 and +0.9 V (vs. SCE) at 5 sec intervals ($t_a = t_c = 5$ sec) in 0.1 M n-Bu$_4$N$^+$Br$^-$/CH$_3$CN containing 5 mM TTF. The indicated time corresponds to the length of time the electrode was held at -0.2 V. (b) (upper) the TTFBr$_{0.7}$ model used for the simulation and (lower) simulated frequency response for double potential step experiments with $t_a = t_c = 5$ sec, N = 1 x 10^4, r_0 = 5 nm, h_0 = 100 nm, $k_a^S = 5$ x 10^{-7}, $k_c^S = 4.98$ x 10^{-7}, $k_a^L = 3.33$ x 10^{-8}, $k_c^L = 3.03$ x 10^{-8} mol cm^{-2}sec^{-1}. The subscripts a and c refer to the anodic and cathodic potential step, respectively; the superscripts S and L refer to the short and long face of the crystal, respectively.

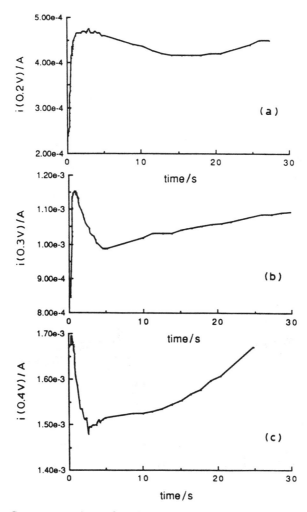

Figure 4. Current transients for electrocrystallization of TTFBr$_{0.7}$ at gold electrodes upon stepping from 0.0 V to (a) 0.2 (b) 0.3 and (c) 0.4 V (vs. SCE) in 0.1 M n-Bu$_4$N$^+$Br$^-$/CH$_3$CN containing 5 mM TTF.
Reproduced with permission from reference 13, copyright 1989 Elsevier Sequoia SA.

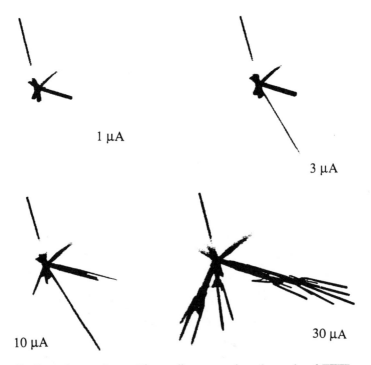

Figure 5. Crystal growth at a 10 μm diameter microelectrode of $TTFBr_{0.7}$ in 0.1 M n-$Bu_4N^+Br^-/CH_3CN$ containing 5 mM TTF. The frames were taken in sequence in the order of increasing current as indicated.

Adsorption of impurities on the short face may also be responsible for termination of growth in this direction. It should be noted that at 10 uA, the current density at a tip of a 5 μm diameter crystal (estimated from the video frames) is 50 A cm^{-2}. Although the current is distributed over several crystal growth sites, it is not unreasonable that oxidation of TTF at the tip of this crystal becomes diffusion limited. The resulting large overpotential results in a sufficiently large Gibb's free energy, resulting in the observed dendritic instability, similar to that discussed above for TTFBr$_{0.7}$ crystals grown at constant potential. The morphology observed at 30 μA is particularly dramatic, clearly showing the dendritic instabilities that result from the large overpotential. The dendrite formation appears to occur primarily at the crystal tips (where the greatest excess free energy is present). Dislocations along the (100) or (110) planes may also be responsible for the observed behavior. This is suggested in the frame taken at 10 μA which indicates growth parallel to these faces along the [001] direction.

Growth of the low temperature superconductor (TMTSF)$_2$PF$_6$ at high overpotentials also affords rather dramatic crystalline morphologies. Figure 6 illustrates that growth of this compound via oxidation of TMTSF at a 10 μm microelectrode held at a potential of +0.7 V (vs. SCE). In this case, fractal-like deposits result; preliminary estimates of the fractal dimension suggest values in the range 1.5 - 1.6.

TMTSF

These preliminary videomicroscopy experiments indicate some key points concerning electrocrystallization of low-dimensional organic conductors. It is clear that the single crystals of TTFBr$_{0.7}$ behave as extensions of the electrode. That is, their conductivity allows electron transport from the base metal electrode, with actual electron transfer to molecular species in solution taking place at the crystal surfaces. The electron transfer rates at the different crystal faces is expected to differ, and this may be the source of the change in anisotropy at different applied currents. These experiments also indicate that careful attention to electrochemical growth conditions is imperative if single crystal materials with low defect concentrations are desirable. If these materials are to be used to form interconnects in nano-fabrication, further understanding of the effect of various experimental parameters on morphology needs to be achieved.

Experimental Section

The experimental apparatus for EQCM experiments comprised a 5 MHz AT-cut quartz crystal (Valpey-Fisher or McCoy Electronics) and a homemade oscillator designed to drive the crystal at its resonant frequency. Gold electrodes (2000 Å thick) were deposited on chromium underlayers (200 Å thick) on both sides of the crystal using evaporative techniques. The frequency of the EQCM was monitored with a Hewlett Packard 5384A frequency counter. Commercially available potentiostats (Princeton Applied Research 173 or 273) were used for electrochemical experiments. The experimental apparatus and quartz crystal electrode format have been described in greater detail elsewhere(16). Video microscopy was performed with a Wild M5-APO stereomicroscope in conjunction with a Panasonic video camera and a Panasonic A6-6010s time-lapse videocassette recorder. Single frame images of

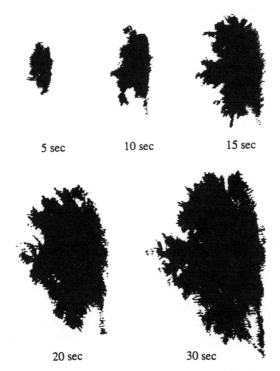

Figure 6. Crystal growth at a 10 μm diameter microelectrode of $(TMTSF)_2PF_6$ in 0.1 M $n\text{-}Bu_4N^+PF_6^-/CH_2Cl_2$ containing 5 mM TMTSF.

the crystals were exported to a Macintosh IIx via a Data Translation Quick Capture board and were analyzed using Image, a public domain image analysis program.

Literature Cited

[1]Ward, M. D. *Electroanalytical Chemistry*, Vol 16, **1989**, Bard, A. J., Ed.; Marcel Dekker, New York, p 181.

[2]Ward, M. D. *Inorg. Chem.* **1986**, *25*, 4444.

[3](a) Miller, J. S. *Science* **1976**, *194*, 189. (b) Miller, J. S.; Epstein, A. J. *Prog. Inorg. Chem.* **1976**, *20*, 1.

[4](a) Jerome, D; Mazaud, A.; Ribault, M.; Bechgaard, K. *J. Phys.* (Paris) Lett. **1980**, *41*, L95. (b) Bechgaard, K.; Jacobsen, C. S.; Mortensen, K.; Pedersen, H. J.; Thorup, N. Solid State Comm. **1980**, *33*, 1119.

[5](a) Williams, J. M.; Carneiro, K. *Adv. Inorg. Chem. Radiochem.* **1985**, *29*, 249. (b) Williams, J. M.; Beno, M. A.; Wang, H. H.; Reed, P. E.; Azevedo, L. J.; Schirber, J. E. *Inorg. Chem.* **1984**, *23*, 1790.

[6](a) Fleischmann, M.; Thirsk, H. R. *Advances in Electrochemistry and Electrochemical Engineering*, Vol 3; Delahay, P., Ed.; Wiley-Interscience, New York, 1963, p 123. (b) Hillman, A. R., Mallen, E. F. *J. Electroanal. Chem.* **1987**, *220*, 351. (c) Bockris, J. O'M; Razumney, G. A. *Fundamental Aspects of Electrocrystallization*, Plenum, New York, 1967.

[7]Williams, D. F. *Science*, **1977**, *197*, 1194.

[8]Berkovitch-Yellin, Z.; *J. Amer. Chem. Soc.* **1985**, *107*, 8239.

[9](a) Ward; M. D.; Johnson, D. C. *Inorg. Chem.* **1987**, *26*, 4213. (b) Ward, M. D.; Fagan, P. J.; Calabrese, J. C.; Johnson, D. C. *J. Amer. Chem. Soc.* **1989**, 111, 1719.

[10]Lamache, M.; Kacemi, K. E. *Mol. Cryst. Liq. Cryst.* **1985**, *120*, 255.

[11]Chiang, L. Y.; Johnston, D.C.; Stokes, J. B.; Bloch, A. N.; *Synth. Met.* **1987**, *19*, 697.

[12]Anzai, H.; Tokumoto, M.; Saito, G. Mol. Cryst. Liq. Cryst. 1985, 125, 385.

[13]Ward, M. D. *J. Electroanal. Chem.* **1989**, *273*, 79.

[14](a) Buttry, D. A. Electroanalytical Chem.; Bard, A. J., Ed.; Marcel Dekker, New York, 1991, Vol 17, p 1. (b) Ward, M. D.; Buttry, D. A. *Science*, **1990**, *249* 1000.

[15]La Placa, S. J.; Corfield, P. W. R.; Thomas, R.; Scott, B. A. *Solid State Commun.* **1975**, *17*, 635.

[16]Ward,M. D.; *J. Phys. Chem.* **1988**, *92*, 2049.

RECEIVED January 16, 1992

Chapter 18

Electrodeposition of Nanoscale Artificially Layered Ceramics

Jay A. Switzer, Richard J. Phillips, and Ryne P. Raffaelle

Graduate Center for Materials Research, University of Missouri—Rolla, Rolla, MO 65401

Ceramics are generally viewed as highly insulating refractory materials with coarse microstructures. They are both processed and used at high temperatures. We have recently shown, however, that it is possible to electrochemically deposit $Tl_aPb_bO_c/Tl_dPb_eO_f$ superlattices with individual layers as thin as 2 nm. The superlattices were deposited from a single aqueous solution at room temperature, and the layer thicknesses were galvanostatically controlled. Substitution of Tl_2O_3 into PbO_2 appears to stabilize a face-centered cubic structure with an average lattice parameter of 0.536 nm. The lattice parameters for the TlPbO oxides vary by less than 0.3% when the atomic percent Pb is changed from 46% to 88%. X-ray diffraction evidence is presented for the heteroepitaxial growth of the PbTlO oxides. Because the modulation wavelengths are of electron mean free path dimensions, this new class of degenerate semiconductor metal-oxide superlattices may exhibit thickness-dependent quantum optical, electronic, or optoelectronic effects.

A superlattice is a multilayer structure with coherent stacking of atomic planes (*1*). An idealized superlattice structure with square-wave modulation of composition and/or structure is shown in Figure 1. The thicknesses of the layers are not necessarily equal, as long as the structure is periodic. When the modulation wavelength is in the nanometer range, each layer is only a few unit cells thick. Quantum confinement of carriers in nanomodulated materials often leads to technologically important optical and electrical properties which are intermediate between those of discreet molecules and extended network solids.

The major device emphasis in nanomodulated materials has been on semiconductor-based systems. The superlattices are grown using semiconductors which are lattice-matched, so that coherent interfaces can be formed by epitaxial growth. Although the crystal structures of the component semiconductors are quite similar, the band structures of the layers are designed to vary as widely as possible. Quantum confinement is achieved by modulating either the bandgap or doping level of the semiconductor (*2-4*). Band diagrams for compositional and doping superlattices are shown in Figure 2. An example of a compositional superlattice would be GaAs/GaAlAs, in which nanometer-scale layers of small bandgap GaAs are sandwiched between layers of larger-bandgap GaAlAs. Superlattices of this type are also known as multiple quantum wells.

0097–6156/92/0499–0244$06.00/0

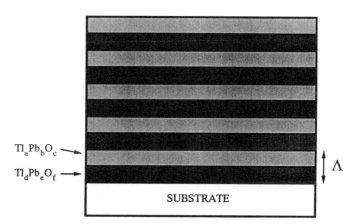

Figure 1. Idealized superlattice structure. Equal thicknesses of the $Tl_aPb_bO_c$ and $Tl_dPb_eO_f$ layers are not required, as long as the structure is periodic.

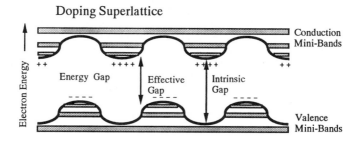

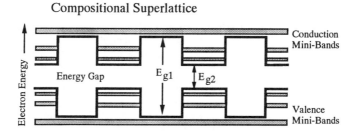

Figure 2. Band structures of doping and compositional superlattices.

Quantum effects, such as resonant tunneling, enhanced carrier mobility (two-dimensional electron gas), bound states in the optical absorption spectrum, and nonlinear optical effects (e.g., intensity-dependent refractive indices) have been observed in semiconductor multiple quantum wells (2-4). Examples of devices based on these structures include tunnel diodes, fast optical and optoelectronic switches, high electron mobility transistors, and quantum well lasers.

We have recently demonstrated that it is possible to electrodeposit nanoscale ceramic superlattices based on the TlPbO system (5). The idea of electrochemically depositing nanomodulated superlattices is not new, but it has not been applied previously to the deposition of nonmetallic materials. Several research groups have shown that compositionally modulated metallic alloys can be electrochemically deposited from a single plating bath by cycling either the potential or current (6-9). The interest in nanomodulated metallic systems stems from their enhanced mechanical and magnetic properties (7,10).

We have chosen the $Tl_aPb_bO_c/Tl_dPb_eO_f$ system for our study for several reasons: (1) during our previous work on the electrochemical and photoelectrochemical deposition of Tl_2O_3 films, we found that it was possible to deposit highly oriented films (11-13), (2) the deposition of PbO_2 (14,15) and $Pb_8Tl_5O_{24}$ (16,17) are well documented, (3) these oxides deposit directly at room temperature and require no heat treatment, (4) there is a nearly isomorphous series of mixed thallium/lead oxides that grow epitaxially, and (5) the device applications of these types of materials have not been studied previously. The generalized double galvanostatic pulse that we used to deposit the superlattices is shown in Figure 3. The layer thickness is proportional to the product of current and time.

The end members of the series, PbO_2 and Tl_2O_3, are both degenerate n-type semiconductors. They have the high electrical conductivity of metals, with the optical properties of semiconductors. Because of their high majority carrier concentrations ($>10^{20}/cm^3$), they also have high reflectivity in the near-IR. Thallium(III) oxide, for instance, is a degenerate n-type semiconductor with a bandgap of 1.4 eV and a room temperature resistivity of only 70 μohm-cm (18). Lead(IV) oxide has a larger bandgap of approximately 1.8 eV (14). Transmission optical spectra of 2 μm thick films of thallium(III) oxide and lead(IV) oxide are shown in Figure 4. These materials are, therefore, metal oxide analogs of semiconductors such as GaAs and AlGaAs, and they may function as multiple quantum wells when the modulation wavelength is in the nanometer range (19,20).

Experimental

Electrochemical depositions were performed using either a Princeton Applied Research (PAR) Model 273A potentiostat/galvanostat, or a system consisting of a Stonehart BC 1200 potentiostat/galvanostat, PAR Model 175 universal programmer, and PAR Model 379 digital coulometer. Superlattices were deposited onto polycrystalline 430 stainless steel disks. The disks were sealed in epoxy or mounted in a PAR Model K105 flat specimen holder, so that only the front surface was exposed to the solution. The exposed electrode area ranged from 0.78 to 1.96 cm^2. The final polish on the stainless steel disks was 0.05 μm alumina. Films were deposited from aqueous solutions of 0.005 M $TlNO_3$ and 0.1 M $Pb(NO_3)_2$ in 5 M NaOH. The strong base is necessary to dissolve the $Pb(NO_3)_2$, which precipitates in 1 M NaOH as PbO. (**CAUTION:** Thallium salts are extremely toxic.) The superlattices were deposited by galvanostatically pulsing between 6.2 mA/cm^2 and a lower current density such as 0.06 mA/cm^2. Each layer thickness was proportional to the product of current density and time. Elemental analysis was done by EDS on a JEOL 35CF scanning electron microscope. X-ray diffraction patterns were run on a Scintag XDS-2000 diffractometer. Optical spectra were run on a CARY 5 UV-Vis-NIR spectrophotometer.

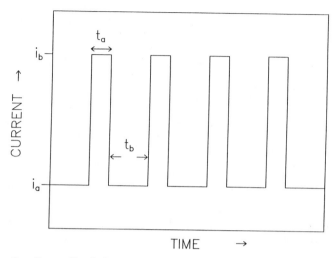

Figure 3. Generalized double galvanostatic pulse used to electrodeposit superlattices. The superlattices could also be deposited by modulating either the potential or mass-transport.

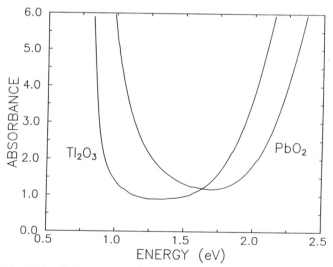

Figure 4. Transmission optical spectra of 2 μm thick films of thallium(III) oxide and lead(IV) oxide.

Results and Discussion

Since Tl_2O_3 and PbO_2 can both be electrodeposited from alkaline solution and there is considerable overlap of their deposition iV curves, it is possible to electrodeposit $Tl_aPb_bO_c$ films with compositions that are a function of applied potential or current density (5). We have carefully chosen the reactant concentrations so that it is possible to vary the Pb/Tl ratio over a wide range. Since the TlNO₃ concentration is only 0.005 M, and the $Pb(NO_3)_2$ concentration is 0.1 M, the deposition becomes mass-transport-limited in (Tl⁺) when the current density is raised over about 1 mA/cm². The atomic percent Pb was varied from 46% to 88%, when the current density was increased from 0.06 mA/cm² to 6.2 mA/cm² (see Table I).

The x-ray diffraction patterns of the $Tl_aPb_bO_c$ films are very similar to the Tl_2O_3 pattern. An important difference between the patterns is that mixed-index reflections such as (211) are not observed for the mixed oxide, but they are present for Tl_2O_3. The systematic absence of mixed index reflections is consistent with an fcc structure. This is in agreement with the assignment of Sakai et al. (16), of electrodeposited $Pb_8Tl_5O_{24}$ as an fcc fluorite-type structure with a cubic lattice parameter of 0.53331 nm. For comparison, Tl_2O_3 has a bcc bixbyite structure with a=1.05434 nm, and PbO_2 has either an orthorhombic or tetragonal structure (21). The lattice parameters that we measure for our mixed oxides vary by only 0.3% over the entire composition range listed in Table I. Hence, these oxides are ideal for the epitaxial growth of nanomodulated superlattices.

Table I. Composition and Cubic Lattice Parameters for Electrodeposited $Tl_aPb_bO_c$ Films as a Function of Applied Current Density.

Applied Current Density (mA/cm²)	Measured Potential (mV vs. SCE)	Atomic % Pb	Cubic Lattice Parameter (nm)
0.06	60	46	0.536
0.62	120	66	0.535
1.24	140	72	0.537
6.20	270	88	0.535

A modulated structure was prepared for SEM studies by alternately pulsing the electrode at 0.06 mA/cm² for 8772 seconds and at 6.20 mA/cm² for 88 seconds. The layers were intentionally made quite thick, so that they could be easily imaged in the SEM. The cross section of a fractured film is shown in Figure 5. The micrograph clearly shows that the composition is modulated.

Samples with much shorter modulation wavelengths and layer thicknesses were prepared by using shorter dwell times during the double-pulse galvanostatic deposition. The superlattices with modulation wavelengths in the nanometer range were characterized by x-ray diffraction. The periodicity of the superlattice manifests itself in two ways in the x-ray diffraction pattern. Reflections are observed at low angles which correspond to the modulation wavelengths. These peaks at low angles would be observed even with a modulated amorphous material. At high angles the periodicity is seen as satellites around the main Bragg reflections for the material. The analysis of our superlattices was done at high angles. The x-ray diffraction pattern for a superlattice with a 6.8 nm modulation wavelength is shown in Figure 6. The modulation wavelength, Λ_x, can be calculated from Equation 1, where λ is the x-ray wavelength used, L is the order of the reflection, and θ is the diffraction angle. In Equation 1, L = 0 for the Bragg reflection, while the first satellite at low angle has the value L = -1, and the first satellite at higher angle has the value L = +1.

$$\Lambda_x = \frac{(L_1 - L_2)\lambda}{2(\sin\theta_1 - \sin\theta_2)} \tag{1}$$

The modulation wavelength can also be estimated from Equation 2, which is derived from Faraday's law:

$$\Lambda_F = \frac{1}{nFA} \left(\frac{i_a t_a M_a}{\rho_a} + \frac{i_b t_b M_b}{\rho_b} \right) \tag{2}$$

where n is the number of electrons transferred, F is Faraday's number, A is the electrode area, M is the formula weight, and ρ is the density. For the calculation, we assumed that both layers had the same density (10.353 g/cm³) and formula weight (228 g/mol). This is only an approximation, since the density and formula weight varies somewhat with composition. We have found, however, that for all of the compositions listed in Table I, approximately 1.1 to 1.2 µm of material are deposited per coulomb/cm² of anodic charge that is passed. There is remarkably good agreement between Λ_F and Λ_x in Table II considering the approximation used for Λ_F.

Table II. Comparison of modulation wavelengths calculated from Faraday's law (Λ_F) and from x-ray satellite spacings (Λ_x) for ceramic superlattices deposited at various current densities (J) and dwell times (t). λ represents the individual layer thickness calculated from Faraday's law. The electrode area ranged from 0.78 to 1.96 cm².

J_a (mA/cm²)	t_a (s)	J_b (mA/cm²)	t_b (s)	λ_{Fa} (nm)	λ_{Fb} (nm)	Λ_F (nm)	Λ_x (nm)
0.036	50	4.95	0.5	2.0	2.8	4.8	5.0
0.063	40	6.31	0.4	2.9	2.9	5.8	5.9
0.036	70	4.95	0.7	2.8	3.9	6.7	6.8
0.064	50	6.41	0.5	3.7	3.7	7.4	7.4
0.060	60	6.02	0.6	4.1	4.1	8.2	8.0
0.061	70	6.17	0.7	4.9	4.9	9.8	9.9
0.060	90	5.95	0.9	6.1	6.1	12.2	12.9

A fundamental question to be addressed in these complex oxides, is how much lattice mismatch a system can tolerate before switching from epitaxial growth to the nucleation of a distinct second phase. Since superlattice satellites are observed in the x-ray diffraction pattern of nanomodulated samples made by pulsing the current density between 0.036 mA/cm² and 4.95 mA/cm², it is quite likely that these compositions grow epitaxially and the interfaces are coherent. A general definition of epitaxy is the growth of crystals on a crystalline substrate that determines their

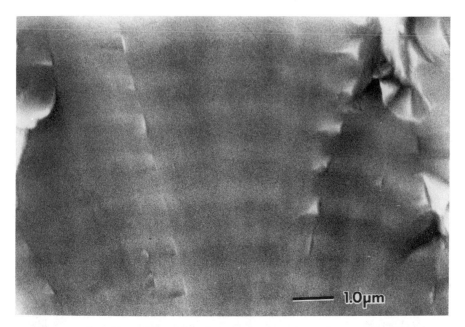

Figure 5. Scanning electron micrograph of a modulated $Tl_aPb_bO_c/Tl_dPb_eO_f$ film. A cross-sectional view of a fractured film is shown. The film was deposited by cycling between 0.05 mA/cm^2 and 5 mA/cm^2 in a solution of 0.005 M TlNo$_3$ and 0.1 M Pb(NO$_3$)$_2$ in 5 M NaOH. Layer thicknesses are intentionally quite thick ($\sim 0.5\,\mu$) so that SEM imaging is possible.

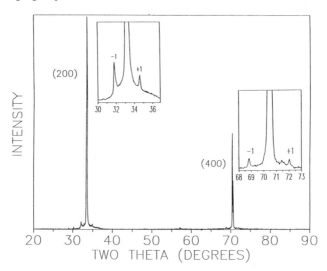

Figure 6. X-ray diffraction pattern for the $Tl_aPb_bO_c/Tl_dPb_eO_f$ superlattice with a modulation wavelength of 6.8 nm. Insets: Satellites marked by +1 and -1 around the (200) and (400) reflections at 34° and 71°.

orientation (22). Hence, one way to determine whether two materials grow epitaxially is to grow one composition of a known and unique orientation, and then determine by x-ray diffraction whether an overlayer with a different composition follows the same orientation.

The oxides in our work are nearly ideal for this type of experiment, since it is relatively easy to deposit unique orientations of the oxides depending on the applied current or film composition. For example, both the TlPbO mixed oxides and thallium(III) oxide deposit in a near random orientation when deposited at low current densities such as 0.036 mA/cm^2, but develop strong preferred orientations when deposited at higher current densities such as 4.95 mA/cm^2.

The x-ray diffraction patterns for various prelayers and overlayers are shown in Figures 7 and 8. A film of TlPbO grown at 0.036 mA/cm^2 develops the random orientation previously described by Sakai et al. (16). The (111) reflection has the highest intensity in this pattern. If the TlPbO film is deposited at 4.95 mA/cm^2, however, the (220) reflection is the most intense. A TlPbO film that is then grown at 0.036 mA/cm^2 (the random orientation) onto a [110] oriented film that was predeposited at 4.95 mA/cm^2, follows the [110] orientation of the predeposit (Figure 7). Similarly, a film of thallium(III) oxide deposited at 4.95 mA/cm^2 has a strong [100] orientation, and this orientation is followed by the 0.036 mA/cm^2 overlayer of TlPbO (Figure 8). These studies indicate that the two pairs TlPbO (0.036 mA/cm^2)/TlPbO (4.95 mA/cm^2) and TlPbO (0.036 mA)/Tl$_2$O$_3$ (4.95

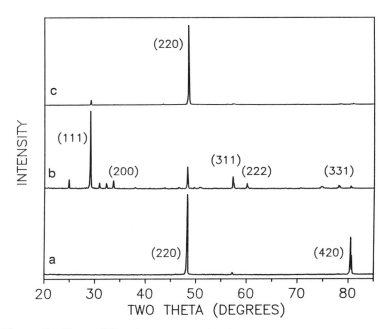

Figure 7. X-ray diffraction patterns showing the epitaxial growth of one TlPbO composition onto another TlPbO composition. (a) Oriented TlPbO film deposited at 4.95 mA/cm^2 onto 430 stainless steel. (b) Near-random TlPbO film deposited at 0.036 mA/cm^2 onto 430 stainless steel. (c) Oriented TlPbO film deposited at 0.036 mA/cm^2 onto an oriented predeposit of TlPbO that was deposited at 6.95 mA/cm^2.

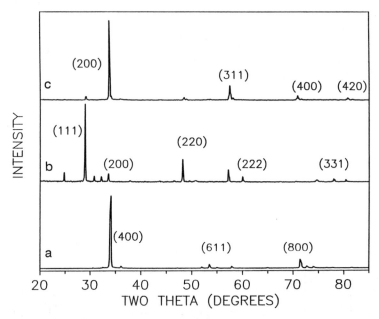

Figure 8. X-ray diffraction patterns showing the epitaxial growth of TlPbO onto Tl$_2$O$_3$. (a) Oriented Tl$_2$O$_3$ film deposited at 4.95 mA/cm^2 onto 430 stainless steel. (b) Near-random TlPbO film deposited at 0.036 mA/cm^2 onto 430 stainless steel. (c) Oriented TlPbO film deposited at 0.036 mA/cm^2 onto an oriented predeposit of Tl$_2$O$_3$ that was deposited at 4.95 mA/cm^2.

mA/cm2) grow epitaxially. We are presently exploring the growth of orthorhombic lead(IV) oxide onto both of the mixed oxides and onto thallium(III) oxide. We do not expect to see epitaxial growth in these cases.

The effect of quantum confinement on the optical and electrical properties of these materials is unknown. Quantum confinement should produce thickness-dependent blue-shifting of the optical absorption spectra. Since the materials are non-reflective in the 1.0 - 1.3 µm range, but have a sharp plasma edge (i.e., high reflectivity) at wavelengths greater than 1.3 µm, we are exploring their possible application as fast optical shutters, mirrors, or waveguides in the near-IR. Another important area of future research will be to study epitaxial growth in these systems by measuring nucleation and growth rate constants by potential-step transient experiments for varying degrees of lattice mismatch. These *in situ* studies should provide valuable information to architects of future superlattice structures.

Acknowlegments

This work was supported in part by the Division of Materials Research of the National Science Foundation under Grant No. DMR-9020026, the Office of Naval Research under Grant No. N0001491-J-1499, and the Mitsubishi Corporation through a Mitsubishi Kasei Faculty Development Award.

Literature Cited

1. Schuller, I.K.; Homma, H. *MRS Bulletin*, **1987**, *18*.
2. Esaki, L. *IEEE J. of Quantum Electronics*, **1986**, *QE-22*, 1611.
3. Döhler, G.H. *ibid*, **1986**, *QE-22*, 1682.
4. Harris, J.J.; Pals, J.A.; Woltjer, R. *Rep. Prog. Phys.*, **1989**, *52*, 1217.
5. Switzer, J.A.; Shane, M.J.; Phillips, R.J. *Science*, **1990**, *247*, 444.
6. Cohen, U.; Koch, F.B.; Sard, R. *J. Electrochem. Soc.*, **1983**, *130*, 1987.
7. Tench, D.; White, J. *Metall. Trans. A*, **1984**, *15*, 2039.
8. Yahalom, J. et al. *J. Mater. Res.*, **1989**, *4*, 755.
9. Lashmore, D.S.; Daniel, M.P. *J. Electrochem. Soc.*, **1988**, *135*, 1218.
10. Bennett, L.H. et al. *J. Magn. Magn. Mater.*, **1987**, *67*, 239.
11. Phillips, R.J.; Shane, M.J.; Switzer, J.A. *J. Mater. Res.*, **1989**, *4*, 923.
12. Switzer, J.A. *Am. Ceram. Soc. Bull.*, **1987**, *66*, 1521.
13. Switzer, J.A. *J. Electrochem. Soc.*, **1986**, *133*, 722.
14. Mindt, W. *J. Electrochem. Soc.*, **1969**, *116*, 1076.
15. Thomas, J.C.G.; Wabner, D.W. *J. Electroanal. Chem.*, **1982**, *135*, 243.
16. Sakai, M.; Sekine, T.; Yamazaki, Y. *J. Electrochem. Soc.*, **1983**, *130*, 1631.
17. Tillmetz, W.; Wabner, D.W. *Z. Naturforsch. Teil B*, **1984**, *39*, 594.
18. Shukla, V.N.; Wirtz, G.P. *J. Am. Ceram. Soc.*, **1977**, *60*, 253.
19. *Interfaces, Quantum Wells, and Superlattices*; Leavens, C.R.; Taylor, R., Eds.; NATO Series B: Physics; Plenum: New York, New York, **1988**; Vol. 179.
20. *Physics and Applications of Quantum Wells and Superlattices*; Mendez, E.E.; von Klitzing, K., Eds.; NATO Series B: Physics; Plenum: New York, New York, **1987**; Vol. 170.
21. Moseley, P.T.; Hutchison, J.L.; Bourke, M.A.M. *J. Electrochem. Soc.*, **1982**, *129*, 876.
22. Bauer, E.G. et al. *J. Mater. Res.*, **1990**, *5*, 852.

RECEIVED February 19, 1992

THREE-DIMENSIONAL FRAMEWORKS AND AMORPHOUS NETWORKS

Chapter 19

Crystal Engineering of Novel Materials Composed of Infinite Two- and Three-Dimensional Frameworks

Richard Robson, Brendan F. Abrahams, Stuart R. Batten, Robert W. Gable, Bernard F. Hoskins, and Jianping Liu

Inorganic Section, School of Chemistry, University of Melbourne, Parkville 3052, Victoria, Australia

A general approach to the construction of new types of infinite frameworks based on a number of simple structural prototypes is described. Whereas $Cu(4,4',4'',4'''$-tetracyanotetraphenylmethane$)BF_4.xC_6H_5NO_2$ contains a single diamond-like framework which generates huge intra-framework spaces filled with essentially fluid nitrobenzene, $Cu(1,4$-dicyanobenzene$)_2BF_4$ contains five independent diamond-like frameworks which interpenetrate leaving no space for solvent. Simple mixing of the components NMe_4^+, Zn^{2+}, Cu^+ and CN^- leads to the spontaneous assembly of the intended diamond-related array of composition $[NMe_4][ZnCu(CN)_4]$. A new simple structural prototype is provided by $Cd(CN)_2.2/3H_2O.tBuOH$ which contains an infinite honeycomb-like framework consisting of interconnected square planar and tetrahedral centres in 1:2 proportions; $Cd(CN)_2.1/3$hexamethylenetetramine has a geometrically very different framework structure which nevertheless has an identical connectivity or topology. A number of metal-4,4'-bipyridine derivatives consist of two perpendicular stacks of 2D square grid sheets which interpenetrate to give an unprecedented 3D concatenation. $M[C(CN)_3]_2$ crystals (M=Zn, Cu, Cd, Ni, Co, Mn) consist of two independent, interpenetrating rutile-like frameworks. A new type of 3D net is produced by interconnecting 5,10,15,20-tetra(4-pyridyl)-21H,23H-porphine palladium building blocks via Cd centres. A PtS-related framework can be deliberately constructed by interconnecting square planar $Pt(CN)_4^{2-}$ building blocks by tetrahedral Cu^I centres.

It is possible that unusual materials could be made by building infinite frameworks based on a number of structural prototypes in which each atom of the parent net has been replaced by a stereochemically appropriate molecular building block and each bond of the parent has been replaced by an appropriate molecular connection. It is probably wise in early attempts at this framework construction to use the simplest available structural prototypes such as diamond and Lonsdalite (tetrahedral centres), α–polonium (octahedral centres), NbO (square planar centres), PtS (equal numbers

0097–6156/92/0499–0256$06.00/0

of tetrahedral and square planar centres) and rutile (trigonal and octahedral centres in 2:1 proportions) but others that exist or that can be envisaged could be put to similar use. Given the wide range of molecular building blocks and connectors that can be conceived for these purposes each prototype net in principle affords a whole family of related frameworks. Such solids would be of fundamental structural interest and may have useful properties.

The working hypothesis we have adopted is that self-assembly to yield ordered infinite frameworks may occur spontaneously in reaction mixtures containing appropriately functionalised building blocks and connecting units. In ideal cases one could imagine devising reaction systems having no option but to condense into the intended infinite array.

Channels, Cavities and Interpenetration

Examination of models and consideration of the sorts of molecular building blocks that might realistically be used for framework construction leads to the realisation that, for many simple nets, if the connecting units have an element of rod-like rigidity they need have only modest length (by normal molecular standards) to provide structures with relatively large cavities, windows and channels. As a consequence, these materials may have interesting and useful properties with applications in areas such as ion exchange, molecular sieves and zeolite-like catalysis. Appropriate functionalisation of the components, either before framework construction or afterwards may make possible the synthesis of a range of permeable solid catalysts each tailor-made for a particular chemical transformation. Compared with zeolites these frameworks in principle offer bigger channels, more facile and more widely variable functionalisation for the introduction of catalytic sites and better substrate access to those sites.

An aspect of these frameworks which is becoming more apparent the more of them we study, is their tendency to yield remarkable structures in which two or more entirely independent giant molecules are intimately entangled, but in an ordered fashion. In such cases the channels and cavities generated by one framework are very neatly filled by the other(s). A few interpenetrating networks have been recognised for many years. The discovery of the first example, Cu_2O, consisting of two independent, interpenetrating diamond-like nets dates back to the very early days of structural analysis (1). Most known cases of interpenetration involve varying numbers of diamond-related nets eg. two independent nets (2), three nets (3), five nets (4) and six nets (5). Rare examples of interpenetrating 3D nets not derived from diamond are provided by the silicate mineral neptunite, in which two essentially 3-connected nets interpenetrate (6) and β-quinol (7). Rare examples of interpenetrating sheet structures are provided by $Ag[C(CN)_3]$ (8), benzene-1,3,5-tricarboxylic acid (9) and certain of its inclusion compounds (10) all of which involve sheets resembling hexagonal mesh chicken wire and Hittorf's violet phosphorus (11). Several of these examples involve rather weakly H-bonded networks where the distinction from collections of discrete molecules is not clear cut and examples of interpenetrating frameworks that are strongly bonded internally remain relatively rare. It does seem likely at this stage that work along the lines described below will greatly increase the range of known types of interpenetration.

Concatenated arrays are of interest not only at a fundamental structural level as new and special geometrical and topological types but also because they may show unusual properties stemming from their unusual structures. For example, the ordered entanglement may lead to unusual mechanical properties. It may also provide a means of positioning and orienting sub-units of separate frameworks into unusually close contact; if the sub-units have appropriate delocalised electronic π-systems this oriented close contact may afford an approach to the development of organic superconductors with higher T_c's than than those currently known.

Diamond-Related Frameworks

4,4',4",4'''-tetrasubstituted tetraphenylmethanes and 1,3,5,7-tetrasubstituted adamantanes are natural choices as building blocks for the construction of diamond-related frameworks, a number of general approaches to which have been considered elsewhere (12). Replacement of the acetonitrile ligands in $Cu^I(CH_3CN)_4^+$ by 4,4',4",4'''-tetracyanotetraphenylmethane led to the assembly of a diamond-like array (12). Part of the structure of the positively charged framework is represented in Figure 1, which highlights an adamantane-like unit, a fundamental structural component (see Figure 2) of the diamond net. The structure consists of alternating tetrahedral C and Cu^I centres, inter-connected by $C-C_6H_4-CN-Cu$ rods 8.856(2) Å in length. The figure gives a visual impression of the large relative size of the intraframework space which accounts for approximately two thirds of the volume of the crystal and which is occupied by large amounts of fluid nitrobenzene (at least 7.7 $C_6H_5NO_2$ per Cu) together with mobile BF_4^- ions. A crystal two thirds of which is fluid is quite extraordinary.

When the acetonitrile ligands of $Cu^I(CH_3CN)_4BF_4$ are substituted by 1,4-dicyanobenzene under conditions identical to those used above for the tetraphenylmethane system, the outcome is very different. The crystals formed contain no solvent and have composition $Cu^I(1,4\text{-dicyanobenzene})_2BF_4$. Diamond-like frameworks are indeed formed ($Cu-NC-C_6H_4-CN-Cu$ rods 11.76Å in length) but the structure contains five independent frameworks which interpenetrate as represented in Figures 3a and 3b. An adamantane unit has three 2-fold axes of symmetry one of which is indicated in Figure 2a; the view looking almost directly down a 2-fold axis is shown in Figure 2b. The framework structure of $Cu(1,4\text{-dicyanobenzene})_2BF_4$, showing only the metal centres and their connectivity is represented in Figure 3a, as seen from a viewpoint slightly off the 2-fold axis analogous to that in Figure 2b. As can be seen, the adamantane units are distorted. The heavy connections represent one framework. The interpenetrating arrangement is such that the adamantane units of the 2nd, 3rd, 4th and 5th frameworks can be imagined generated from the first by a translation along the 2-fold axis common to all of them by 1/5s, 2/5s, 3/5s and 4/5s respectively (see Figure 2a for s). Channels of rhombic cross-section are generated in which the BF_4^- ions are located as shown in Figure 3b.

A study of the way the structure varies, in particular whether or not interpenetration occurs, as counter ion and solvent are varied, would clearly be valuable with these Cu(I) derivatives of 1,4-dicyanobenzene and 4,4',4",4'''-tetracyanotetraphenylmethane.

$Zn(CN)_2$ and $Cd(CN)_2$ each contain two independent diamond-related frameworks with tetrahedral metal centres and MCNM rods, one neatly filling the spaces generated by the other (12). As an exercise in the deliberate manipulation of framework assembly we attempted to devise ways of creating a single, non-interpenetrating, diamond-like structure consisting of tetrahedral metal centres inter-connected by linear cyanide bridges. One strategy explored was to attempt the construction of a negatively charged single framework by substituting every other Zn^{2+} in a $Zn(CN)_2$ diamond net with Cu^+, the required counter cations thus making interpenetration impossible. Models suggested NMe_4^+ would fit snugly into the adamantane cavities of a single framework and the overall framework charge required that only half the cavities would need to be occupied by cations. Bringing together in aqueous solution the components Zn^{2+}, Cu^+, CN^- and NMe_4^+ under extremely simple conditions led to the spontaneous assembly of the intended array (12). Iwamoto has shown independently that in the presence of CCl_4, $Cd(CN)_2$ can be

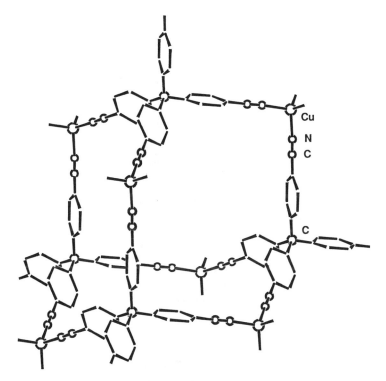

Figure 1. Framework structure of
Cu(4,4',4",4'''-tetracyanotetraphenylmethane)BF$_4$.xC$_6$H$_5$NO$_2$
highlighting an adamantane-like unit.

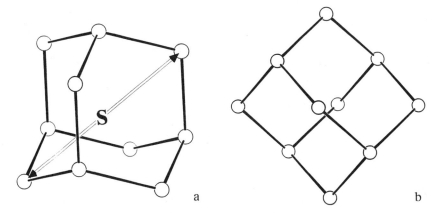

Figure 2. (a) An adamantane-like component of the diamond net showing one of
the 2-fold axes. (b) View of adamantane unit looking slightly offset from the 2-
fold axis.

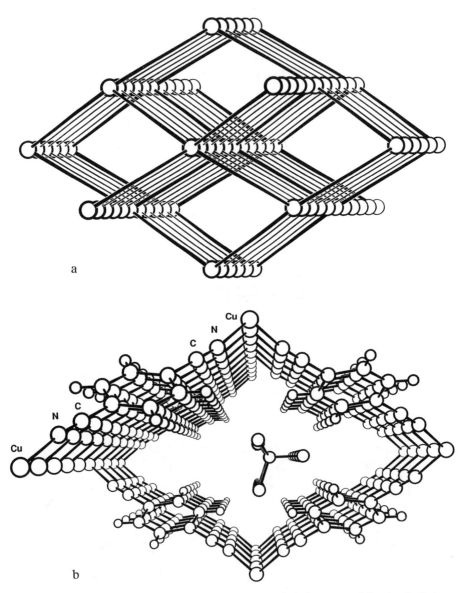

Figure 3. (a) Arrangement of copper centres and their connectivity in Cu(1,4-dicyanobenzene)$_2$BF$_4$ from a viewpoint similar to that in Figure 2b. Heavy connections indicate one particular framework. (b) Perspective view down one of the rhombic channels of Cu(1,4-dicyanobenzene)$_2$BF$_4$.

induced to crystallise as a single framework with CCl_4 molecules in the adamantane cavities (13).

A New Prototype Honeycomb Net Involving Square Planar and Tetrahedral Centres in 1:2 Proportions

$Cd(CN)_2$ crystallises from aqueous tBuOH as $Cd(CN)_2.2/3H_2O.tBuOH$ with the novel honeycomb stucture shown in Figure 4 (14). Disordered tBuOH occupies the channels. The infinite $[Cd(CN)_2]_n$ framework consists essentially of square planar and tetrahedral centres in 1:2 proportions. Upon exposure to the atmosphere tBuOH and H_2O are lost and the crystals collapse to a collection of microcrystals of the normal interpenetrating form of $Cd(CN)_2$ as revealed by powder X-ray diffraction. When $Cd(CN)_2$ is crystallised in the presence of hexamethylenetetramine (HMTA) crystals of $Cd(CN)_2.1/3(HMTA)$ are obtained with an infinite 3D $[Cd(CN)_2]_n$ framework which at first sight appears very different from that in $Cd(CN)_2.2/3H_2O.tBuOH$ but which in fact is topologically identical (15). The geometrical relationship between the two is shown in Figures 5a and b. The linear hexagonal channels present in $Cd(CN)_2.2/3H_2O.tBuOH$ (Figure 5a) are deformed, the original connectivity remaining unbroken, to produce in $Cd(CN)_2.1/3HMTA$ pronounced zig-zags (Figure 5b) which allows the four N donors of each HMTA to become coordinated.

This new, simple 3D net, seen in its geometrically most regular form in $Cd(CN)_2.2/3H_2O.tBuOH$ may provide the prototype (analogous to the diamond prototype in the systems above) for a whole family of potentially interesting solids; e.g. it may be possible to construct frameworks in which the planar ribbons of edge-sharing square units apparent in Figures 4 and 5a have been replaced by banks of coplanar porphyrins or similar plate-like building blocks. The hexagonal channels, which with such large building blocks would be of correspondingly large dimensions, would then provide access for substrate molecules to the catalytic metal-porphyrin sites lining the channels.

Square Grid 2D Sheets and their Interpenetration to form a Novel 3D Network

Mixtures of $Cd(PF_6)_2$ and 4,4'-bipyridine (bipy) in 1:4 ethanol - water yield an initial product in the form of thin plates which is then gradually replaced by a second product consisting of more chunky crystals. The structure of the second type of crystal is presently under investigation. The initial crystals of formulation $[Cd(H_2O)_2(bipy)_2](PF_6)_2.2bipy.4H_2O$ consist of stacks of parallel infinite square grid cationic sheets of composition $[Cd(H_2O)_2(bipy)_2]_n^{2n+}$ with the structure shown in Figure 6. Each Cd is essentially octahedral with two trans water ligands and four bipy N donors. Uncoordinated bipy molecules project through the square holes in the sheets. Only minor modification of the reaction conditions, namely, increasing the EtOH:H_2O ratio to 5:2, yields crystals of $[Cd(H_2O)(OH)(bipy)_2]PF_6$. We have not yet completed a full structural analysis of this material but it has the same space group (P4/ncc) and almost the same unit cell dimensions (a,10.992(5); c,17.608(8) Å) as those of the compound $[Cd(H_2O)_2(bipy)_2]SiF_6$ (P4/ncc; a,11.016(2); c,17.586(3) Å) for which a full analysis has been carried out. The latter compound is obtained from aqueous methanolic solutions containing $Cd(ClO_4)_2$, Na_2SiF_6 and bipy and it is isostructural with the corresponding Zn (16) and Cu derivatives, consisting of two perpendicular and equivalent stacks of square grid $[Cd(H_2O)_2(bipy)_2]_n^{2n+}$ sheets which interpenetrate to give the remarkable 3D arrangement shown in Figure 7. Any particular sheet has an infinite number of perpendicular ones enmeshed in it. On the evidence above it is almost certain that

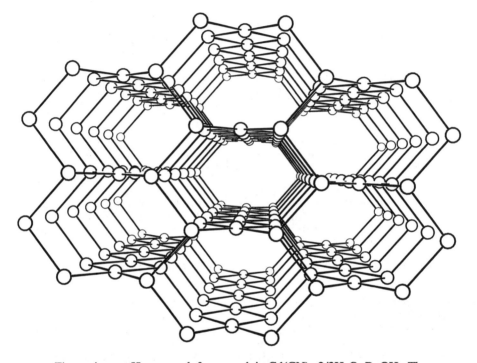

Figure 4.   Honeycomb framework in $Cd(CN)_2.2/3H_2O.tBuOH$. The centres represented are all Cd's and connections are all of the type CdCNCd. C and N atoms are not shown, nor are the water molecules coordinated above and below the apparently square planar Cd's. Reproduced with permisson from ref. 14. Copyright 1990 Royal Society of Chemistry.

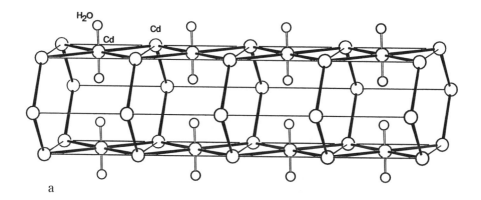

a

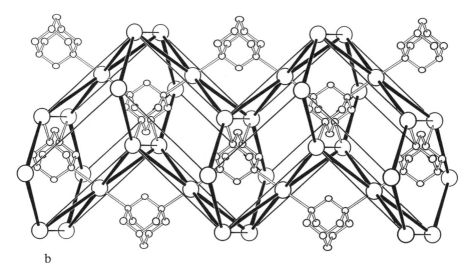

b

Figure 5. (a) Side-view of one of the hexagonal channels of Cd(CN)$_2$.2/3H$_2$O.tBuOH. Heavy connections represent the cyanide linkages between pairs of Cd atoms. Lighter lines indicate only geometrical relationships. The smaller circles represent oxygen atoms of coordinated water. (b) Side-view of zig-zag hexagonal channel in Cd(CN)$_2$.1/3(C$_6$H$_{12}$N$_4$) analogous to the linear hexagonal channel in Cd(CN)$_2$.2/3H$_2$O.tBuOH shown in (a). Heavy and light lines have the same significance as in (a). Reproduced from ref. 15. Copyright 1991 American Chemical Society.

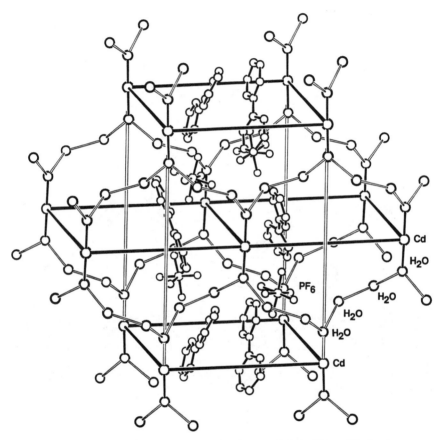

Figure 6.   Structure of [Cd(H$_2$O)$_2$(bipy)$_2$](PF$_6$)$_2$.2bipy.4H$_2$O where bipy = 4,4'-bipyridine. Heavy lines represent the Cd.bipy.Cd rod-like components of the square grid sheets.

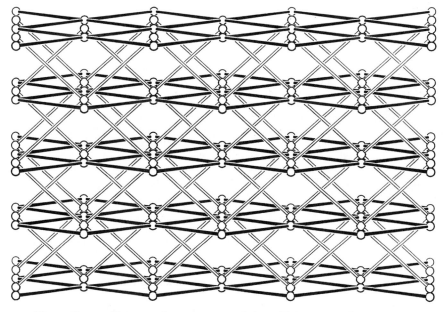

Figure 7.   Framework structure consisting of interpenetrating square grid sheets common to [Cd(H$_2$O)(OH)(bipy)$_2$]PF$_6$ and[M(H$_2$O)$_2$(bipy)$_2$]SiF$_6$ (M = Zn, Cu, Cd). Only metal centres and bipy connections are shown.   Reproduced with permisson from ref. 16. Copyright 1990 Royal Society of Chemistry.

[Cd(H$_2$O)(OH)(bipy)$_2$]PF$_6$ has a nearly identical interpenetrating sheet structure in which one of the coordinated water molecules has lost a proton. These Cd-bipy compounds, then, provide the unusual situation where almost identical square grid sheets in one case are stacked one on top of the other to give a non-interpenetrating structure and in another case interpenetrate to yield an unprecedented type of concatenated 3D network.

4,4'-Bipyridine coordinated through both nitrogens to metal cations may have an electronic configuration resembling that in the methyl viologen dication (the 1,1'-dimethyl-4,4'-bipyridinium or paraquat dication) a species undergoing ready reduction to the stable mono-cation-radical and consequently much used in electron transfer studies. Moreover, bipy is one of a number of bridging species incorporated into mixed valence Ru^{II}/Ru^{III} complexes of the famous Creutz-Taube type which mediate in facile electron transfer (17). Infinite framework solids constructed from mixed valence metal centres inter-connected by bipy may therefore show unusual electrical properties.

Rutile-Related Frameworks

The rutile prototype consists of trigonal and octahedral centres in 2:1 proportions. One of the simplest imaginable potential trigonal connectors is the tricyanomethanide ion, C(CN)$_3$⁻. X-ray structural analyses of Zn[C(CN)$_3$]$_2$ (18) and Cu[C(CN)$_3$]$_2$ reveal that both consist of two independent rutile-related frameworks which interpenetrate in the manner shown, for the Zn compound, in Figure 8. Cell dimensions and space group determination indicate that the corresponding Cd^{II}, Ni^{II}, Co^{II} and Mn^{II} compounds have the same structure. Tetragonal elongation of two Cu-N bonds is observed in the Cu compound but otherwise the structure is very similar to that of the Zn compound. As can be seen by inspection of Figure 8, each framework in Zn[C(CN)$_3$]$_2$ contains "4-membered"[Zn$_2$C(3)$_2$] and "6-membered" [Zn$_3$C(3)$_3$] rings. The nature of the interpenetration is such that every 6-membered ring of one framework has a ZnNCC rod of the other framework passing through it, but projection of rods through 4-membered rings does not occur. Parts of the independent frameworks are forced by the interpenetration into unusually close contact eg. C(1)...C(1) = 3.135(4) and 3.138(4) Å; C(2)...N(1) = 3.193(3) Å. This highlights the general point made above that interpenetration may lead to enforced proximity and thence to unusual properties.

By linking together C(CN)$_3$⁻ units with cations having preferences for geometries other than octahedral, it should be possible to generate a number of new geometrical and topological types of 3D structure; the highly unusual interpenetrating sheet structure in Ag[C(CN)$_3$] (8) in which Ag^I acts as a 3-connecting centre illustrates this potential for the generation of new structural types. Trigonal building blocks larger and more elaborate than C(CN)$_3$⁻ can readily be envisaged and may provide whole families of new materials. Wells (19) has surveyed 3-connected and (3,n)-connected nets, many of which remain purely hypothetical. These trigonal building blocks may yield frameworks providing real examples of hitherto hypothetical nets.

A New Type of 3D Net Involving 4-Connected Porphyrin Building Blocks

Solutions of 5,10,15,20-tetra(4-pyridyl)-21H,23H-porphine palladium (Pd.py.porph) with excess Cd(NO$_3$)$_2$ in water-methanol-ethanol gave crystals of (Pd.py.porph).2Cd(NO$_3$)$_2$.hydrate whose structure (20) is represented in Figures 9a and b. All porphyrin units are equivalent and are attached via their pyridyl nitrogens to four Cd's (Figure 9a). Two diametrically opposed Cd(2)'s are coordinated by two

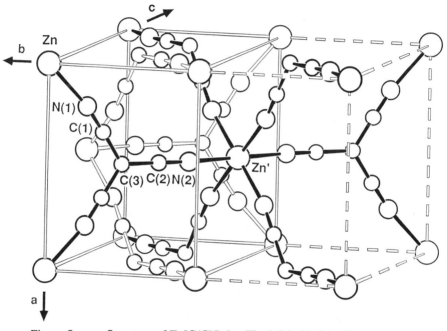

Figure 8.   Structure of Zn[C(CN)₃]₂. The left half of the figure represents the actual unit cell. One framework is omitted from the right half so that the relationship to rutile of the framework extending into both halves is easily recognised.   Reproduced with permisson from ref. 18. Copyright 1991 Royal Society of Chemistry.

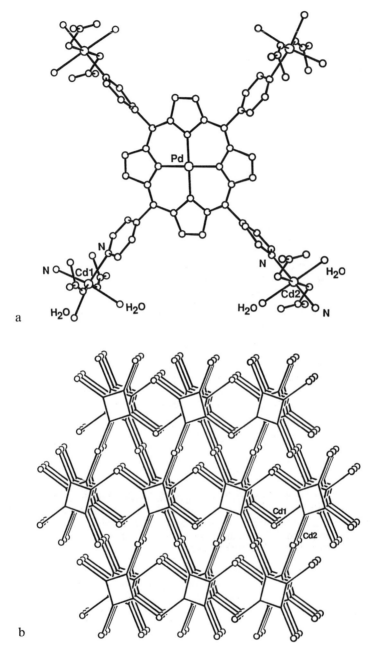

Figure 9. (a) Environment of the porphyrin unit in (Pd.py.porph).2Cd(NO$_3$)$_2$.hydrate. (b) Extended 3D framework in (Pd.py.porph).2Cd(NO$_3$)$_2$.hydrate. Only the cadmiums and the porphyrin meso(5,10,15,20) carbons are indicated, the latter occupying the corners of the squares seen here obliquely. Reproduced from ref. 20. Copyright 1991 American Chemical Society.

trans pyridines, two trans H_2O's and two trans monodentate NO_3^-'s. With regard to the framework structure these Cd(2)'s can be regarded as "linear"connectors. The two Cd(1)'s associated with a particular porphyrin are "bent" connectors because the two coordinated pyridines in these cases are cis. Linear polymeric strips consisting of Pd.py.porph units connected together by linear Cd(2)'s can be seen in Figure 9b, half of them being parallel with each other, the others being parallel to a line at 59° to the first set. Attached to every linear strip via its pendent bent Cd(1)'s is an infinite number of other strips all at 59° to the first; half of these pass over and half pass under the first strip. At these points of overlap essentially parallel Pd.py.porph units make face to face contact (Pd...Pd, 4.68Å) the PdPd vector making an angle of 30° with the normal to the porphyrin planes; in this way stacks of porphyrins are generated with all Pd's co-linear. To the best of our knowledge this is a new geometrical and new topological type of infinite 3D network.

On the basis of this preliminary study the prospects look good for constructing many new 2D and 3D nets using porphyrin and phthalocyanin building blocks.

PtS-Related Frameworks

The PtS prototype consists of equal numbers of inter-connected square planar and tetrahedral centres (Figures 10a and b). In an experiment similar to the generation of the diamond-related $[CuZn(CN)_4][NMe_4]$ described above, an attempt was made to link together stable, square planar $Pt(CN)_4^{2-}$ building blocks via tetrahedral Cu^I centres. Bringing together Cu^+, $Pt(CN)_4^{2-}$ and NMe_4^+ led to the assembly of the $[CuPt(CN)_4]_n^{n-}$ framework as the NMe_4^+ derivative which did have the PtS-related structure (21) (Figure 11). Three sorts of channels can be seen. Hexagonal channels of large dimensions run in both the a and b directions, the largest Pt...Pt separation across the channel being 13.50 Å. Also running parallel to a and b are smaller channels of roughly square cross-section. The cations which are disordered over four equivalent orientations are located in the hexagonal channels. A second type of square channel (edge 7.61 Å) runs parallel to c; in crystals sealed together with the aqueous mother liquor in Lindemann tubes these channels are occupied by very disordered water molecules. The framework in this case, in contrast to that in $Cd(CN)_2.2/3H_2O.tBuOH$, survives loss of solvent, the channels parallel to c being vacant after exposure of the crystals to the atmosphere.

The PtS prototype provides a very attractive model for the construction of permeable solids with catalytic potential. If one imagines replacing the planar PtS_4 units apparent in Figures 10a and b with flat, rigid, 4-connecting building blocks larger than the $Pt(CN)_4^{2-}$ used above, such as porphyrins and phthalocyanins, the potential offered by this net for structures with large channels giving access to banks of catalytic sites can be appreciated.

Conclusions and Future Prospects

The exploratory work reported here provides considerable encouragement that the idea of using simple 3D nets as prototypes for the construction of new infinite frameworks is not only feasible but potentially extremely fruitful. Deliberate crystal engineering along these lines to produce 2D and 3D frameworks with planned structural features begins to appear a realistic objective. This approach has already uncovered new examples of the little studied phenomenon of interpenetration and the discovery of further types can be anticipated; increasing knowledge in this area may allow, in future materials, the juxtapositioning of structural components for specific applications eg. in the area of electrical properties. The use of appropriately functionalised, large, rigid building blocks such as porphyrins and phthalocyanins

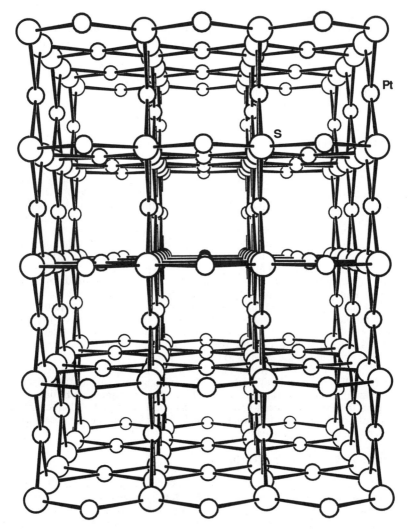

Figure 10a.   Perspective view of the PtS structure down the tetragonal
axis.

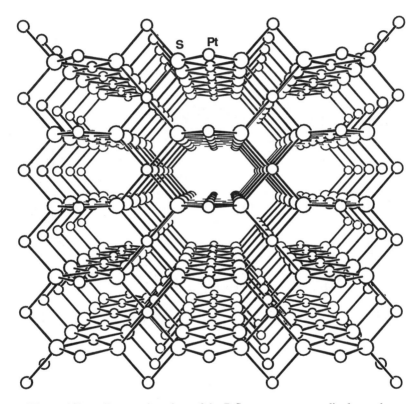

Figure 10b. Perspective view of the PtS structure perpendicular to the tetragonal axis.

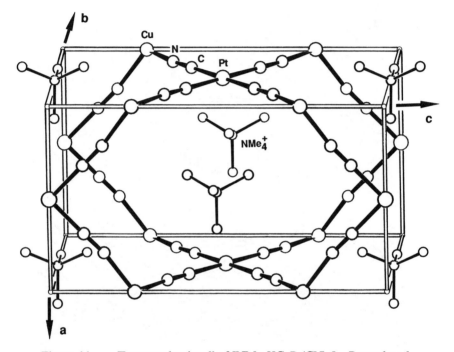

Figure 11.   Tetragonal unit cell of [NMe$_4$][CuPt(CN)$_4$].   Reproduced
with permisson from ref. 21. Copyright 1990 Royal Society of Chemistry.

may afford materials with correspondingly large channels and cavities with potentially useful applications in areas such as catalysis.

Acknowledgement

We are very grateful for support from the Selby Scientific Foundation.

Literature Cited

1. Niggli, P. Z. Krist., 1922, 57, 253.
2. Zhdanov, H.S. C.R.Acad. Sci., 1941, 31, 352. Shugam, E.; Zhdanov, H.S. Acta Physiochim. URSS, 1945, 20, 247. Kamb, B.; Davis, B.L. Proc. Natl. Sci. U. S. A., 1964, 52, 1433. Kamb, B. Science (Washington, D.C.), 1965, 150, 205. Kamb, B. J. Chem. Phys., 1965, 43, 3917. Brown, A.J.; Whalley, E. J. Chem. Phys., 1966, 45, 4360.
3. Ermer, O.; Eling, A. Angew. Chem.(Int. Edn.), 1988, 27, 829. Ermer, O.; Lindenberg, L. Helv. Chim. Acta, 1988, 71, 1084.
4. Ermer, O. J. Amer. Chem. Soc., 1988, 110, 3747.
5. Kinoshita, Y.; Matsubara, I.; Higuchi, T.; Saito, Y. Bull. Chem. Soc. Japan, 1959c, 32, 1221.
6. Cannillo, E.; Mazzi, F.; Rossi, G. Acta Cryst., 1966, 21, 200.
7. Powell, H.M. J. Chem. Soc., 1950, 300.
8. Konnert, J.; Britton, D. Inorg. Chem., 1966, 5, 1193.
9. Duchamp, D.J.;Marsh, R.E. Acta Cryst., 1969, B25, 5.
10. Herbstein, F.H.; Kapon, M.; Reisner, G.M. Proc. Roy. Soc., London, Ser. A, 1981, 376, 301. Davis, J.E.D.; Finocchiaro, P.; Herbstein, F.H. in "Inclusion Compounds"; Atwood, J.L.; Davis, J.E.D.; MacNicol, D.D., Eds.; Academic Press : New York, 1984; Chapter 11, P. 407.
11. Thurn, H.; Krebs, H. Acta Cryst. 1969, B25, 125.
12. Hoskins, B.F.; Robson, R. J. Amer. Chem. Soc., 1990, 112, 1546.
13. Kitazawa, T.; Nishikiori, R.; Kuroda, R.; Iwamoto, T. Chem. Lett., 1988, 1729.
14. Abrahams, B.F.; Hoskins, B.F.; Robson, R. J. Chem. Soc., Chem. Commun., 1990, 60.
15. Abrahams, B.F.; Hoskins, B.F.; Liu, J.; Robson, R. J. Amer. Chem. Soc., 1991, 113, 3045.
16. Gable, R.W.; Hoskins, B.F.; Robson, R. J. Chem. Soc., Chem. Commun., 1990, 1677.
17 Creutz, C. Progress in Inorg. Chem., 1983, Vol. 30, p.1.
18. Batten, S.R.; Hoskins, B.F.; Robson, R. J. Chem. Soc., Chem. Commun., 1991, in press.
19. Wells, A.F. "Three-Dimensional Nets and Polyhedra", (Wiley, 1977), Chs. 5-8.
20. Abrahams, B.F.; Hoskins, B.F.; Robson, R. J. Amer. Chem. Soc., 1991, 113, 3606.
21. Gable, R.W.; Hoskins, B.F.; Robson, R. J. Chem. Soc., Chem. Commun., 1990, 762.

RECEIVED January 16, 1992

Chapter 20

Zeolite Inclusion Chemistry
Clusters, Quantum Dots, and Polymers

Thomas Bein

Department of Chemistry, Purdue University, West Lafayette, IN 47907

An overview on zeolite structures and properties is given, followed by a discussion of various strategies for the intrazeolite encapsulation of metal clusters and organometallics, quantum size semiconductor clusters, and polymer filaments. Several examples are described in more detail. Palladium clusters of low nuclearity have been prepared in the cavities of X zeolites via ion exchange with $Pd(NH_3)_4Cl_2$, oxidative dehydration, and subsequent reduction of the dry Pd(II)zeolites with hydrogen at 295 K. EXAFS analysis of Pd K-edge data shows that intrazeolite Pd_{2-4} clusters are formed by partial occupation of SI' and SII' positions of the sodalite subunits of the zeolite. Ensembles of CdSe have been synthesized within the cage system of zeolite Y via ion exchange with Cd(II) and subsequent treatment with H_2Se. Se,O bridged cadmium dimers and Cd_4O_4 cubes are formed in the sodalite unit. The anchoring chemistry of $Me_3SnMn(CO)_5$ in acid forms of zeolite Y was studied with X-ray absorption spectroscopy (Sn, Mn edge EXAFS) and in-situ FTIR/TPD-MS techniques. The compound attaches to the zeolite framework at the oxygen rings of the supercage. The attachment of the molecule occurs through the Sn moiety by loss of CH_4 gas while the Sn-Mn bond and the CO ligand sphere are still intact. Different degrees of substitution of the methyl groups by the acidic oxygen framework are observed. Intrazeolite polyacrylonitrile (PAN) was formed from preadsorbed acrylonitrile in zeolite Y and mordenite on reaction with radical initiators. Chain length analysis with gel permeation chromatography revealed a peak molecular weight of 19,000 for PAN in NaY, and about 1,000 for the polymer in mordenite. When intrazeolite PAN is pyrolyzed under nitrogen, black encapsulated material results that has lost the nitrile groups and hydrogen. After recovery from the zeolite hosts, the pyrolyzed polyacrylonitrile shows electronic DC conductivity at the order of 10^{-5} Scm^{-1}.

Zeolites Inclusion Chemistry: Structure and Properties of the Host

Few classes of inorganic solids have gained as much importance for host/guest chemistry as the microporous aluminosilicates called zeolites.[1,2,3] While many layered materials are known, including aluminosilicate clays, group 4 phosphates, graphite,

0097–6156/92/0499–0274$06.00/0

and metal chalcogenides, zeolites are the archetypical three-dimensional "molecular sieves". One major difference between these materials is that structural integrity of the zeolites upon encapsulation or intercalation of guest molecules is maintained, while the layered structures change the interlayer spacing and interlayer orientation when intercalation takes place. There are several other groups of crystalline microporous structures of growing importance, for example molybdenum phosphates,[4] or metal sulfides,[5] which are not subject of this discussion.

Zeolites are open framework oxide structures (classically aluminosilicates with hydrophilic surfaces) with pore sizes between 0.3 and 1.2 nm and exchangeable cations. The fundamental building blocks of the classical zeolite structure are the SiO_4 and AlO_4 tetrahedra that are conceptually viewed as joining up in "secondary building units" such as single six-rings (S6R, which are in fact twelve-rings consisting of six metals and six alternating oxygens arranged much like a crown ether), double-six-rings (D6R) and complex 5-1 structures. Other important structural units are the "sodalite cage" (a truncated octahedron) and the "supercage" (a unit with 12.5 Å internal diameter). These units are then joined to form open pore structures with channels based on rings containing 6 to 20 (in a gallophosphate) metal and oxygen atoms. Not only channels but also cage structures can be formed. There are now more than seventy different structure types known and new ones continue to emerge. Zeolites have remarkable temperature stability; survival of structures at 1000 K is not uncommon. The presence of Al atoms in the silica framework results in one negative charge per Al atom such that cations are required to balance the charge. The zeolitic pores constitute a significant fraction of the crystal volume (up to about 50%) and are usually filled with water. In hydrated zeolites, the cations have a high mobility giving rise to ion exchange capability, and the water molecules can be removed at elevated temperature. A general formula for zeolite-type materials based on 4-connected networks is the following[6]:

$$M_x M'_y N_z [T_m T'_n..O_{2(m+n+..)}-e(OH)_{2e}](aq)_p \cdot qQ$$

with tetrahedral T-atoms Be, B, Al, Ga, Si, Ge, P, transition metals, and M, M' exchangeable and nonexchangeable cations, N nonmetallic cations (normally removable on heating), (aq) chemically bonded water and Q sorbate molecules which need not be water. The essential part in the square brackets represents the 4-connected framework which is usually anionic. A simpler formula used for many zeolites is:

$$(M^{n+})_{x/n} [(AlO_2)_x (SiO_2)_y] \cdot zH_2O$$

Structures of representative zeolites.[7] The following Table 1 presents a brief description of important zeolite structure types. Structural building blocks and cage topologies of those zeolites are shown in Figure 1. Zeolite A (LTA) consists of cubooctahedra (sodalite cages) linked together through double four-rings (D4R) to form a central cage (alpha-cage), accessible through six S8R with a free diameter of 4.1 Å.

Zeolite Y is isostructural with zeolite X (FAU). It differs from the latter in its higher Si/Al ratio which reduces the number of cations present in the framework. Zeolite X contains typically about 88 cations per unit cell, while Y contains about 56. The framework consists of sodalite cages, connected through four D6R that enclose a supercage with 12.5Å internal diameter. The supercages are accessible through four S12R windows with a free aperture of 7.4Å. Relevant cation locations are the center of the hexagonal prism (site SI) and in the six-membered rings connecting into the sodalite (SI'; SII') and into the supercage (SII). The ions are often slightly displaced from the center plane of the rings.

In mordenite (MOR), complex 5-1 units form two major intersecting channels, one defined by windows with oxygen 12-rings (6.5Å x 7.0Å) and one defined by windows with oxygen 8-rings. Mordenite is thermally very stable due to the large number of energetically favored five-membered rings in the framework.

Silicalite (MFI), $Si_{96}O_{192}$, belongs to the pentasil family of which the isostructural ZSM-5 is the most prominent. It has two intersecting pore systems, one consisting of zig-zag channels of near-circular cross section (5.3 x 5.6 Å) and another of straight channels of elliptical shape (5.1 x 5.5 Å) perpendicular to the first. All the intersections in silicalite are of the same size. The heat of adsorption and the room temperature uptake of water is small compared to the Al-containing zeolites.

Table 1. Representative Zeolite Structure Types.

Name	Unit Cell/Composition	Cage Type Channels/Å[a]	Main
LTA, Linde A	$Na_{12}[(AlO_2)_{12}(SiO_2)_{12}]$ 27 H_2O	a, ß	4.1 ***
FAU, Faujasite	$Na_{58}[(AlO_2)_{58}(SiO_2)_{134}]$ 240 H_2O	ß, 26-hedron(II)	7.4 ***
RHO, Rho	$Na_{12}[(AlO_2)_{12}(SiO_2)_{36}]$ 44 H_2O	a, D8R	3.9x5.1 ***
MOR, Mordenite	$Na_8[(AlO_2)_8(SiO_2)_{40}]$ 24 H_2O	complex 5-1	6.5x7.0* <-> 2.6x5.7*
MFI, ZSM-5	$Na_n[(AlO_2)_n(SiO_2)_{96-n}]$ 16 H_2O	complex 5-1	{5.3x5.6 <-> 5.1x5.5} ***
AFI, AlPO-5	$AlPO_4$	12R	7.3 *

[a] The number of stars (*) at the channel description indicates the dimensionality of channel connections. The cage types are depicted in Figure 1.

Zeolite synthesis.[8] A typical synthesis involves an aluminosilicate gel prepared from silicate and aluminate sources, hydroxide, and water, which is crystallized under hydrothermal conditions. Many zeolite preparations also contain organic cations such as tetramethyl ammonium hydroxide.[9] The organic cations are thought to be structure-directing templates, at least to a certain degree, because their presence changes the resulting zeolite phase and because they are often incorporated in the pores during synthesis. For example, tetrapropylammonium (TPA) ions are sited at the intersections of four channels in ZSM-5 with each of the four propyl groups directed along individual channels.[10] The SiO_2/Al_2O_3 ratio in these gels influences the final framework composition of the product, and usually all the aluminium available is incorporated into the final zeolite composition. The H_2O/SiO_2 and OH^-/SiO_2 ratios strongly influence the nature of the polymeric species present, and the rate of conversion of these species to three-dimensional zeolite frameworks.

Adsorption and acidity in zeolites. Intrazeolite water can be removed from the pores by heating and/or evacuating, resulting in a large, accessible intracrystalline surface area. Access of other species is limited to molecules having effective diameters small enough to enter the pores. However, the zeolite framework has a certain flexibility such that the pore size expands at elevated temperature. A classical example is the exclusion of nitrogen in NaA zeolite at cryogenic temperature (77 K), while at 180 K, the windows have expanded enough to adsorb 5 molecules of nitrogen in each cage. Other important factors influencing adsorption are polarizability of the adsorbate, wettability of the zeolite surface, and electric field gradients near cations. Zeolites

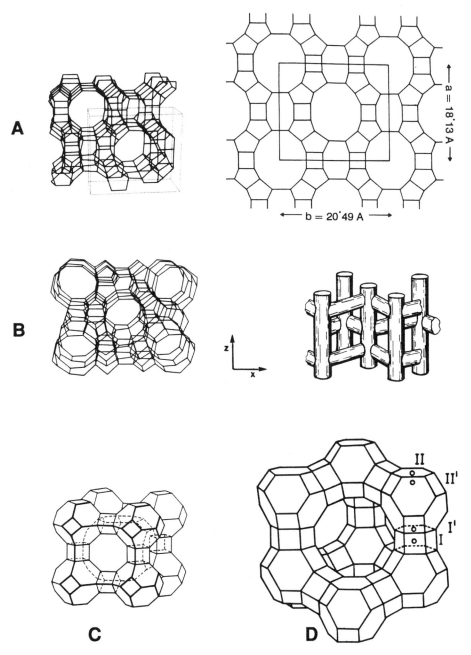

Figure 1.   The structures of zeolites. A, Mordenite, B, Silicalite, C, zeolite A, and D, Faujasite.

exhibit type I adsorption isotherms, where the initial rise at low partial pressure is associated with micropore filling.

Moderate acid will replace the original cations by hydronium ions;[11] the material can be heated to form hydrogen zeolites:

$$H_3O^+ZO^- \rightarrow H^+ZO^- + H_2O$$

Bronsted acidity can also be introduced by hydrolysis of hydrated transition metal ions, and ammonium exchange with subsequent removal of ammonia under heating.[12]

The variety of zeolites and zeolite-related materials has grown enormously in recent years with the incorporation of transition metals into the framework,[13] and the discovery of metal aluminophosphate sieves.[14] Structural analogs to the zeolites as well as new structures with $AlPO_4$ elemental composition[15] include one example with a pore size greater than that of zeolite Y (VPI-5, pore size ~ 12Å).[16] Silicon has been incorporated into many of the $AlPO_4$ structures resulting in negatively charged frameworks.[17] New developments include the titanosilicates[18] containing octahedral Ti and the gallophosphate "cloverite" with impressive clover-like channels formed by 20-membered rings.[19]

Zeolites Inclusion Chemistry: Encapsulation Strategies

The crystalline microporosity and well-defined internal surfaces of zeolites, in addition to their great chemical variety, make these materials very attractive hosts for many areas of inclusion chemistry.[20] A brief overview of encapsulation strategies follows.

The two most obvious and common ways to incorporate matter into zeolite pores are (i) by adsorption and diffusion from gas or liquid phase, and (ii) by ion exchange. A third, usually more difficult approach involves incorporation of species during hydrothermal synthesis. The zeolite pores can be viewed as nanometer size reaction chambers that accommodate numerous organic and inorganic molecules with enough space for chemical conversions. It follows that there is a vast potential for ingenious chemistry in these hosts.

Supported metal clusters are important for catalytic applications and have therefore attracted attention for many years. Noble metals have been introduced into zeolites by ion exchange of complex cations such as $Pd(NH_3)_4^{2+}$. Degassing in oxygen eliminates premature autoreduction,[21] and subsequent hydrogen reduction forms very small metal clusters stabilized by the zeolite structure. (see example of Pd zeolites below). Other, more powerful reducing agents including H-atoms and sodium vapor have been used for metals such as Ni or Fe. Often a particle size distribution with additional external phase is obtained.

Metals can also be introduced by adsorption of the elemental vapor or melt, for instance in the case of mercury or alkali metals. Adsorption of molecular "precursors" such as carbonyls of iron, cobalt, nickel and molybdenum, and subsequent thermal or photochemical decomposition has become an important approach for metals that are difficult to reduce. Other ligands such as alkyls or acetylacetonates have also been used for this purpose. In all these cases, thermal decomposition carries the risk of excessive mobility of the precursors or intermediates such that agglomeration and particle formation at the external surface of the zeolite crystals can occur. Barrer has described the synthesis of salt-bearing zeolites including the famous dry synthesis of ultramarin in 1828, which is sodalite containing intercalated Na-polysulphides.[22] Adsorption of numerous non-ionic and salt species into zeolites was also described, either as such or as precursors for oxides, hydroxides, or metals.

Examples in four different areas of zeolite inclusion chemistry will be discussed in the following. Noble metal and semiconductor clusters, organometallics, and intrazeolite polymer filaments are objects of new and continued research activity.

Palladium Ensembles in Zeolites. Small metal particles or clusters have attracted great interest during the last decade. The optical, electronic and catalytic characteristics of clusters are expected to change from 'bulk' properties to 'molecular' properties within a certain size-range.[23,24] This change is represented by the transition of the electronic band structure of a crystal to the molecular orbital levels of species few atoms in size. Since the cluster size determines the relative population of coordination sites[25] and possibly its molecular symmetry, it is thought to be responsible for modified selectivities in a number of catalytic reactions.[26] Controlled synthesis of stable clusters with defined size is of particular interest, because this would potentially allow to fine-tune the properties of electronic materials and catalyst systems.

The particle size of Pd and Pt phases encapsulated in zeolite hosts can be controlled by chosing appropriate reduction conditions for the ion exchanged, dehydrated precursor form.[27,28,29] Based on x-ray diffraction studies, a bidisperse distribution of Pd particles appears to be typical: In addition to particles 10-20 Å in size, a second Pd phase was assigned to Pd atoms in the cage systems.[30,31] In the following we describe the synthesis and characterization of Pd ensembles, consisting of 2 to 4 correlated atoms, which are stabilized at room temperature in an open, chemically accessible zeolite matrix.[32] Characterization of the clusters was done by means of x-ray absorption spectroscopy (EXAFS).[33,34]

$Pd(NH_3)_4^{2+}$ was ion exchanged into zeolite NaX. Fourier transformed EXAFS data of Pd-X zeolite samples after different treatments are shown in Figure 2: after oxygen pretreatment at 625 K (PdXO, Figure 2.A); and after subsequent H_2 exposure at room temperature (PdX, Figure 2.B). In sample PdXO, the coordination of Pd^{2+} to oxygen with a Pd-O distance of 2.07 Å (1.5 Å uncorrected for phase shifts; fit results give the true distances) is detected as the main peak. This corresponds exactly to results from x-ray diffraction studies of similar palladium exchanged zeolites. The coordination of Pd to the zeolite is indicated in the EXAFS spectrum by the concomitant appearence of a peak at about 3.0 Å (uncorrected), typical for Si/Al scatterer of the zeolite framework.

If the sample PdXO is exposed to hydrogen at 295 K (PdX), drastic changes are visible in the corresponding EXAFS data. The data do not show any sign of remaining cationic Pd-zeolite-oxygen coordination between 1 to 2 Å, the range of the Pd-O contribution in sample PdXO. Instead, a very small, well-resolved contribution from Pd-Pd scattering is present. No outer palladium shells are visible, indicating an extremely small cluster size.

Analysis of the corresponding EXAFS data shows that under carefully chosen conditions it is possible to achieve a total reduction of zeolite-supported palladium cations to *metal clusters of molecular size at room temperature*. This is concluded from the fact that no remaining ionic Pd-oxygen coordination at the original distance of 2.07 Å (sample PdXO) is detected after reduction in sample PdX. A new, disordered Pd-O coordination at 2.76 Å in sample PdX is resulting from a weak interaction between the reduced metal atoms and the support which gives rise to an increased static disorder. Similar long metal-oxygen bond distances have been reported for a palladium phase in Y zeolites (Pd-O = 2.74 Å after H_2 reduction at 420 K)[22], and for silver clusters in zeolite Y (Ag-O = 2.67 and 2.79 Å after H_2 reduction at 348 K). The average coordination number of Pd-Pd in the zeolite is 1.5. This could point to the formation of dimeric and trimeric clusters. On average each Pd atom is surrounded by 3 oxygens and six Si/Al atoms.

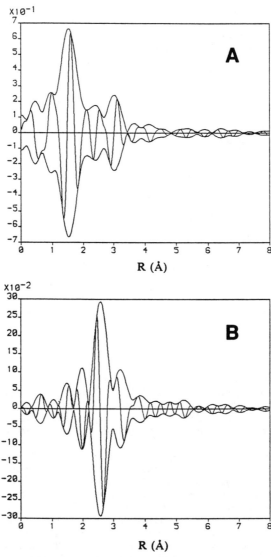

Figure 2. Fourier transformed EXAFS data of A, Sample PdXO; B, Sample PdX.

A model arrangement of four palladium atoms occupying the sodalite unit as shown in Figure 3 corresponds to an average Pd-Pd coordination of N = 1.5. Each palladium atom is located in front of one hexagonal window in dislocated SI' or SII' positions within the sodalite unit (SI'* and SII'*). The SI' sodalite position is surrounded by three nearest SII' positions and vice versa. Our proposed model of Pd$_2$ to Pd$_4$ molecular clusters is comparable to the structure of silver clusters in sodalite units, where the participating Ag atoms were located in SI'* and SII'* positions. Furthermore, x-ray diffraction studies of Ca(II)-exchanged and dehydrated zeolite Y and of Cd(II)-exchanged zeolites X were interpreted with very similar arrangements in SI'* positions, which are stabilized by OH or H$_2$O in SII' positions.[35,36]

The intrazeolite stabilization of the unusually small Pd clusters is possible through the interplay of two types of interaction: (1) The zeolite sodalite cage offers a stabilizing oxygen-coordination sphere for the reduced Pd atoms. (2) In addition, the different sets of SI'* and SII'* cage positions provide *templates* for the arrangement of palladium ensembles with sufficient Pd-Pd bonding overlap. These well-defined 'molecular clusters' with low nuclearity stabilized at ambient temperature in an open pore structure are attractive model systems to study changes in electronic, optical and vibrational characteristics, and to explore their catalytic properties under reaction conditions.

Quantum Size Cadmium Selenide Clusters in Zeolite Y. The preparation of well-defined semiconducting clusters with homogeneous morphology and size distribution is a prerequisite for understanding the physical origin of quantum size effects and exciton formation.[37,38,39] Classical preparation methods for these clusters include wet colloidal techniques,[40,41,42,43,44] and growth in dielectric glassy matrices,[45] or in polymers.[46] However, particle sizes obtained with these techniques are usually non-uniform and agglomeration of individual particles often occurs.[47] Zeolites are excellent hosts for nanometer size semiconductor clusters,[48] as demonstrated with Se species and CdS clusters.[49,50] A structural study of related cadmium selenide clusters stabilized in zeolite Y is discussed below.[51] Colloidal cadmium selenide is of interest for photosensitized electron-transfer reactions utilized for solar energy conversion and photocatalysis.[52,53,54]

Synthesis of intrazeolite CdSe was achieved by ion exchange with Cd(II), subsequent heating in oxygen to 673 K, and exposure to H$_2$Se gas at 298 K. Cluster size and geometrical arrangements could be determined by comprehensive analysis of Cd- and Se-edge EXAFS data as well as synchrotron x-ray powder diffraction and model calculations.

A comparison between EXAFS data of the cadmium-exchanged zeolite CdY and the samples treated with hydrogen selenide gas shows that major changes occur. From EXAFS data analysis and x-ray diffraction data of Cd(II) zeolite Y it was found that four Cd^{2+} ions occupy SI' cation positions in the sodalite cages, and that they are bridged by oxygen in adjacent SII' six-ring sites, thus forming small Cd$_4$O$_4$ cubes. On exposure to H$_2$Se it is expected that the selenium reacts with cadmium and the sample is expected to differ from the Cd(II) zeolite in its overall appearance primarily by an additional selenium shell. If the Fourier transforms of both samples are compared, a new peak at about 2.3 Å (uncorrected) in the CdSe-Y sample appears due to the selenide in this system. This assignment is further justified by the striking similarity in phase function and location in R-space to that of a reference sample of bulk CdSe.

X-ray powder data show that the Cd^{2+} in this sample is located primarily at SI'. This site, within the sodalite unit, is preferred by multivalent cations because some of the high charge density can be compensated by hydroxyl anions at SII', also in the sodalite unit. The cadmium at SI' has a pseudo octahedral coordination resulting from trigonal coordination to framework oxygens of the six-ring (which is part of the double

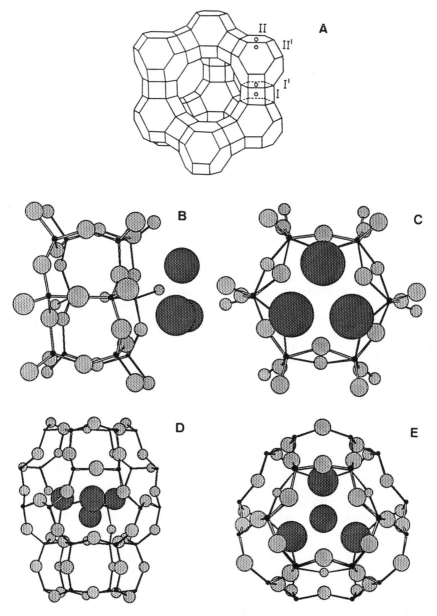

Figure 3.   Geometric arrangements of palladium ensembles in a faujasite subunit. A: display of a partial faujasite unit cell with cation positions SI, SI', SII and SII'. B,C: geometric arrangements of a palladium trimer located in front of a double six-ring (side and top view; Pd atoms dark-shaded). D,E: four Pd atoms located in adjacent SI' and SII' windows in a sodalite cage. Radii of atoms are sized as follows: $Si^{4+} = 0.42$ Å , $O^{2-} = 1.32$ Å, $Pd(0) = 1.37$ Å. Si^{4+} and O^{2-} atoms are shown with 30% of their ionic radii, $Pd(0)$ with 60% of its metallic radius.

From ref. 32. Copyright 1989 American Chemical Society.

six-ring interconnects of the zeolite Y structure) with a bond length of Cd(1)-O(3) = 2.41 Å, and the three SII' atoms, Cd(1)-O(5) = 2.42 Å. O(5) represents a site jointly occupied by Se and oxygen and hence the Cd(1)-O(5) bond length is a weighted average of Cd-O (EXAFS data: 2.26 Å) and Cd-Se (EXAFS data: 2.60 Å).

In the CdSe-Y sample, the major fraction is composed of cadmium cations in adjacent SI' positions bridged on average by two extra-framework oxygen in SII' sites and one additional selenium ion in SII' sites. The second phase consists of Cd ions, present in the twelve-ring windows of the large cavities (SIII) coordinated to extra framework oxygen and one selenium.

The EXAFS-derived coordination number for Cd-O(5) at 2.26 Å is about half of that expected for a complete cube arrangement (expected for 70% Cd located in sodalite cage: N = 3 x 0.7 = 2.1; found: N = 1.4). In addition, the small Cd-Cd interaction (N = 0.4) is not sufficient to account for a complete cube arrangement According to the XRD results, the cadmium cations are located in SI' and the O^{2-} ions in SII' positions. This indicates the formation of oxygen-bridged dimers Cd_2O_2. Based on the form factor for oxygen, the occupancy per unit cell was calculated to be 28.9. Since Cd_2O_2 dimers would account for only 11.4 oxygens, the remainder is assigned to selenium ions (0.4 Se per sodalite cadmium), detected by EXAFS with a bond distance Cd-Se of 2.60 Å. Thus, on average the sodalite units contain two cadmium ions occupying SI' positions, bridged by two oxygen in SII'. An additional selenium atom in SII' is bound to one cadmium atom (see Figure 4). This arrangement can be considered as a modified fragment of the Cd_4O_4 cube found in CdY zeolite.

The formation of Cd-Se ensembles in sodalite subunits must involve diffusion of H_2Se from the supercage through zeolite six-ring windows with a typical pore opening of 2.6 Å. The diameter of H_2Se (2.34 Å) clearly allows the diffusion to take place. The intrazeolite Cd-Se bond distance derived from this study is close to that of bulk CdSe and indicates some covalent character of these species. It is likely that HSe^- fragments replace the original (O5) species in the precursor $Cd_4(O)_4$ cube arrangement, forming Cd_2O_2Se fragments occupying the sodalite units. Additional CdO_2Se fragments at SIII positions are also detected. Non-coordinated helical selenium chains and a small fraction of CdSe are formed primarily in samples with a higher Se/Cd ratio.

Reactivity of Trimethyltin Manganesepentacarbonyl in Zeolite Cavities.

The immobilization of organometallic catalysts on heterogeneous supports has attracted growing attention,[55,56,57] because the high selectivity of many molecular catalysts[58] could be combined with the facile product separation and catalyst recovery inherent to heterogeneous systems. Zeolite hosts might offer additional features such as diffusional or transition state selectivity. Different strategies for the deposition of catalytically active organometallics into zeolites include physisorption of neutral metal carbonyls with only weak framework interactions,[59,60,61,62,63] diffusional blocking ("ship in the bottle" concept) of phthalocyanine (Pc) and other chelate complexes,[64,65,66] and ligation at transition metal cations. Intrazeolite Ru,[67] Ir,[68] and particularly Rh carbonyl complex cations have been studied in great detail.[69,70,71] Migration of intrazeolite Rh-carbonyl species and eventual formation of extrazeolite Rh(0) particles appears to occur under experimental conditions used for catalytic hydroformylation reactions[72].

Higher stability of the intrazeolite species should be expected from surface-attached complexes that utilize the bridging zeolite hydroxyls for anchoring reactions, such as the reaction of $Rh(allyl)_3$ with partially proton-exchanged X and Y type zeolite.[73,74,75,76]

Immobilization concepts for organometallic fragments in microporous solids are being developed in this laboratory. If two metal centers are present as in

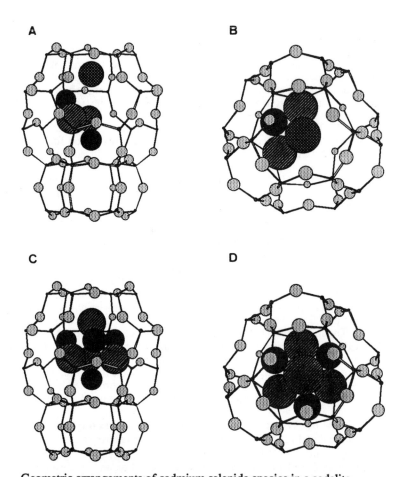

Figure 4.   Geometric arrangements of cadmium selenide species in a sodalite
subunit of the zeolite framework, viewed from the side and from top.
Framework oxygens: light shaded, extra framework oxygens: dark
striped, Cd cations black, Se atom light dotted.
Radii R = 1.32 Å for O^{2-}, R = 0.97 for Cd^{2+}, R = 1.17 for covalent Se.
A, B: $Cd(O_2)CdSe$ bridged fragment.
C, D: Cd_4O_4 cube. One sodalite unit incorporating the Cd moieties and
one of the four interconnecting double six rings of the faujasite
framework are presented. All framework atoms are sized 25% of their
radii. For clarity, the cube ensembles are shown with 75% of the radii.
From ref. 51. Copyright 1989 American Chemical Society.

heterobimetallic complexes, the complex can be 'anchored' to the support via one appropriately chosen, oxophilic metal, whereas the catalytic reaction may proceed at the second metal center. We could recently demonstrate the new bimetallic approach by utilizing the intrazeolite attachment chemistry of Ge-transition metal complexes.[77] Framework attachment of $Cl_2(THF)GeMo(CO)_5$ and $Cl_2(THF)GeW(CO)_5$ was possible in both the Na form of zeolite Y and in the proton-exchanged form. Both compounds react with the acidic zeolite at elevated temperatures under removal of chloride ligands.

As a recent example, we discuss the surface chemistry and stability of $Me_3SnMn(CO)_5$ in similar zeolite supports.[78] The precursor is expected to react with the internal surface hydroxyl groups of the zeolites by loss of methane gas whereas the Sn-Mn bond and the coordination sphere of Mn should remain intact. EXAFS and in situ FTIR data were used to characterize these systems.

EXAFS: Mn-absorption edge. After adsorption of 0.5 complexes $Me_3SnMn(CO)_5$ per supercage from hexane at room temperature into acidic Y zeolite, a full carbonyl coordination shell remained at the Mn atom: 5.2 carbon atoms at 1.87 Å and 6.0 oxygen atoms (Sn backscattering included) at 2.98 Å. At 373 K, the carbonyl coordination remains stable with 4.9 carbon atoms and 6.4 oxygen atoms (Sn modulation included) at 1.84 and 2.97 Å, respectively. At 423 K, dramatic changes are observed. The carbon and oxygen shells of the CO ligands have disappeared. The EXAFS data indicate the formation of small Mn clusters with Mn-Mn distances between about 2.4 and 3.7 Å. At 523 K, the Fourier transform presents a Mn-O bond as a first shell (N = 3.4 atoms, and R = 2.21 Å) and Mn-Si as an outer shell (N = 4.7 atoms, and R = 3.37 Å).

EXAFS: Sn-absorption edge. Fit results of the data of the intrazeolite precursor at 298 K, stripped of the Sn-Si shell show 2.4 atoms of C or O at 2.13 Å and the Sn-Mn bond at 2.56 Å (N = 1.2 atoms). The Sn-Mn bond length shrinks significantly compared to the Sn-Mn bond of the precursor. This effect was also found for the Sn-Mn bond (average distance = 2.59 Å) in $Cl_3SnMn(CO)_5$[79]. Most likely, the replacement of methyl groups by electronegative ligands increases electron density for the Sn-Mn π interaction by decreasing π back-bonding in the Mn-C bonds. This is also supported by the elongated Mn-CO distance in this sample (1.83 Å in NaY versus 1.87 Å in the acidic Y at RT).

From combined EXAFS, FTIR, TPD-MS and separate MS results of $Me_3SnMn(CO)_5$ in acidic Y zeolite, conclusions can be drawn as shown in Figure 5. Already at room temperature, the compound attaches to the zeolite framework via the oxygen rings of the supercage. The attachment of the molecule occurs through the Sn moiety by loss of CH_4 gas while the Sn-Mn bond and the CO ligand sphere are still intact. The substitution of methyl ligands by the oxygen atoms of the zeolite framework is supported by the EXAFS results which indicate that the electronegative oxygen atoms increase the Sn-Mn bond strength of the attached precursor in acidic Y zeolite. The resulting weakened Mn-CO interaction is confirmed by the higher CO-stretching frequencies of the precursor in acidic Y. Based on IR data, both mono- and di-substitution of methyl groups by the framework oxygen atoms occurs. The mono-substituted species, $(Oz)Me_2SnMn(CO)_5$, is the major product at room temperature while the di-substituted species, $(Oz)_2MeSnMn(CO)_5$, is formed at higher temperature, *ca.* 373 K. The attached species decompose at 423 K by loss of CO ligands and cleavage of the Sn-Mn bond. One methyl ligand remains on the zeolite-attached tin atom. Unidentified Mn species are left inside the zeolite cavities as a result of the decomposition.

These results show that the heterobimetallic compound $Me_3SnMn(CO)_5$ can be anchored into the cages of acidic zeolites at room temperature under retention of the Sn-Mn bond. The methyl ligands are good leaving groups in acidic Y but do not react in

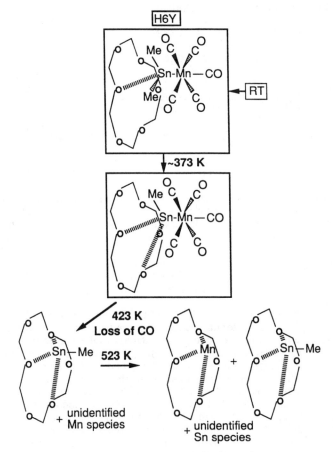

Figure 5.   The chemistry of Me₃SnMn(CO)₅ in acid Y zeolite.

NaY. The thermal stability of the intrazeolite, attached species permits further studies on their chemical reactivity and potential catalytic activity.

Polyacrylonitrile Chains in Zeolite Channels: Polymerization and Pyrolysis. The quest for ever smaller and faster devices for information processing aims at the control of electronic conductivity at nanometer dimensions. Conducting organic polymers are interesting candidates for this purpose. Studies of the conduction mechanism of conjugated polymers[80,81,82,83,84] would benefit substantially if the low-dimensional structures were available as decoupled, structurally well-defined entities. We have recently demonstrated the encapsulation of conjugated polymers such as polypyrrole, polyaniline and polythiophene in zeolite channels.[85] Precursor monomers are introduced into the zeolite host and are subsequently polymerized by appropriate oxidants in the pore system. Methylacetylene gas reacts with the acid sites in zeolites L, Y, and others to form reactive, conjugated oligomers.[86] Short-chain oligomers of polythiophene were prepared, oxidatively doped to the conducting state and stabilized in Na-pentasil zeolites[87]. The synthesis of these and related systems is a step towards the design of oriented *"molecular wires"*, that could in principle permit to process signals or to store information inside channel systems with nanometer dimensions.

As a last example of zeolite inclusion chemistry, we discuss the assembly of polyacrylonitrile (PAN) strands in different large-pore zeolites, zeolite Y and mordenite, and explore the pyrolysis reactions of the encapsulated polymer (Figure 6).[88]

Acrylonitrile vapor was adsorbed in the degassed (670 K, 10^{-5} Torr) zeolite crystals at a vacuum line for 60 min at 298 K. To an aqueous suspension of the acrylonitrile-containing zeolite were added aqueous solutions of potassium peroxodisulfate and sodium bisulfite as radical polymerization initiators.[89] The zeolite frameworks could be dissolved with HF to recover the intrazeolite polyacrylonitrile (PAN). IR and NMR data show no damage to the polymers after this treatment. For pyrolysis, the zeolite/PAN adducts were heated under nitrogen or vacuum to 920 and 970 K for extended periods.

On saturation, zeolite Y contains 46 and mordenite 6 molecules of acrylonitrile per unit cell. The polymer recovered from the zeolite hosts is identical to bulk PAN as shown by NMR spectroscopy.[90] The ^1H NMR spectra show two bands at a ratio of 2 to 1, one at 2.0 ppm corresponding to the methylenic group in the polymer, and the other at 3.1 ppm confirming a methine group. The ^{13}C NMR data of the zeolite-extracted PAN show CH_2 (27.5 ppm), CH (32.7 ppm), and -CN (120.1 ppm), identical to the bulk material.

Infrared spectra of the zeolite/polymer inclusions and of PAN extracted from the zeolites show also peaks characteristic of the bulk polymer,[91,92] including methylenic C-H stretching vibrations of the backbone (2940 cm^{-1} and at 2869 cm^{-1}), and a band at 2240 cm^{-1} due to the pendant nitrile group. The spectra of the extracted intrazeolite polymers are indistinguishable from the spectrum of the bulk polymer. We conclude that the polymer formed in the zeolites is polyacrylonitrile.

Gel permeation chromatography was used to determine the molecular weight of the polymer extracted from the zeolites, relative to a PAN broad standard (M_w = 86,000; M_n = 23,000). For PAN extracted from NaY, the main molecular weight distribution peaks at 19,000, corresponding to 360 monomer units or about 0.2 µm for an extended chain. For PAN extracted from Na-mordenite, a bimodal molecular weight distribution is observed; a small fraction peaks at 19,000 and a large fraction at about 1,000, corresponsing to a 0.01 µm chain in a fully extended form. The shorter chain length in mordenite might indicate crystal defects or diffusional constraints for the polymerization reaction.

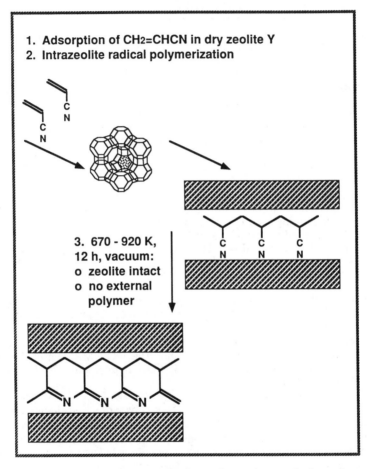

Figure 6.   Intrazeolite polymerization and tentative pyrolysis product of
             acrylonitrile.

Pyrolysis changes the color of the PAN-containing zeolites from white to grey-black.[93] The electronic absorption spectrum of pyrolyzed PAN extracted from NaY shows the appearance of a feature at about 350 nm, as in the bulk, probably due to conjugated carbon-nitrogen double bonds in the ladder polymer.[94] The rest of the spectrum is practically structureless and resembles the absorption spectrum of graphite. The methylenic stretching vibrations and the nitrile band of the intrazeolite PAN main chain disappear. New bands appear in the 1400-1600 cm^{-1} region which have been assigned to C=C and C=N double bonds in the pyrolyzed bulk PAN, as well as the tail of the electronic excitation corresponding to free carrier absorption. A sample of bulk PAN heated to 800 K instead of 920 K shows more defined features in the 1400-1600 cm^{-1} region, similar to the case of pyrolyzed PAN (920 K) extracted from the zeolite. This suggests an early stage of graphitization for the intrazeolitic material. Zeolite/PAN samples pyrolyzed at different temperatures, times, and atmospheres, (nitrogen or vacuum) show spectra similar to the spectrum of bulk PAN pyrolyzed to 800 K.

Samples of NaY with pyrolyzed (920 K) PAN show no measurable dc conductivity. This is not surprising because the polymer is encapsulated completely within the insulating zeolite, and because no external polymer coats the zeolite crystal surfaces. However, the conductivity of the pyrolyzed PAN *extracted* from the zeolite is at the order of 10^{-5} Scm^{-1}, almost identical with that of bulk PAN pyrolyzed at 800 K, and five orders of magnitude smaller than that of the bulk sample pyrolyzed at 920 K. In contrast to the bulk polymer, pyrolysis treatment conditions above about 870 K have little effect on the resulting conductivity of the extracted intrazeolite samples. We conclude that the spatial limitations within the zeolite channels prevent the formation of more extended, graphitized structures with higher conductivity.

This study demonstrates the inclusion synthesis of polyacrylonitrile in the channel systems of NaY and Na-mordenite zeolites, and its pyrolysis to yield a conducting material consisting of nanometer size carbon filaments. These and related systems are promising candidates for low-field conductivity at nanometer scale dimensions.

Conclusion. The examples discussed above demonstrate the versatility and great potential of zeolite molecular sieves to encapsulate and stabilize often unusual forms of matter. The zeolite pores form nanometer size reaction chambers that permit ready access to the external world via adsorption and ion exchange. The regular arrangement of the cages and their variable connectivity is of great interest for superlattice assembly of electronic and optical materials.

So far, research in this area has emphasized metal clusters, organometallics, and quantum size semiconductor clusters and superlattices. As experimental techniques become available, the dynamics of intrazeolite reactions, such as catalysis, ligand exchange,[95] electron transfer and radical reactions, and polymerizations will be explored in more detail. As molecular sieves with ever larger pores are being discovered, the future potential to assemble and understand supramolecular structures is enormous.

Acknowledgments. The author wishes to thank the students in our group who have worked on the above projects for their important contributions (A. Borvornwattananont and P. Enzel). The fruitful collaboration with our coauthors is also gratefully acknowledged (M. M. Eddy, N. Herron, D. C. Koningsberger, K. Moller, and G. D. Stucky). We appreciate funding from the Sandia University Research Program (DOE), the U. S. Department of Energy (DE-FG04-90ER14158), Sprague Electric Company, and ONR (G. D. S.). Acknowledgment is also made to the Donors of the Petroleum Research Fund, administered by the American Chemical

290 SUPRAMOLECULAR ARCHITECTURE

Society, for partial funding of this research. The operational funds for NSLS beamline X11A are supported by DOE grant # DE-ASO580ER10742.

Literature cited

1   Breck, D.W. "Zeolite Molecular Sieves", R.E. Krieger Publishing Co., Malabar, FL,1984
2   "New Developments in Zeolite Science and Technology", Murakami, Y; Iijima, A; Ward, J.W., eds., Kodansha, Tokyo, 1986
3   R. Szostak, Molecular Sieves. Principles of Synthesis and Identification, Van Nostrand Reinhold, New York, 1989.
4   Haushalter, R. C.; Strohmaier, K. G.; Lai, F. W. Science 1989, 246, 1289
5   Bedard, R. L.; Wilson, St. T.; Vail, L. D.; Bennett, J. M.; Flanigen, E. M. Studies Surf. Science Catal. 49, Elsevier,1989, 375
6   Meier, W.M. "New Developments in Zeolite Science and Technology", Murakami, Y; Iijima, A; Ward, J.W., eds., Kodansha, Tokyo, 1986, 13
7   Atlas of Zeolite Structure Types. Meier, W. M.; Olson, D. H. 2nd Ed., Butterworths, London, 1987
8   (a) Barrer, R. M. "Hydrothermal Synthesis of Zeolites", Academic Press, London, 1982. (b) Szostak, R. "Molecular Sieves. Principles of Synthesis and Identification", Van Nostrand Reinhold, 1989
9   B. M. Lok, T. R. Cannan, and C. A. Messina, Zeolites, 1984, 4, 289.
10   R. J. Argauer, and G. R. Landolt, US Patent 3,702,886, 1972.
11   P. A. Jacobs, Carboniogenic Activity of Zeolites, Elsevier, Amsterdam, 1977.
12   J. W. Ward, J. Catal., 1968, 11, 238.
13   Ball, W.J.; Dwyer, J.; Garforth, A.A.; Smith, W.J. "New Developments in Zeolite Science and Technology", Murakami, Y; Iijima, A; Ward, J.W., eds., Kodansha, Tokyo, 1986, 137
14   Messina, C.A.; Lok, B.M.; Flanigen, E.M. U. S. Patent 4,544,143
15   S. T. Wilson, B. M. Lok, C. A. Messina, T. R. Cannan, and E. M. Flanigen, J. Am. Chem. Soc., 1982, 104, 1146.
16   M. E. Davis, C. Saldarriaga, C. Montes, J. Garces, and C. Crowder, Zeolites, 1988, 8, 362.
17   B. M. Lok, C. A. Messina, R. L. Patton, R. T. Gajek, T. R. Cannan, and E. M. Flanigen, J. Am. Chem. Soc., 1984, 106, 6092.
18   Chemical & Engineering News, Sept. 30, 1991, 31
19   Kessler, H. Mat. Res. Soc. Symp. Proc. Vol. 233, 1991, 47
20   (a) Ozin, G. A.; Kuperman, A.; Stein, A. Angew. Chem. Int. Ed. Engl., 1989, 28, 359. (b) Ozin, G. A.; Gil, C. Chem. Rev. 1989, 89, 1749.
21   Sachtler, W. H. M., Stud. Surf. Sci. Catal. 49, Elsevier, 1989, 975
22   Barrer, R. M. "Hydrothermal Synthesis of Zeolites", Academic Press, London, 1982, 306
23   "Contribution of Clusters Physics to Materials Science and Technology", Ed. Davenas, J., Rabette, P. M. ; NATO ASI Series, E104 ; Martinus Nijhoff: Dordrecht, 1986
24   "Metal Clusters", Ed. M. Moskovits; Wiley: New York, 1986
25   Van Hardeveld, R.; Hartog, F. Surf. Sci. 1969, 15, 189
26   Somorjai, G. A., in "Catalyst Design", Hegedus, L. L., Ed.; Wiley: New York, 1987, 11
27   Exner,D.; Jaeger, N.; Moller, K.; Schulz-Ekloff, G. J. Chem. Soc., Faraday Trans. 1, 1982, 78, 3537

28   Exner, D.; Jaeger, N.; Moller, K.; Nowak, R.; Schrubbers, H.; Schulz-Ekloff, G.; Ryder, P. Studies in Surf. Sci, Vol. 12, Ed. Jacobs, P.A.; Elsevier:Amsterdam, 1982, 205

29   Bergeret, G.; Gallezot, P.; Imelik, B. J. Phys. Chem. 1981, 85, 411

30   Gallezot, P.; Imelik, B. Adv. Chem. Ser. 1973, 121, 66

31   Bergeret. G.; Tran Manh Tri; Gallezot, P. J. Phys. Chem. 1983, 87, 1160

32   Moller, K.; Koningsberger, D. C.; Bein, T. J. Phys. Chem., 1989, 93, 6116

33   Teo, B. K. "EXAFS: Basic Principles and Data Analysis"; Springer: Berlin, 1986

34   "X-Ray Absorption: Principles, Applications, Techniques of EXAFS, SEXAFS and XANES", Koningsberger, D. C.; Prins, R., Ed.; Wiley: New York, 1987

35   Costenoble, M. L.; Mortier, W. J.; Uytterhoeven, J. B. J. Chem. Soc., Faraday Trans. 1 1978, 74, 466

36   Calligaris, M.; Nardin, G.; Randaccio, L.; Zangrando, E. Zeolites 1986, 6, 439

37   Brus, L. J. Phys. Chem. 1986, 90, 2555

38   Nedeljkovic, J. M.; Nenadovic, M. T.; Micic, O. I.; Nozik, A. J. J. Phys. Chem. 1986, 90, 12

39   Nozik, A. J.; Williams, F.; Nenadovic, M. T.; Rahj, T.; Micic, O. I. J. Phys. Chem. 1985, 89, 397

40   Henglein, A. Ber. Bunsen Ges. Phys. Chem. 1984, 88, 969

41   Ramsden, J. J.; Webber, S. E.; Grätzel, M. J. Phys. Chem. 1985, 98, 2740

42   Dannhauser, T.; O'Neil, M.; Johansson, K.; Whitten, D.; McLendon, G. J. Phys. Chem. 1986, 90, 6074

43   Tricot, Y.-M.; Fendler, J. H. J. Phys. Chem. 1986, 90, 3369

44   Variano, B. F.;Hwang, D. M.; Sandroff, C. J.; Wiltzius, P.; Jing, T. W.; Ong, N. P. J. Phys. Chem. 1987, 91, 6455

45   Ekimov, A. I.;Efros, Al. L.; Onushchenko, A. A. Solid State Comm. 1985, 921

46   Wang, Y.; Mahler, W. Opt. Comm. 1987, 61, 233

47   Chestnoy, N.; Hull, R.; Brus, L. E. J. Chem. Phys. 1986, 85, 2237

48   Stucky, G. D.; Mac Dougall, J. E. Science, 1990, 247, 669

49   Parise, J. B.; MacDougall, J. E.; Herron, N.; Farlee, R.; Sleight, A. W.; Wang, Y.; Bein, T.; Moller, K.; Moroney, L. M. Inorg. Chem. 1988, 27, 221

50   Herron, N.; Wang, Y.; Eddy, M.; Stucky, G. D.; Cox, D. E.; Moller, K.; Bein, T.; J. Am. Chem. Soc., 1989, 111, 530

51   Moller, K.; Eddy, M. M.; Stucky, G. D.; Herron, N.; Bein, T. J. Am. Chem. Soc., 1989, 111, 2564

52   Bard, A. J. J. Phys. Chem 1982, 86,172

53   Grätzel, M. Acc. Chem. Res. 1982, 15, 376

54   Darwent, J. R. J. Chem. Soc., Faraday Trans. 1 1984, 80, 183

55   Lamb, H. H.; Gates, B. C.; Knözinger, H. Angew. Chem. Int. Ed. Engl. 1988, 27, 1127

56   Basset, J.M.; Choplin, A J. Mol. Catal. 1983, 21, 95

57   Bailey, D.C.; Langer, S.H. Chem. Rev. 1981, 81, 109

58   Jardin, F.H. Prog. Inorg. Chem. 1981, 28, 63

59   Ballivet-Tkatchenko, D.; Coudurier, G.; Mozzanega, H.; Tkatchenko, I.; Kinnemann, A. J. Mol. Catal. 1979, 6, 293

60   Suib, S.L.; Kostapapas, A.; McMahon, K.C.; Baxter, J.C.; Winiecki, A.M. Inorg. Chem. 1985, 24, 858

61   Bein, T.; McLain, S.J.; Corbin, D.R.; Farlee, R.F.; Moller, K.;
      Stucky, G.D.; Woolery, G.; Sayers, D.   J. Am. Chem. Soc. 1988,
      110, 1801
62   Bein, T.; Schmiester, G.; Jacobs, P.A.   J. Phys. Chem. 1986, 90,
      4851
63   Yang, Y.S.; Howe, R.F. "New Developments in Zeolite Science and
      Technology", Murakami, Y.; Iijima, A.; Ward, J.W., eds., Kodansha, Tokyo,
      1986, 883
64   Herron, N. Inorg. Chem. 1986, 25, 4714
65   Diegruber, H.; Plath, P.J.; Schulz-Ekloff, G.   J. Mol. Catal. 1984,
      24, 115
66   Herron, N.; Stucky, G.D.; Tolman, C.A.   J.C.S., Chem. Comm.
      1986,521
67   Verdonck, J.J.; Schoonheydt, R.A.; Jacobs, P.A.   J. Phys. Chem.
      1983, 87, 683
68   Gelin, P.; Naccache, C.; Ben Taarit, Y.; Diab, Y.   Nouv. J. Chim.
      1984, 8, 675
69   Bergeret, G.; Gallezot, P.; Gelin, P.; Ben Taarit, Y.; Lefebvre, F.;
      Naccache, C.; Shannon, R.D.   J. Catal. 1987, 104, 279
70   Davis, M.E.; Schnitzer, J.; Rossin, J.A.; Taylor, D.; Hanson, B.E.
      J. Mol. Catal. 1987, 39, 243
71   Shannon, R.D.; Vedrine, J.C.; Naccache, C.; Lefebvre, F.   J. Catal.
      1984, 88, 431
72   Rode, E.J.; Davis, M.E.; Hanson, B.E.   J. Catal. 1985, 96, 574
73   Huang, T.N.; Schwartz, J.; Kitajima, N.   J. Mol. Catal. 1984, 22,
      38
74   Huang, T.N.; Schwartz, J.   J. Am. Chem. Soc. 1982, 104, 5244
75   Corbin, D.R.; Seidel, W.C., Abrams, L.; Herron, N.; Stucky, G.D.;
      Tolman, C.A. Inorg. Chem. 1985, 24, 1800
76   Taylor, D. F.; Hanson, B. E.; Davis, M. E. Inorg. Chim. Acta 1987,
      128, 55
77   (a) Borvornwattananont, A.; Moller, K.; Bein, T. J. Chem. Soc.,
      Chem. Commun. 1990, 28. (b) Borvornwattananont, A.; Moller, K.;
      Bein, T., Synthesis/Characterization and Novel Applications of Molecular Sieve
      Materials, R. L. Bedard, T. Bein, M. E. Davis, J. Garces, V. A. Maroni, G.
      D. Stucky, Eds., Mat. Res. Soc. Symp. Proc. 233, 1991, p. 195.
      (c) Borvornwattananont, A.; Bein, T., submitted.
78   Borvornwattananont, A.; Bein, T., submitted.
79   Onaka, S., Bull. Chem. Soc. Japan, 1975, 48(1), 319.
80   Proceedings of the International Conference on Science and Technology of
      Synthetic Metals, ICSM '88 and '90; Synth. Metals 1988, 28 (1-3) and 29(1),
      and Synth. Metals 1991, 41-43.
81   Handbook of Conducting Polymers; T. A. Skotheim, Ed.; Marcel Dekker,
      New York, Vol. 1, 1986.
82   Conducting Polymers. Special Applications; L. Alcacer, Ed.; D. Reidel,
      Dordrecht , 1987.
83   Hopfield, J. J.; Onuchic, J. N.; Beratan, B. N. Science 1988, 241, 817
84   (a) Molecular Electronic Devices; F. L. Carter, Ed., Marcel Dekker, New
      York, 1982.
      (b) Molecular Electronic Devices II; F. L. Carter, Ed., Marcel Dekker, New
      York, 1987.
85   (a) Enzel, P.; Bein, T.   J. Phys. Chem. 1989, 93 , 6270.

(b) Enzel, P.; Bein, T. J. Chem. Soc., Chem. Commun. 1989, 1326.
(c) Bein, T.; Enzel, P. Angew. Chem., Int. Ed. Engl. 1989, 28, 1692.
(d) Bein, T.; Enzel, P. Mol. Cryst. Liq. Cryst., 1990, 181, 315.
(e) Bein, T.; Enzel, P. ; Beuneu, F.; Zuppiroli, L.
"Inorganic Compounds with Unusual Properties III. Electron Transfer in Biology and the Solid State", M. K. Johnson et al., Eds., ACS Adv. Chem. Ser., No. 226, 1990, p. 433

86   Cox, S. D.; Stucky, G. D. J. Phys. Chem., 1991, 95, 710.
87   Caspar, J. V.; Ramamurthy, V.; Corbin, D. R. J. Am. Chem. Soc., 1991, 113,   600.
88   Enzel, P.; Bein, T., submitted.
89   At molar ratios acrylonitrile: peroxodisulfate: bisulfite of 1:0.0027:0.0035.
90   G. Svegliado, G. Talamini, and G. J. Vidotto, Polym. Sci. A-1, 1967, 5, 2875.
91   M. M. Coleman, and R. J. Petcavich, J. Polym. Sci. Phys. Ed., 1978, 16, 821.
92   T. -C. Chung, Y. Schlesinger, S. Etemad, A. G. MacDiarmid, and A. J. Heeger, J. Polym. Sci. Phys. Ed., 1984, 22, 1239.
93   X-ray powder diffraction data demonstrate that the zeolite framework remains intact even after pyrolysis treatments at 970 K for NaY. Scanning electron micrographs show no apparent external bulk polymer coating the zeolite crystals indicating that most of the polymer chains reside in the interior of the zeolite crystals.
94   C. L. Renschler, A. P. Sylwester, and L. V. Salgado, J. Mat. Res. 1989, 4, 452.
95   Ozin, G. A.; Özkar, S.; Pastore, H. O.; Poë, A. J.; Vichi, E. J. S. J. Chem. Soc., Chem. Commun., 1991, 141.

RECEIVED April 7, 1992

Chapter 21

Three-Dimensional Periodic Packaging

Sodalite, a Model System

G. D. Stucky, V. I. Srdanov, W. T. A. Harrison, T. E. Gier, N. L. Keder,
K. L. Moran, K. Haug, and H. I. Metiu

Department of Chemistry, University of California,
Santa Barbara, CA 93106–9510

The 3-d periodicity of molecular sieve surfaces coupled with the ability to vary pore size, topology and chemical potential of the framework permits considerable latitude in the assembly of confined atomic and molecular arrays. Sodalite, one of the simplest zeolite analogue structures with a 60 atom cage can be synthesized with a broad range of different atoms to give effective cage charges varying from 0 to -6. Non-hydrogen atom clusters of up to 9 atoms and hydrogen atom containing clusters with as many as 17 atoms can be assembled within the cage during synthesis or by gas phase or ion exchange inclusion. The optical and structural properties of the included clusters can be systematically modified by changes in the cage dimensions and framework electric field. The structure of both the frameworks and the clusters within the cages of sodalite structural analogues can be precisely determined. In addition to new framework compositions with the sodalite structure, approaches to synthesize new classes of materials consisting of semiconductor, metal or molecular clusters confined within open framework structures are discussed.

Recent progress towards the goal of nanostructure photonic and electronic components has evolved into the development of commercial devices which are currently in the range of 0.5 to 1 μm in size. 1-d confinement of atomic or molecular monolayers (~5 Å) by molecular beam epitaxy (MBE), electrochemical, atomic layer epitaxy (ALE) and Langmuir Blodgett approaches has been well documented. However, generation of 2-d and 3-d confined structures by using tilted superlattices or related approaches has so far given only mixed results,

0097–6156/92/0499–0294$06.00/0

primarily because of lateral resolution limitations (~100Å). Another dimension of sophistication is required to precisely fabricate periodic 3-d structures with confinement sizes below 100 Å. In this size regime, where the volume to surface area ratio of the bulk material rapidly decreases, one finds the transition from extended band structure to quantum confined structure, i.e. the interface between solid state and molecular inorganic chemistry. This fascinating area of cluster chemistry is currently being intensively investigated by synthetic chemists, theorists and device engineers.

There are several important requirements related to the properties of an ideal 3-d heterostructure. These include:

- size and topographical uniformity
- 3-d periodicity
- tunability with respect to atomic modification of
  - topography
  - cluster dimensions
  - surface states defined by the cluster/packaging interface
  - intercluster coupling
- thermal and optical stability of both substrate and the 3-d clusters
- optical transparency of the 3-d enclosure surface

Currently, there are few examples of structurally defined 3-d periodic arrays of packaged clusters. As a packaging medium, molecular sieves have the potential to be used to generate 3-d heterostructures consisting of ordered assemblies of various clusters with sizes between 6-13 Å. In this family, the sodalite structure has the simplest cage and 3-d periodic geometry with exceptional crystallinity. As such, it provides an excellent opportunity to investigate how one might use host composition to control cluster structure and electronic properties.

An Inorganic 60 Atom Cage

Figure 1 shows the truncated octahedron of the sodalite structure and for comparison the truncated icosahedron found for C_{60} buckminsterfullerene(1). The truncated octahedron is constructed with tetrahedrally coordinated metal atoms (Si,Al) which are linked by corner sharing oxygen atoms. The upper right hand part of Figure 1 shows only the 24 metal atoms which make up the sodalite cage. The 36 connecting oxygen atoms (lower part of Figure 1) make up the remainder of the 60 atoms in the polyhedron. A hypothetical organic analog would be a saturated polyether. The pore openings are typically described as "6 rings" (metal atoms only) which are in fact 12 atom ring openings with alternating metal and oxygen atoms, while the "4 rings" correspondingly define 8 atom ring openings. In the remainder of this paper we will retain the usual zeolite molecular sieve convention of denoting these pores by the metal atoms only (i.e. as 6 rings and 4 rings). This simple cage structure is an important fundamental building block in molecular sieve and zeolite chemistry and can be used to generate open zeolite structures by structural architecture based on putting together "clusters of cages" as in zeolite A, zeolite X or Y, and the hexagonal form of zeolite Y.

There are a large number of atomic and molecular clusters that can be synthesized within the sodalite cage. An example is the eight atom cubane-like cluster formed by cadmium sulfide in the sodalite (ß) cages of zeolite Y (Figure 2) *(2)*. Na_4ClO_3 is an 8 atom cluster found in the sodalite cage with a different geometry, i.e. a cubane type structure with one oxygen atom missing from a corner site and a chlorine atom at the center.*(3)* The five atom M_4X cluster (Figure 3) is illustrated for Zn_4S in $Zn_4S(BeSiO_4)_3$*(4)* and discussed in more detail below. One can have at least 9 non-hydrogen atoms, or as many as 17 atoms if hydrogen atoms are included, within a sodalite cage (Table 1).

The term "empty cage" is used to refer to structures which do not contain atoms at the center of the cage. This is denoted in empirical formulae by [], e.g $Na_3[](AlSiO_4)_3$ for an "empty" cage containing $[Na_3]^{3+}$, versus $Na_4Br(AlSiO_4)_3$ for a "filled" cage containing $[Na_4Br]^{3+}$. The sodium atoms of the $[Na_4Br]^{3+}$ cluster are tetrahedrally located at four of the 6 ring windows as illustrated for $[Zn_4S]^{6+}$ in Figure 3. Removal of the Br⁻ atom from the center of the cage and one sodium atom from the $[Na_4]^{3+}$ tetrahedron along with some displacement of the sodium atoms gives the topography of the $[Na_3]^{3+}$ cluster. Empty cage structures can be made by direct synthesis or by the reactions indicated below (i.e. starting with a sodalite which has a hydroxide group at the center of the cage) in which sodium hydroxide is removed by extraction to give "empty cages" containing three sodium atoms*(5)*.

$$Na_4[OH](AlSiO_4)_3 \quad \xrightarrow{\text{soxhlet}} \quad Na_3[] \, (AlSiO_4)_3 + NaOH$$

$$Na_4[OH]_x[Br]_{1-x}(AlSiO_4)_3 \rightarrow Na_{4-x}[][Br]_{1-x}(AlSiO_4)_3 + xNaOH$$

Alternatively, a specified number of empty and filled cages can be synthesized by starting with a material which has some cages containing hydroxide and others containing the desired atoms. Hydroxide extraction then will leave the desired fraction of cages filled with the cluster surrounded by empty cages.

Numerous other cage geometries and charges are accessible such as the 4-9 combination, $Na_8[SO_4][](AlSiO_4)_6$*(6)* (an equal number of cages containing four atom $[Na]_4^{4+}$ and nine atom $[Na_4SO_4]^{2+}$ clusters), the 5 atom mixed cluster found in $[Zn_3GaAs](BO_2)_{12}$*(7)* and clusters designed for ternary metal atom cages as in $[(CH_3)_4N](MgAl_2P_3O_{12})$*(8)*.

The sodalite crystal structure is usually a cubic close packed array of truncated octahedral cages, however, lower crystallographic symmetries including tetragonal, hexagonal, and orthorhombic can be obtained by appropriate framework atom substitutions. Using different atomic group compositions also modifies the cage size, cage electric field and dielectric properties. Examples of sodalites are known with all of

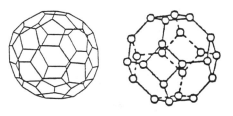

Truncated Icosahedron   Truncated Octahedron

M$_{60}$ Polyhedra

Figure 1.   60 Atom truncated octahedron of the sodalite structure with oxygen atoms included (bottom) and excluded (upper right) compared with C$_{60}$ truncated icosahedron.

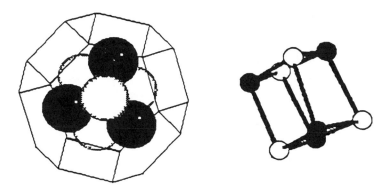

Figure 2.   M$_4$X$_4$ cubane-like cluster in sodalite structure cage.

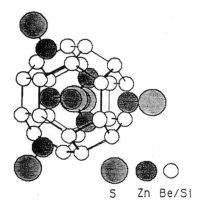

S   Zn   Be/Si

Figure 3.   Five atom M_4X sodalite cage cluster along with MX
fragments from surrounding cages.

Table 1.   Examples of the number of atoms that can be put in a single
sodalite cage.

# Atoms	Examples of Cluster Types
3	$[Na_3]^{3+}$ "Empty Cage"
4	$[Na_4]^{3+}$; $[Na_4]^{4+}$
5	$[Cd_4Te]^{6+}$
6,7	$[Na_4S_x]^{n+}$ ultramarines
8	$[Cd_4S_4]^{6+}$
9	$[Na_4MnO_4]^{3+}$
13	$[Na_4Al(OH)_4]^{3+}$
15	$[Na_3(H_2O)_4]^{3+}$
17	$[(CH_3)_4N]^+$

the group combinations shown in Table 2. Note that the formal sodalite cage charge varies from 0 to -6.

The cages can be considered as potential wells with barriers between the cages dependent upon framework dielectric properties (i.e. framework charge and atomic composition). In the following discussion, we will consider how the cage electric field, the cage geometry and the intercage separation influence cluster properties.

Optical Spectra and the Cage Electric Field

One of the simplest, but most intriguing clusters is synthesized by gas phase deposition and consequent diffusion of sodium atoms into an "empty" cage $[Na_3]^{3+}$ to give a four sodium atom cluster, $[Na_4]^{3+}$, with an unpaired delocalized electron within the cage (9),(10),(11),(12). The UV-VIS absorption spectrum of the Na_4^{3+} sodalite is dominated by an electronic transition between the internal Stark effect broadened ground and first excited state of the Na_4^{3+} cluster. EPR studies of this sodalite "electride" have been discussed previously (10,11,12); however, no quantitative measurements have been previously reported for the optical properties of this phase.

In a recent experiment(13) the sodium vapor emerging from a temperature controllable source inside a high vacuum apparatus for metal vapor deposition(14) was deposited onto the surface of the $Na_3[](AlSiO_4)_3$ sodalite at a background pressure of $P=1x10^{-7}$ torr. A series of samples was prepared with different concentrations of $[Na_4]^{3+}$ clusters formed by diffusion of sodium atoms into the "empty" $Na_3[](AlSiO_4)_3$ sodalite cages. The diffuse reflectance spectra of the samples containing approximately one $[Na_4]^{3+}$ cluster per 50, 10 and 4 empty cages are shown in Figure 4.

The spectrum corresponding to the lowest concentration ratio (1:50) is shown on an expanded scale in Figure 6. This defines the absorption spectrum of an isolated Na_4^{3+} color center. An additional band appears in the UV region of the spectrum ($\sim$38,000 cm^{-1}) as soon as the increasing concentration of sodium atoms leads to the formation of Na_4^{3+} clusters in adjacent cages. This band expands into the IR region as more of the 14 nearest neigbor (eight via 6-rings and 4 via 4-rings) cages are filled around a given Na_4^{3+} cluster cage, leading at the end to a black metallic(9) material.

Time dependent first order perturbation theory calculations to determine the absorption cross section for the Na_4^{3+} clusters in $Na_{3+x}[]_{1-x}[e^-]_x(AlSiO_4)_3$ can be carried out rigorously for this one electron problem(15) using the cage potential field and the cluster geometry as input parameters. The hypothetical four atom Na_4^{3+} cluster in free space would show a single absorption line at 3.2 ev. This line is split into a multiplet by the sodalite cage electric field and by higher energy transitions to the framework states.

Several semiempirical calculations (16),(17),(18),(19). have been carried out to determine the cage and framework potential of aluminosilicates, with widely varying results for the implied framework atom charges. The optical properties are clearly a sensitive function of

Table 2.   Formal charge variation with framework composition. The
             corresponding total formal charge variation within the
             sodalite cage is from 0 to -6.

T_d Corner Shared Molecular Sieve Compositions

	Groups		Charge	
IV-V	SiO_2	SiO_2	0	Silicalite
III-V	AlO_2^-	PO_2^+	0	ALPO
III-IV-V	AlO_2^- $(SiO_2)_\delta$ $(PO_2^+)_{1-\delta}$		δ	SAPO
III-IV	$(AlO_2^-)_x$	$(SiO_2)_y$	x-	Zeolite[*]
IIb-V	ZnO_2^{2-}	PO_2^+	1-	ZIPO
II-V	BeO_2^{2-}	AsO_2^+	1-	BEASO
III-III	BO_2^-	BO_2^-	2-	Borate
IIb-IV	ZnO_2^{2-}	GeO_2	2-	ZIGE

[*] e.g. Offretite (zeolite) = $(K_2Ca)_{2.5}[(AlO_2)_5(SiO_2)_{13}]$

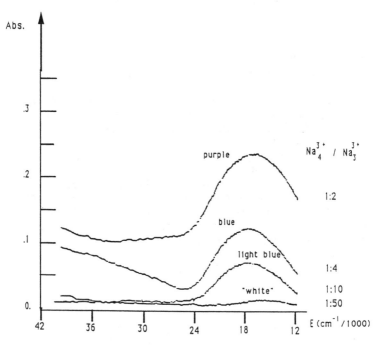

Figure 4.   Absorption spectra (absolute scale) for Na_4^{3+} centers in
              sodium sodalite at Na_4^{3+}/Na_3^{3+} loadings of 1:2, 4, 10 and 50.

charge as shown in Figure 5 for theoretical results obtained using recently published cage electric field parameters(20),(21),(22),(23) with a fixed polarization direction and zeolite orientation. The spectrum is also sensitive to Na-Na distances, which are presently being characterized. The best agreement between theory and experiment is shown in Figure 6. The theoretical model is for the orientationally averaged, high resolution spectrum with the following framework charges: Si=+1.5, Al=0.85, O=-0.84, and Na=+1. These are in close agreement with the model of Leharte (21). The isolated cage results will be extended to theoretical modelling of the intercluster interactions which contribute to the infrared absorption with increasing sodium atom loading, and which define the fully loaded cage and cluster band structure.

Framework Atom Substitution and Inter/Intracluster Geometry

An obvious way to tune the cluster electro-optic properties within a given structure is to change the cage and channel framework atomic composition. The net empirical charge is kept constant by isovalent substitution as exemplified in an isostructural molecular sieve series containing Al/Si, Ga/Si, Al/Ge, and Ga/Ge framework atoms.(24) This substitution does several things. It changes the average framework electronegativity, framework potential, band structure of the framework and consequently intercage coupling. It also modifies the inherent framework polarizability along with the linear and nonlinear optical properties. This has been supported by our recent studies of the nonlinear optical response of the four noncentrosymmetric sodalites with the above compositions (25). Second harmonic generation (SHG) measurements performed on the crystalline powders using Nd-Yag 1064 nm radiation showed that substitution of silicon by germanium in $Na_3[](AlSiO_4)_3 \cdot 4H_2O$ and $Na_3[](GaSiO_4)_3 \cdot 4H_2O$ increases SHG efficiency by factor of 3 while substitution of aluminum by gallium in $Na_3[](AlSiO_4)_3 \cdot 4H_2O$ and $Na_3[](AlGeO_4)_3 \cdot 4H_2O$ increases SHG by only 30%. Evaluation of the P43m point group $\chi_{(123)}$ nonzero polarizability tensor elements using the SHG measurements for this series is in progress.

The geometry changes associated with isovalent framework substitution also can dramatically affect both cluster geometry and diffusion properties through the pores. Structural parameters are given in Figure 7 for (1) a series of "empty" cage sodalite analogue structures filled with water, (2) an example of a dehydrated empty cage and (3) an anhydrous sodalite with a hydroxyl group at the center of the cage. Various framework compositions are indicated (26),(27),(28),(29). Several key points arise from these data.

Framework atomic radii in $Na_3[](ZnAsO_4)_3 \cdot 4H_2O$ (27), (30) are 0.60Å (Zn^{2+}) and 0.34Å (As^{5+}) (cf 0.47Å (Ga^{3+}) and 0.39Å (Ge^{4+})) so that one might expect to obtain for a given zeolite structural analogue the largest known pores and channels with the zinc arsenates. However, increasing the atomic radii of the framework metal atoms does not necessarily give either a larger cage or pore opening. This is indirectly evident from the lattice parameters for the cubic unit cells of the sodalite

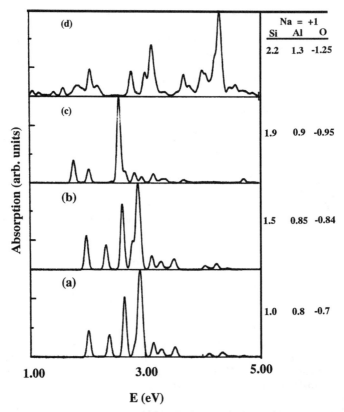

Figure 5.   Calculated absorption cross section of Na_4^{3+} cluster in a
              sodalite cage at high dilution with framework and sodium
              atom charges indicated, a) an interpolation between the
              charges of Si = 0.0, Al = 0.0, O = -0.25, and Na = 1.0 *(22)* and
              the charges given by Leherte *(21)*; b) reference *(21)*;
              c) reference *(20)*; d) an interpolation between the charges of
              Van Genechten *(20)* and the much stronger charge model of
              Skorczyk *(23)* which places a charge of 3.03 on Si, 2.45 on Al,
              -1.62 on O.

EXPERIMENT AND THEORY

Framework charges from the model (c)

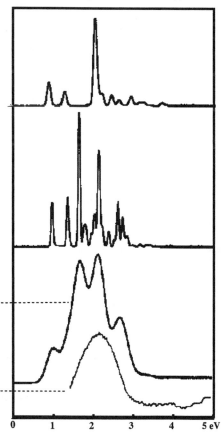

* High resolution , single
orientation absorption
spectrum of the solvated
electron in sodalite.

* High resolution
orientationally averaged
absorption spectrum

* Low resolution
orientationally averaged
absorption spectrum

* Experiment -
polycrystalline sample

Figure 6.   Theoretical orientationally averaged high resolution
spectrum of a single Na_4^{3+} cluster in a sodalite cage at high
dilution with atomic charges of Si=+1.5, Al=0.85, O=-0.84
and Na=+1 (top) and observed absorption spectrum
$Na_4^{3+}:Na_3^{3+}$ = 1:50.

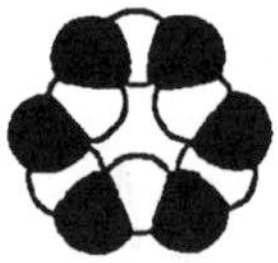

Composition	a(Å)	Θ(T-O-T') °	A (Å^2)	δ Na-ring (Å)	Ref.
Na$_3$(AlSiO$_4$)$_3$ • 4H$_2$O	8.848(1)	136.2(3)	5.575	1.53(1)	(26)
Na$_3$(ZnPO$_4$)$_3$ • 4H$_2$O	8.8281(1)	126.1(3)	5.280	1.66(1)	(27)
Na$_3$(AlGeO$_4$)$_3$ • 4H$_2$O	8.964(1)	129.6(2)	5.399	1.61(1)	(28)
Na$_3$(GaGeO$_4$)$_3$ • 4H$_2$O	9.003(1)	125.5(2)	5.172	1.68(1)	(28)
Na$_3$(ZnAsO$_4$)$_3$ • 4H$_2$O	9.02 73(1)	123.8(3)	4.741	1.73(1)	(27)
Na$_3$(AlSiO$_4$)$_3$	9.122(1)	156.3(6)	8.288	0.24(2)	(26)
Na$_4$(AlSiO$_4$)$_3$[OH]	8.734(1)	132.9(3)	5.37	1.38(1)	(29)

Figure 7.   Sodalite framework composition modification of cage
structure for hydrated empty cages.

analogue systems, and specifically is reflected in the decrease in the T-O-T' angles as the atomic radii increase (Figure 7).*(31)* A is the area of the triangle of oxygen atoms which define the six ring pore openings of the sodalite cage. A and the pore openings *decrease* in size with increasing atomic radii.

δ is the distance of the sodium atom from the plane of the six ring tetrahedral atoms. If $\delta = 0$, the sodium atoms no longer have a single cage identity, and are equally shared between adjacent cages. At that point, as far as the sodium atoms are concerned, the structure is an expanded lattice with no definable sodium atom clusters. Upon dehydration of $Na_3[](AlSiO_4)_3 \cdot 4H_2O$ there is a dramatic framework displacement of the 60 atom cage. The area of the pore opening increases 80% so that the structure becomes much more permeable. The T-O-T' angle increases 20° in going from $Na_3[](AlSiO_4)_3 \cdot 4H_2O$ to $Na_3[](AlSiO_4)_3$ and the sodium atom moves to within 0.24 Å of the center of the pore opening. 0.24 Å is the displacement required for the Na_3^{3+} cluster identity to be lost and the structure to be transformed into an expanded Na^+ lattice. The sodium atoms can be pulled back into a given cage by placing within the cage a charged species such as the hydroxyl ion. The consequence of this is a smaller T-O-T' angle and pore opening.

An example of the precision with which one can determine cluster and packaging structural properties in the sodalite systems is provided by our recent synthesis and characterization of the Ga/Ge sodalite structural analogue, $Na_3[](GaGeO_4)_3 \cdot 4H_2O$. The site ordering of the framework atoms has a profound effect on the potential distribution, cluster geometry and electronic properties. From an X-ray diffraction point of view, one is asking to differentiate between two atoms which differ by one electron (Ga(31) and Ge(32)). MASNMR spectroscopy could partially resolve this question, but would not provide the structural details needed for the definition of optical-structural relationships. If indexed by "black box" software routines currently used, X-ray data for both polycrystalline and single crystal samples of $Na_3[](GaGeO_4)_3 \cdot 4H_2O$ give a body centered lattice, space group I43m. Single crystal structure refinement in this space group (isotropic, no hydrogen atoms) converges to R = 0.016 and $R_w = 0.019$ with an average T-O distance of 1.791(1)Å for a disordered array of gallium and germanium atoms. A more careful examination of the single crystal diffraction data shows several reflections which are an exception to the body centered assignment with intensities < 2σ(background). Including these additional reflections in the space group P43n gives a refinement with R = 0.012 and $R_w = 0.014$ (isotropic, no hydrogen atoms). More important than this decrease in R factors, are the changes in the Ga-O (1.840(1)Å) and Ge-O (1.745(1)Å) bond distances, consistant with an ordered gallium/germanium framework. These results have been recently supported by [71]Ga MASNMR.

Other ways of varying the framework geometry are shown in Figure 8, for example, placing a relatively large atomic group such as MnO_4^-*(32)* at the center of the cage increases the pore size opening and moves the sodium atoms towards the center of the six ring opening. The metal atoms at the six ring sites are also closer to the expanded lattice

Composition	a(Å)	Θ(T-O-T') °	A (Å^2)	δ M-ring (Å)
Na$_4$(AlSiO$_4$)$_3$ [OH][1]	8.734(1)	132.9(3)	5.37	1.18(1)
Na$_4$(AlSiO$_4$)$_3$ [MnO$_4$][2]	9.0992(7)	146.5(7)	6.941	0.50(1)
Zn$_4$(B O$_2$)$_6$ [O] [3]	7.4659(3)	127.2(3)	3.65	1.01(1)
Zn$_4$(B O$_2$)$_6$ [S] [4]	7.6404(6)	132.0(4)	4.05	0.87(1)
Zn$_4$(B O$_2$)$_6$ [Se] [4]	7.6800(7)	133.4(3)	4.15	0.81(1)
Li$_4$(BePO$_4$) [Br] [5]	8.0823(4)	128.8(3)	4.33	0.60(1)

1 Felsche, J.; Luger, S. Ber. Bunsen-Ges. Phys. Chem. 1986, 90(8), 731-6.

2 Srdanov, V; Gier, T.E.; Harrison, W.T.A.; Stucky, G. D. 1991

3 Smith-Verdier, P.; Garcia-Blanco, S. Zeit. Krist. 1980, 151, 175-177.

4 Moran, K. L.; Harrison, W. T. A.; Gier, T. E.; MacDougall, J. E.; Stucky, G. D.
Mater. Res. Soc. Symp. Proc., 164(Mater. Issues Microcryst1990, Semicond.), 123-8.

5 Gier, T.E.; Harrison, W.T.A., Stucky , G.D. Angew. Chem. Int. Ed. 1991, in press.

Figure 8.   Use of cage packaging to control cluster and intercluster
geometries.

positions in the sodalite analog structures which have small cages because of small atomic radii (0.11 Å (B^{3+}), 0.27 Å (Be^{2+}) and 0.17 Å (P^{5+})).

In this connection, it is of interest to note that a comparison of the optical spectrum of Ag_4Br in $Ag_4[Br](AlSiO_4)_3(33),(34),(35)$ and $Ag_4[Br](BePO_4)_3(36)$ reveals a distinct red shift (~70 nm) in the optical spectrum in the latter. This is consistent with the cages being closer together (cage center − cage center 7.328(1)Å versus 7.17(1)Å); and, as noted above, with the silver atoms in the smaller BePO cages being closer to the center of the six rings and therefore to the expanded lattice configuration. These considerations may be secondary to changes in the framework electric field which are obtained by substituting Be for Al and P for Si. This is an example of the substitution shown in Table 2, giving in this case larger local gradients in the cage electric field. Additional experiments and theoretical modelling are currently underway in order to resolve the relative importance of the above possibilities.

A variety of five atom cluster combinations can be made with II-V, II-VI and III-V atoms packaged in the sodalite structure (Table 3).(37) The tetrahedral geometry of the first coordination sphere around the group VI atom in these structures is the same as that in the corresponding bulk semiconductor. The geometry of this coordination sphere can be substantially modified by the framework packaging. For example, the Zn-S distance for the five atom Zn_4S cluster is significantly shorter in the highly constrained boralite cage in $Zn_4S(B_2O_4)_3$ than in the larger beryllosilicate $(Zn_4S(BeSiO_4)_3$ or beryllogermanate $Zn_4S(BeGeO_4)_3$ cages (Table 3). In the latter two larger cage structures, the Zn-S distance approaches that observed in the bulk semiconductor. Framework substitution chemistry also changes the intercage distance and the "expanded lattice" nature of the cage by virtue of the siting of the zinc atoms. The systematic correlation of these structural changes with optical properties is of considerable interest and currently being investigated.

In summary, although the sodalite cages are relatively small and limit the size of the clusters which can be examined, they provide an excellent opportunity to investigate and model 3–d packaging of clusters. One can grow single crystals as large as a centimeter of several of these compositions. Because there is a large structure field with lattice parameters varying by as much as 20%, lattice matching to generate thin films is feasible. In addition, the six ring pores are sufficiently large so that gas and ion phase inclusion chemistry can be used to modify the framework and synthesize clusters. As noted above, these are noncentrosymmetric crystal structures with second order NLO properties. In the most common space group (P43n) for this structural field there is only one susceptibility tensor element, $\chi_{(123)}$, which can be determined directly from powder data and used to evaluate structure/property relationships. The high optical density has already been demonstrated to give exceptional sensitivity and resolution in cathodo-chromic device applications based on F-centers in halogen and Ge-doped $Na_4Br(AlSiO_4)_3$ (38). Structurally it is possible to determine atomic positions and site occupancies precisely. The large accessible compositional structural field also makes it possible to vary the framework band structure and charge over a wide range.

Other 3-D Packaging Considerations

The above approach for sodalite structure analogues is currently being extended to other periodic 3-d open framework hosts. Some specific goals (in addition to those noted previously) include:

- Stereoselective and orientational defined self assembly
- Larger channels and cage dimensions (> 10Å)
- Definition of order/disorder properties
- Homogeneity and diffusion control

All of these have been sufficiently well demonstrated to confirm the feasibility of ultimately generating nanocomposite devices based on packaging in 3-d periodic hosts. Taking advantage of an organic guest/inorganic host approach, we have used polar molecular sieve channels to self assemble, orient and stabilize molecular arrays of hyperpolarizable organic molecules. Molecular self assembly has been monitored both on the external surface of zeolites by atomic force microscopy(AFM)(39) and within the pores by second harmonic generation. Orthogonal self assembly within the 3-d pores has been demonstrated by co-inclusion of two types of organic guests which have diametrically opposed non-linear optic responses (40),(41),(42). For example, para nitroaniline (NA) has a zero SHG response in the bulk since it is centrosymmetric, but can be oriented by the host channel polar axes of AlPO-5 to give a large SHG response. Conversely, 2 methyl para nitroaniline (MNA) has a large bulk SHG as a noncentrosymmetric phase, but because of the change in stereochemistry upon methyl substitution is not properly oriented in the channels of ALPO-5, giving a zero SHG response at all channel loadings. However NA can be used to co-assemble molecules of MNA and orient them in such a way as to generate as large an SHG response as for NA alone. Yet the two molecules have diametrically opposite (orthogonal) electro-optic properties in both the bulk and the polar channels of ALPO-5.

The primary limitation of the sodalite structures is the relatively small cage and pore size. The clusters per se have dimensions well below the exciton radius for the corresponding bulk semiconductors. The 120 atom 26-hedron cages of zeolites RHO and Y increase the possible cluster diameter to ~11-13Å. The development of new larger pore molecular sieves is being actively pursued by numerous groups using Hoffman and 3-d linked molecular rod complexes(43),(44) for the building blocks of open framework structures, new approaches to molecular sieve synthesis and open frameworks containing both four and six coordinate metal ions. We will close with some recent results related to the last two items.

By carrying out molecular sieve synthesis at lower temperatures one can hope to accomplish several goals: 1) obtain more open structural phases, 2) use the solvent as an effective structural template and 3) intercept metastable phases which have a short half-life at higher temperatures. The precedent for solvent templating is indicated by the work of Jeffrey in the late 1960's(45) who demonstrated that one could isolate and structurally characterize gas and ion-pair hydrates at low temperatures. An example is tetra(i-pentyl)ammonium fluoride which crystallizes at 31.2°C with an empirical formula (isoC$_5$H$_{11}$)$_4$NF•38H$_2$O. In addition to the utility of tetra-alkyl ammonium fluorides as co-solvent mineralizers in zeolite synthesis(46),(47), their templating ability should be increasingly effective with decreasing temperature.

Using beryllium or zinc as framework metal atoms, we have isolated a number of open framework structures at room temperature or lower(27),(30),(48),(49) well within the regime of solvent clathrate templating. The molecular sieve chemistry of these new framework compositions is extensive. Using only the sodium ion as a template over a pH range from 1 to 13 gives nine different phases within a narrow temperature range of which five are open framework and one contains (ZnO)$_2$PO 3-ring configurations. Zeolite X can be synthesized at -18°C overnight using mixed templates (e.g. sodium and tetramethyl ammonium halide) and nonaqueous cosolvents. Organic template phases are equally prolific. 1,4-Diazabicyclo[2.2.2]octane (DABCO) gives seven distinct crystalline phases of which five are open framework structures. The zinc open framework structural field is expanded further when one considers the organophosphonates and phosphites which have been described by Clearfield(50),(51) and Mallouk.(52)

Nature has provided numerous intriguing examples of open framework structures containing both four and six coordinate metal atoms. Using a non aqueous solvent approach with organometallic precursors, we were able to synthesize hureaulite and alluaudite, two open framework iron phosphates.(53) More recently Haushalter has described a large number of reduced Mo(IV)-Mo(V) open framework phosphates.(54) Two other interesting members of the iron phosphate and arsenate mineral family are the 30 ring channel (free pore diameter 14.2Å) cacoxenite ([Al(Al,Fe)$_3$Fe$_{21}$O$_6$(OH)$_{12}$(PO$_4$)$_{17}$(H$_2$O)$_{24}$] •51H$_2$O)(55) and the smaller eight ring channel pharmacosiderite, (KFe$_4$(OH)$_4$(AsO$_4$)$_3$•8H$_2$O).(56) The latter is another example of a cluster of cages, the cages in this case being the M$_4$X$_4$ cubane structure *(57)*, and has been synthesized as M$_4$Ge$_7$O$_{16}$ (M= Li, Na, K, Rb, Cs, H); M$_4$(TiO)$_4$(SiO$_4$)$_3$ (M= K, Na, H); M$_4$(TiO)$_4$(GeO$_4$)$_3$ (M= K, H) and (K,H)$_5$(FeO)$_4$(AsO$_4$)$_3$.(58) In all cases there are 7-8 water molecules associated with the empirical formula. The germanium isostructure is shown in Figure 9.

Summary

3-d packaging with crystalline periodicity gives well defined structures with the flexibility to fine tune optical and electronic properties. This, along with recent advances in 3-d surface synthesis offer the opportunity to construct wires and cluster morphologies not readily accessible via 1-d layer confinement (1-d refers to the dimensionality of the

Table 3. II-VI, III-V, and II,III-V Clusters

Quantum Lattice	M-X Distance (Å)	Cage to Cage Distance (Å)
Zn_4S B	2.260(3)	6.61
Zn_4S H	2.346(2)	7.03
Zn_4S HG	2.345(3)	7.16
Cd_4S H	2.471(3)	7.31
Cd_4S HG	2.508(4)	7.46
Zn_4Se B	2.368(3)	6.66
Zn_3GaP B	2.202(2)	6.59
Zn_3GaAs B	2.299(2)	6.64

ZnS	2.34	CdS	2.52	GaP	2.36
ZnSe	2.45	CdSe	2.62	GaAs	2.44

B = Boralite = B_6O_{12} **H** = Helvite = $Be_3Si_3O_{12}$
HG = GeHelvite = $Be_3Ge_3O_{12}$

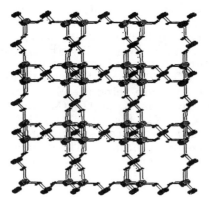

$(Li, Na, K, Rb, Cs, H)_4Ge_7O_{16} \cdot nH_2O$

Figure 9. $K_4Ge_7O_{16}$ channel structure. *(57,58)*

confinement) approaches. The continuing evolution of new 3-d periodic nanocomposite arrays and porous hosts generated in this research should also provide new materials for sorption catalysis and other areas of commercial and academic interest.

Acknowledgments

The authors thank the National Science Foundation (QUEST, Quantized Electron Structure Science and Technology Center, NSF-DMR 88-21499) Grants and the Office of Naval Research for support of this research.

Literature Cited

1. Kroto, H. W.; Heath, J. R.; O'Brien, S. C.; Curl, R. F.; Smalley, R. E. *Nature* (London) **1985**, *318*, 162-163.
2. Herron, N.; Wang, Y.; Eddy, M. M.; Stucky, G. D.; Cox, D. E.; Moller, K.; Bein, T. *J. Am. Chem. Soc.* **1989**, *111(2)*, 530-540.
3. Barrer, R. M.; Cole, J. F. *J. Chem. Soc. A* **1970**, *9*, 1516-1523.
4. Fursenko, D. A. In *Fiz.-Khim. Issled. Mineraloobraz. Sist.*; Godovikov, A. A., Ed.; Akad. Nauk SSSR, Sib. Otd.; Inst. Geol. Geofiz.: Novosibirsk, USSR, 1982, pp. 104-107.
5. Barrer, R. M. In *Hydrothermal Chemistry of Zeolites*, Academic Press: London, UK, 1982.
6. Hassan, I.; Buseck, P. R. *Am. Mineral.* **1989**, *74*, 394-410.
7. (Gier, T. E.; Harrison, W. T. A.; Stucky, G. D. *J. Chem. Mat.*, submitted for publication.)
8. Barrie, J. B. ; Klinowski, J. *J. Phys. Chem.* **1989**, *93*, 5972-5974.
9. Barrer, R. M.; Cole, J. F. *J. Phys. Chem. Solids* **1968**, *29*, 1755-1758.
10. Smeulders, J. B. A. F.; Hefni, M. A.; Klaassen, A. A. K.; DeBoer, E.; Westphal, U.; Geismar, G. *Zeolites* **1987**, 7, 347-352.
11. Geismar, G.; Westphal, U. *Chem.-Ztg.* **1987**, *111*, 277-280.
12. Breuer, R. E. H.; DeBoer, E.; Geismar, G. *Zeolites* **1989**, *9*, 336-340.
13. (Srdanov, V. I.; Haug, K; Metiu, H; Stucky, G. D. *J. Chem. Phys.*, submitted for publication.)
14. (Srdanov, V. I.; Margolese, D.; Saab, A.; Stucky, G. D. *Rev. Sci. Instr.*, submitted for publication.)
15. (Haug, K. and Metiu, H. *J. Phys. Chem.*, submitted for publication.)
16. Masuda, T.; Tsutsumi, K.; Takahashi, H. *J. Colloid Interface Sci.* **1980**, 77, 238-242.
17. Barrachin, B.; CohendeLara, E. *J. Chem. Soc., Faraday Trans. 2* **1986**, *82*, 1953-1966.
18. Ito, T.; Fraissard, J. *J. Chem. Soc., Faraday Trans. 1* **1987**, *83*, 451-462.
19. Beran, S. *J. Mol. Catal.* **1988**, *45*, 225-233.
20. Van Genechten, G.; Mortier, W.J.; Geerlings, P. *J. Chem. Phys.* **1987**, *86*, 5063.
21. Leherte, L. *Chem. Phys. Lett.* **1988**, *145*, 237.
22. Vigne-Meader, F.; Auroux, A. *J. Phys. Chem.* **1990**, *94*, 316.
23. R. Skorcyzk, *Acta Cryst.*, **1976**, *A32*, 447.

24. (Harrison, W. T. A.; Gier, T. E. and Stucky, G. D. *J. Chem. Materials*, submitted for publication.)
25. (Harrison, W. T. A.; Gier, T. E. and Stucky, G. D. *J. Chem. Materials*, submitted for publication.)
26. Felsche, J.; Luger, S.; Baerlocher, C. *Zeolites* **1986**, *6*, 367-372.
27. Nenoff, T. M.; Harrison, W. T. A.; Gier, T. E.; Stucky, G. D. *J. Am. Chem. Soc.* **1991**, 113, 378-379.
28. Keder, N.; Harrison, W. T. A.; Gier, T. E.; Zaremba, C.; Stucky, G. D. submitted for publication.)
29. Luger, S.; Felsche, J.; Fischer, P. *Acta Cryst.* **1987**, *C43*, 1-3.
30. Gier, T.E. and Stucky, G.D. *Nature*, **1991**, 349, 508-509.
31. A similar result has been observed by Newsam and Vaughan for the gallosilicate sodalite; Newsam, J. M.; Jorgensen, J. D. *Zeolites* **1987**, 7, 569-573.
32. (Srdanov, V.; Harrison, W.T.A; Cox, D. and Stucky, G.D., submitted for publication, *Inorg. Chem.*).
33. Stein, A.; Macdonald, P. M.; Ozin, G. A.; Stucky, G. D. *J. Phys. Chem.* **1990**, 94, 6943-6948.
34. Stein, A.; Ozin, G. A.; Stucky, G. D. *J. Am. Chem. Soc.* **1990**, 112, 904-905.
35. Ozin, G. A.; Kirkby, S.; Meszaros, M.; Ozkar, S.; Stein, A.; and Stucky, G. D. in *Materials For Nonlinear Optics: Chemical Perspectives*, American Chemical Society Symposium Series, Marder, S. R., Sohn, J. E. and Stucky, G.D. **1991**, *455*, 554-581.
36. (Gier, T. E.; Harrison, W. T. A. and Stucky, G. D. *Angewandte Chemie*, in press).
37. Moran, K. L.; Harrison, W. T. A.; Gier, T. E.; MacDougall, J. E.; Stucky, G. D. *Mater. Res. Soc. Symp. Proc.*, **1990**, *164 (Mater. Issues Microcryst., Semicond.)*, 123-128.
38. Tranjan, F. M.; Todd, L. T. *J. Electrochem. Soc.* **1988**, *135*, 2288-2291.
39. Weisenhorn, A. L.; MacDougall, J. E.; Gould, S. A. C.; Cox, S. D.; Wise, W. S.; Massie, J.; Maivald, P.; Elings, V. B.; Stucky, G. D.; Hansma, P. K. *Science* **1990**, *247*, 1330-1333.
40. Cox, S. D.; Gier, T. E.; Stucky, G. D.; Bierlein, J. *J. Am. Chem. Soc.* **1988**, *110*, 2986-2987.
41. Cox, S. D.; Gier, T. E.; Stucky, G. D.; Bierlein, J. *Solid State Ionics*, **1988**, *32-33*, 514-520.
42. Cox, S. D.; Gier, T. E.; Stucky, G. D. *Chem. Mater.* **1990**, *2*, 609-619.
43. Iwamoto, T. *Inclusion Compounds, Volume 1*, Atwood, J. L.; Davies, J. D. D.; and MacNicol, D. D. Editors, Academic Press, New York, **1984**, 29-57.
44. Robson, R., "Symposium on Supramolecular Architecture in Two and Three Dimension", American Chemical Society 201st National Meeting, Atlanta, GA (**1991**).
45. Jeffrey, G. A. In *Inclusion Compounds;* Atwood, J. L.; Davies, J. D. D.; MacNicol, D. D., Eds., Academic Press: New York, 1984, Vol. 1, pp. 135-190.
46. Suganuma, F.; Yoshinari, T.; Sera, T. *Jpn. Kokai Tokyo Koho* **1987**, *6*.

47. Xu, Y.; Maddox, P. J.; Couves, J. W. *J. Chem. Soc., Faraday Trans.* **1990**, 86, 425-429.
48. Harrison, W. T. A.; Gier, T. E and Stucky, G. D. *J. Mater. Chem.* **1991**, *1*, 153.
49. Harrison, W. T. A.; Gier, T. E.; Moran, K. L.; Nicol, J. M.; Eckert, H.; Stucky, G. D. *Chem. Mater.* **1991**, *3*, 27-29.
50. Ortiz-Avila, C. Y.; Squattrito, P. J.; Shieh, M.; Clearfield, A. *Inorg. Chem.* **1989**, 28, 2608-2615.
51. Ortiz-Avila, Y.; Rudolf, P. R.; Clearfield, A. *Inorg. Chem.* **1989**, 28, 2137-2141.
52. Cao, G.; Lee, H.; Lynch, V. M.; Mallouk, T. E. *Inorg. Chem.* **1988**, *27*, 2781-2785.
53. Corbin, D. R.; Whitney, J. F.; Fultz, W. C.; Stucky, G. D.; Eddy, M. M.; Cheetham, A. K. *Inorg. Chem.* **1986**, *25*, 2279-2280.
54. Mundi, L.A., Strohmaier, K.G.; Haushalter, R.C. *Inorg. Chem.* **1991**, *30*, 153 and included references.
55. Moore, P. B.; Shen, J. *Nature* **1983**, 306, 356-358.
56. Buerger, M. J.; Dollase, W. A.; Garaycochea-Wittke, I. Z. *Kristallogr., Kristallgeom., Kristallphys., Kristallchem.* **1967**, 125, 92-108.
57. Hauser, E.; Bittner, H.; Nowotny, H. *Monatsh. Chem.* **1970**, 101, 1864-1873. Nevskii, N. N.; Ilyukhin, V. V.; Ivanova, L. I.; Belov, N. V. *Dokl. Akad. Nauk SSSR* **1979**, *245*, 110-111, [Crystallogr.].
58. (Harrison, W. T. A.; Gier, T. E.; Stucky, G. D. *Chem. Comm.*, submitted for publication).

RECEIVED April 7, 1992

Chapter 22

Topotactic Kinetics in Zeolite Nanoreaction Chambers

Geoffrey A. Ozin[1], Saim Özkar[2], Heloise O. Pastore[1,3], Anthony J. Poë[1], and Eduardo J. S. Vichi[3]

[1]Lash Miller Chemical Laboratories, University of Toronto, 80 St. George Street, Toronto, Ontario M5S 1A1, Canada
[2]Department of Chemistry, Middle East Technical University, Ankara, Turkey
[3]Instituto de Quimica, Universidade Estadual de Campinas, Campinas, S.P., Brazil

The first kinetic study is reported for archetypical substitution reactions of PMe_3 and ^{13}CO with the well defined intrazeolite system $Mo(^{12}CO)_6$-$M_{56}Y$ where M = Li, Na, K, Rb and Cs. Excellent isosbestic points and first-order dissociative behaviour are obtained, the activation parameters indicating a highly ordered supramolecular transition-state consisting of activated $Mo(^{12}CO)_6$ and PMe_3 or ^{13}CO anchored to M^+ ions and/or oxygen-framework sites in the α-cage of the host lattice. Generally, an increase in reactant loading produces a decrease in the activating ability of the host lattice but, for reaction with PMe_3, reactivity is enhanced at higher loading and associative attack on $Mo(^{12}CO)_6$ by PMe_3 may be important under some conditions. The global picture that emerges from this investigation is that the α-cages of zeolite Y behave as rigid macrospheroidal multidentate anionic ligands (called zeolates or cavitates) towards extraframework charge-balancing cations to which organometallic and ligand guests can become attached. An approach of this kind can assist with the design of experiments and the understanding of cation and framework anchoring and guest loading effects on the activation parameters and reaction mechanisms compared to gas, solution, surface, and matrix phases where such data are available.

The structure and bonding properties of host-guest inclusion compounds are intimately related to the spatial and topological compatibility of the host and the guest. Knowledge accrued in this field is of fundamental importance to our understanding of chemical systems capable of, for example, size and shape selective catalysis, molecular recognition, or self-assembly into new solid-state materials (1). Charting the dynamical

0097–6156/92/0499–0314$06.00/0

motions and probing the chemical reactivities of an occluded guest not only requires an appreciation of the above, but also consideration of specific interactions, conformational and cooperative effects between the host and guest reactants, transition-states, intermediates and products.

Organic hosts can be of a molecular, macromolecular or supramolecular type, while inorganic hosts are usually of the network variety having different degrees of dimensionality, i.e., tunnel, layer and framework types. Guests can range in character from ions, atoms, organics, inorganics and organometallics to coordination complexes and clusters *(2)*.

Concerning specifically the class of open framework aluminosilicates known as zeolites, a great deal is currently known about the structure and bonding properties of a wide range of zeolite hosts and encapsulated guests *(3)*. Much less, however, is known about the dynamics and virtually nothing about the intimate details of the chemical reactivities of imbibed guests inside zeolite host frameworks.

In this context, the recent impact of molecular mechanics and dynamics calculations in this field has been most impressive *(4)*. They have, for example, been able to successfully delineate the stable structures of different zeolite hosts, predict the distribution of extraframework charge-balancing cations, and map out the coverage dependence of isosteric heats of adsorption, and translational and rotational diffusivities of simple guest molecules. Where corresponding experimental data exist, the agreement with theory is usually good.

Calculations of the above type have not yet been applied to "reactive systems" and there is clearly a perceived need for both quantitative kinetic measurements and complementary molecular mechanics/dynamics simulations for model intrazeolite reactions. In what follows, we describe the first experiments designed to fill this void.

Alkali Metal Cation-Cavitates: Rigid, Macrospheroidal Multidentate, Multisite Zeolate Nanoreaction Chambers: We have recently completed an extensive series of quantitative kinetic measurements in zeolite Y, for archetypical ^{12}CO substitution reactions of $\{Mo^{12}(CO)_6\}$-$M_{56}Y$ (M = Li, Na, K, Rb and Cs) by PMe_3 and ^{13}CO. A key objective of this particular study was to evaluate how the energetics and dynamics of this reaction compared to the situation found in the gas and solution phases. Through such detailed studies of the rates of these reactions as a function of, for example, temperature, pressure, loading and cation type, it was anticipated that one could begin to identify the different ways in which the special environment found within the nanoreaction vessels of a zeolite host could influence the nature of chemical transformations contained therein.

The global picture that emerges from these investigations is that the 13

A nanoreaction chambers of zeolite Y behave as rigid macrospheroidal multidentate anionic ligands (a zeolate or cavitate) towards extraframework charge-balancing cations to which organometallic and ligand guests can become attached. In this context an important point concerns the striking similarity between the framework oxygen 4-ring and 6-ring secondary building units found in zeolite Y and the macrocyclic polyether moieties found in 12-crown-4 and 18-crown-6, respectively. One notes the ability of these kinds of polydentate ligands to selectively coordinate and partially encapsulate alkali metal cations. Lessons from the homogeneous and heterogeneous organometallic chemistry and catalysis literature and the field of zeolite science prepare one for the eventuality that Lewis acid cationic centers of the above type are able to interact with the basic centers of organometallic molecules to cause distortions of structure, changes in reactivity and even alter the mechanism of a chemical transformation. Specifically for metal carbonyls, these effects depend mainly upon the intrinsic basicity of the carbonyl oxygens and the strength of the alkali metal cation-carbonyl interactions.

Recall that the polyhedral tertiary structural building units of zeolite Y (hexagonal prism, β-cage, α-cage), which constitute the final quaternary structure of zeolite Y, are themselves constructed from different topological arrangements of interconnected 4-ring and 6-ring secondary building units.

With the above in mind, the connection to the crown ether (cryptand and spherand) literature is enlightened, the organometallic literature concerning the effects of alkali metal (Lewis acid) anchoring and activating centers is recognized, and the description of the nanoreaction chambers of zeolite Y as alkali-metal cavitates can be appreciated.

Choosing An Archetypical Intrazeolite Reaction: Several factors have to be considered in selecting an intrazeolite reaction for kinetic study. There is a need to obtain enough kinetic data for one to be able to assemble a detailed picture of the effects of the internal surface of the zeolite cavity on the course and rates of the reaction. The internal surface might be modified by the presence of different reactants in various amounts, and such effects will depend on the spatial and electronic properties of the alkali metal cation cavitate. The reactions should therefore ideally satisfy the following criteria:

i)   The intrazeolite reactants, possible intermediates, and products must be structurally and spectroscopically well-defined.

ii)   As complete kinetic information as possible must exist for the same chemical reaction in the gas, solution, surface and/or matrix phases.

iii)   The reaction in all phases must be clean and simple.

A chemical reaction that nicely satisfies most of these criteria involves the thermal dissociative and associative substitution of ^{12}CO in $n\{Mo(^{12}CO)_6\},m\{L\}-M_{56}Y$ by $L = PMe_3$ or ^{13}CO, where M = Li, Na, K, Rb and Cs, and Y refers to the 3-D open-framework host material zeolite Y.

KINETICS

Designing The Experiment: In this system, the establishment of an experimental strategy required a fairly deep appreciation of the properties of the host and guest, as well as host-guest interactions. A complete summary of pertinent knowledge accrued to date, obtained through a "multiprong approach" *(5)* to the characterization of the $n\{Mo(^{12}CO)_6\}$, $m\{L\}-M_{56}Y$ intrazeolite system, is presented in Scheme 1 *(6)*.

Briefly, the spectroscopy, microscopy, diffraction and adsorption methods of analysis listed in Scheme 1 have shown for $M_{56}Y$ saturation loading values of $2Mo(^{12}CO)_6$ and $4PMe_3$ chemisorbed (cation anchoring), with an additional $2PMe_3$ physisorbed (oxygen-framework anchoring) molecules per a-cage. (The latter is indicated by analogy with neutron diffraction data *(7)* for benzene and pyridine in $Na_{56}Y$ and ^{31}P MAS-NMR data *(8)* for PMe_3 in $H_{56}Y$ under similar loading conditions with the PMe_3 probably residing in the 12-ring entrance window to the a-cage of $Na_{56}Y$). These loadings correspond to unit cell formulations of $16\{Mo(^{12}CO)_6\}$-$M_{56}Y$, $32\{PMe_3\}-M_{56}Y$ and $48\{PMe_3\}-M_{56}Y$, respectively. Further, the half saturation loaded $8\{Mo(^{12}CO)_6\}-M_{56}Y$ system can additionally chemisorb up to an average of $2PMe_3$ per a-cage to yield $8\{Mo(CO)_6\}$, $16\{PMe_3\}-M_{56}Y$ and a further 2 PMe_3 per a-cage can be physisorbed to form $8\{Mo(CO)_6\},32\{PMe_3\}-M_{56}Y$. The four site II M^+ ions (a-cage) are tetrahedrally organized in the supercage of $M_{56}Y$ and can trap a single $Mo(^{12}CO)_6$ molecule and two PMe_3 ligands in the arrangement sketched in Scheme 1. Similar cation anchoring schemes are favored for the phosphine-carbonyl complexes $Mo(^{12}CO)_{6-x}(PMe_3)_x-M_{56}Y$ where x = 1, 2, 3. In the case of the $Mo(^{12}CO)_3$ fragment, a 4-ring or 6-ring oxygen framework site is the favored point of attachment to the molybdenum center as illustrated in Scheme 1. For M = Rb and Cs, the extra two site III M^+ ions (a-cage) provide additional Lewis acid centers for the anchoring of occluded guest molecules. Recognition of this distinction between $M_{56}Y$ hosts (denoted Class A: M = Li, Na, K; Class B: M = Rb, Cs) is central to the understanding of the two different patterns of reactivity observed for ^{12}CO substitution of $\{Mo(^{12}CO)_6\}-M_{56}Y$ by PMe_3 and ^{13}CO, to be described in the following sections.

Scheme 1. The Multi Prong Approach to the Characterization of Products in the $n[M(CO)_6]$, $m(L)$-$Na_{56}Y$ System

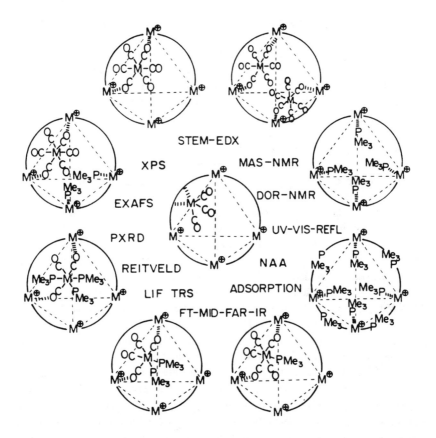

Performing An Experiment: Full details of our combined intrazeolite kinetics cell and FT-MID-IR detection system, as well as of all analytical methods used for the characterization of $n\{Mo(^{12}CO)_6\}, m\{L\}-M_{56}Y$ and related materials will be presented elsewhere *(9)*. For the purpose of this presentation it is more important to appreciate the practical aspects of doing an actual kinetics experiment. One begins with a pressed pellet of the hydrated zeolite Y host material with the unit cell composition $M_{56}[(AlO_2)_{56}(SiO_2)_{136}] \cdot 250H_2O$ and subjects it to the following sequence of pretreatment, impregnation and annealing steps in order to create the desired sample deemed suitable for a kinetics run:

$M_{56}Y$ (Hydrated)

$\quad\big|$ RT $\to \to \to \to$ 450°C, thermal vacuum dehydration
$\quad\downarrow$

$M_{56}Y$ (Dehydrated)

$\quad\big|$ $nMo(CO)_{6(g)}$, RT impregnation
$\quad\downarrow$

$n\{Mo^{12}(CO)_6\}-M_{56}Y$ (1/10 a-cages)

$\quad\big|$ Annealing, 70°C
$\quad\downarrow$

$n\{Mo(CO)_6\}-M_{56}Y$ (Anchored $Mo(CO)_6$)

$\quad\big|$ $mPMe_{3(g)}$, RT impregnation
$\quad\downarrow$

$n\{Mo(CO)_6\}, m\{PMe_3\}-M_{56}Y$ (De-anchored)

$\quad\big|$
$\quad\downarrow$

Kinetics

Loading-Dependent Adsorption Effects: As indicated in the previous section, having prepared a freshly impregnated $n\{Mo(^{12}CO)_6\}-M_{56}Y$ sample, one requires a controlled (50-70°C) thermal post-treatment in order to bring the $Mo(^{12}CO)_6$ guest into its equilibriated (annealed) form. Only under conditions that establish "anchoring site homogeneity" can the "true" symmetry of the $Mo(^{12}CO)_6$ molecule at its cation binding site be realized. In the case of, for example, $Mo(^{12}CO)_6$ in $Na_{56}Y$, this takes the form of a C_{2v} (or lower) symmetry *trans*-ZONa...(OC)Mo(CO)$_4$(CO)...NaOZ anchoring geometry which displays a diagnostic six line ν_{CO} mid-IR pattern (Figure 1). Of special note here is the fact that the structure of this ν_{CO}

multiplet is very sensitive to the loading of $Mo(^{12}CO)_6$, as well as that of other coadsorbed guests, such as PMe_3 and CO. A dramatic illustration of this effect is seen in the virtual collapse of the $Mo(^{12}CO)_6$ ν_{CO} sextet, induced by simply admitting PMe_3 to the sample at room temperature (Figure 1). In essence one is observing a "partial unlocking" of the $Mo(^{12}CO)_6$ guest from its two Na_{II}^+ tethering points through the coadsorption of PMe_3. This kind of "de-annealing" phenomenon can be understood within the experimental and theoretical framework of the loading dependence of the isosteric heats of adsorption (4). On surveying this literature one finds that the integral energy of adsorption due to sorbate-zeolite and sorbate-sorbate interactions decreases monotonically with occupancy and temperature, the effect being most pronounced for the highest ionic potential cations. In the context of molecular dynamics, sorbate-sorbate collisions have two effects. One involves an exchange of energy during a collision, while the other is a reduction of mean free path for motion in the zeolite. The former leads to an increase in the rate of site-to-site barrier crossings whereas the latter leads to a decrease. Temperature and sorbate occupancy control the degree to which these effects contribute to the observed sorbate diffusivity. Increasing the rate of sorbate-sorbate collisions acts to decorrelate molecular motion and serves to increase rotational diffusivity with occupancy (4).

^{12}CO **Substitution Kinetics of** $\{Mo(^{12}CO)_6\}$-$M_{56}Y$ **with** PMe_3 **and** ^{13}CO: Extensive kinetic studies have been completed on the above system for M = Na involving variations in $Mo(^{12}CO)_6$ and PMe_3 loading, and ^{12}CO and ^{13}CO pressure. Similar but less complete studies have also been conducted across the entire alkali-metal cation series Li^+ to Cs^+. These results have permitted us to assemble a comprehensive and quantitative mechanistic picture for these important archetypical intrazeolite metal carbonyl reactions and thereby assess the role of the zeolite environment on the course and dynamics of the chemical transformation. Let us first focus attention on $Mo(^{12}CO)_6$ in the 13Å supercage of $Na_{56}Y$. Thus $n\{Mo(^{12}CO)_6\}$-$Na_{56}Y$ (n < 8) undergoes ^{12}CO substitution reactions in the presence of PMe_3 in $n\{Mo(^{12}CO)_6\}$,$m(PMe_3)$-$Na_{56}Y$, or of ^{13}CO in $n\{Mo(^{12}CO)_6\}$, $m\{^{13}CO\}$-$Na_{56}Y$ to afford cis-$\{Mo(^{12}CO)_4(PMe_3)_2\}$-$Na_{56}Y$ and fully labelled $\{Mo(^{13}CO)_6\}$-$Na_{56}Y$, respectively. No reaction intermediates were detected in the PMe_3 system, as suggested by the excellent isosbestic point in Figure 2. Non-involvement of $Mo(^{12}CO)_5PMe_3$ as an intermediate in the formation of the cis-$Mo(CO)_4(PMe_3)_2$ product was confirmed by direct impregnation of $Mo(^{12}CO)_5PMe_3$, in $Na_{56}Y$, and the demonstration that this reacts much more slowly than $Mo(^{12}CO)_6$. Similar kinetic behaviour was observed for reactions with PMe_3 in the entire series of hosts, $M_{56}Y$, and in all cases these reactions proceed by very well behaved "first-order" processes that

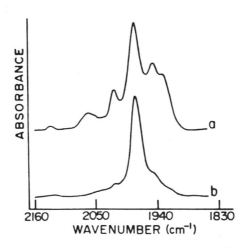

Figure 1: (a) MID-IR spectrum of annealed $4\{Mo(^{12}CO)_6\}$-$Na_{56}Y$; (b) MID-IR spectrum of $4\{Mo(^{12}CO)_6\},48\{PMe_3\}$-$Na_{56}Y$.

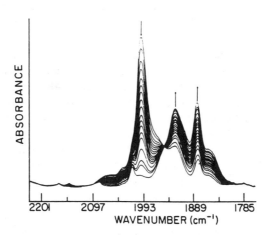

Figure 2: MID-IR spectral traces of a kinetics run for $1.2\{Mo(^{12}CO)_6\},24\{PMe_3\}$-$Na_{56}Y$ at 65.8°C.

involve what we believe to be a supramolecular assembly of $Mo(^{12}CO)_6$ precursor, PMe_3 ligands and extraframework M cations, all housed together in the supercage of $M_{56}Y$. Excellent Eyring plots yield the activation parameters $\Delta H^{\ddagger}$ and $\Delta S^{\ddagger}$ for each of the alkali metal cations listed in Table I. A striking "volcano-shaped" effect is observed for these $\Delta H^{\ddagger}$ values on passing from Li^+ to Cs^+. This is counterbalanced by an "inverse volcano" effect for $-T\Delta S^{\ddagger}$ (Figure 3).

Table I: Activation Parameters For Dissociative Reactions of $Mo(^{12}CO)_6$.

Entering Ligand	Medium	$\Delta H^{\ddagger}/$ kJ mol^{-1}	$\Delta S^{\ddagger}/$ JK^{-1} mol^{-1}
none[a]	gas phase	157	38
^{14}CO[b]	gas phase	126	- 2
PBu_3^n [b]	Decalin	133	28
^{13}CO[c]	$Na_{56}Y$	65	-127
PMe_3	$Li_{56}Y$	49	-182
	$Na_{56}Y$	70	-107
	$K_{56}Y$	88	- 86
	$Rb_{56}Y$	75	-105
	$Cs_{56}Y$	57	-144

[a]Irreversible CO loss induced by pulsed laser pyrolysis technique at 670-760K; ref. 10. [b]Ref. 11. [c]$P(^{13}CO) = 100$ Torr.

The values for $\Delta H^{\ddagger}$ and $\Delta S^{\ddagger}$ are considerably smaller than those found for similar types of reactions in the solution and gas phases (Table I). This dramatic decrease, for what we describe as "intracage" first order dissociative ^{12}CO substitution reactions, is believed to originate in much stronger cation anchoring of the $\{Mo(^{12}CO)_5...(^{12}CO)\}^{\ddagger}$ transition state compared with that of the ground state $Mo(^{12}CO)_6$. This could also account for the large negative values for $\Delta S^{\ddagger}$ since the much more weakly anchored $Mo(^{12}CO)_6$ in the ground state is transformed during ^{12}CO dissociation into the tightly anchored $\{Mo(^{12}CO)_5...(^{12}CO)\}^{\ddagger}$ transition state, this transformation being associated with increased back-bonding in the less highly coordinated intermediate, and the consequently greater negative charge on the oxygen atoms of the carbonyl ligands. The volcano-shape of the alkali metal dependence of the $\Delta H^{\ddagger}$ parameter alerts one to the existence of two classes of reactivity behaviour for these dissociative substitutions in alkali metal cation-cavitate environments. These are denoted Class A for M = Li, Na, K and Class B for M = Rb, Cs (Figure 4). The results indicate that the most strongly activated $Mo(^{12}CO)_6$

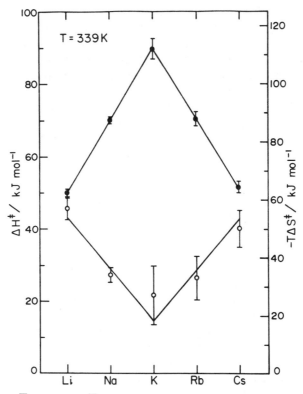

Figure 3: $\Delta H^{\mp}$ and $-T\Delta S^{\mp}$ for the reaction of $n\{Mo(^{12}CO)_6\},m\{PMe_3\}$-$M_{56}Y$ (M = Li, Na, K, Rb and Cs).

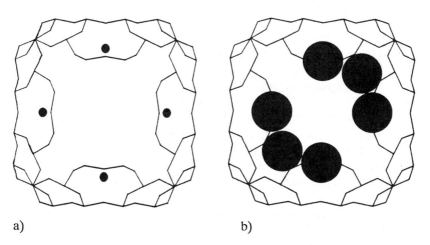

a)   b)

Figure 4: (a) Class A cavitates (M = Li, Na, K) showing 4M site II cations; (b) Class B cavitates (M = Rb and Cs) with 4M site II and 2M site III cations.

precursors have the most well organized $\{Mo(^{12}CO)_5...^{12}CO\}^{\ddagger}$ transition-states (and vice-versa). This kinetic behaviour can be traced to differences that exist in the cation populations, topologies, spatial demands and ionic potentials between the Class A and Class B cavitates (Figure 4). In essence the results imply that ionic potential control of the activation parameters dominates the chemical reactivity of $Mo(^{12}CO)_6$ in the $4M_{II}^+$ cation trap of Class A materials, whereas both ionic potential and spatial demands of the more highly populated $4M_{II}^+ + 2M_{III}^+$ environment, found in Class B materials, play the major role.

Further Mechanistic Studies: The studies reported above involved quite low loadings of $Mo(^{12}CO)_6$, loadings of PMe_3 corresponding to the presence only of chemisorbed PMe_3, and loadings of ^{13}CO corresponding to a pressure of 100 Torr. The results can be understood initially in terms of the sequence of reactions (1)-(4). The fact that the only product of reactions with PMe_3 is *cis*-$Mo(CO)_4(PMe_3)_2$, and that $Mo(CO)_5PMe_3$ is not formed as an intermediate, requires successive dissociation of two ^{12}CO ligands (reactions (1) and (2)) before attack on the $Mo(^{12}CO)_4$ intermediate by PMe_3. In the absence of added ^{12}CO, $k_2 >> k_{-1}[CO]$, $k_{-2}[CO]$, and k_1, and the rate determining step is simply the forward reaction in (1). The formation of the essentially fully labelled $Mo(^{13}CO)_6$ in the reaction with ^{13}CO suggests a similar sequence of reactions to that shown in (1)-(4) but with the added feature that the $Mo(^{12}CO)_4$ intermediate must undergo rapid

$$ZONa...(OC)Mo(CO)_4(CO)...NaOZ \underset{k_{-1}}{\overset{k_1}{\rightleftharpoons}}$$
$$ZONa...(OC)Mo(CO)_3(CO)...NaOZ+CO \qquad (1)$$

$$ZONa...(OC)Mo(CO)_3(CO)...NaOZ \underset{k_{-2}}{\overset{k_2}{\rightleftharpoons}}$$
$$ZONa...(OC)Mo(CO)_2(CO)...NaOZ+CO \qquad (2)$$

$$ZONa...(OC)Mo(CO)_2(CO)...NaOZ + ZONa...PMe_3$$
$$\overset{k_3}{------>} ZONa...(OC)Mo(CO)_2(PMe_3)(CO)...NaOZ \quad (3)$$

$$ZONa...(OC)Mo(CO)_2(PMe_3)(CO)...NaOZ + ZONa...PMe_3$$
$$\overset{k_4}{------>} ZONa...(OC)Mo(CO)_2(PMe_3)_2(CO)...NaOZ \quad (4)$$

replacement of the ^{12}CO by the much more prevalent ^{13}CO before final addition of ^{13}CO to the $Mo(^{13}CO)_4$ in reactions analogous to (3) and (4). This is supported by the similarity of the activation parameters for reaction of $\{Mo(^{12}CO)_6\}$-$Na_{56}Y$ with PMe_3 and ^{13}CO (Table I).

A) PMe$_3$ Loading Effects in the Absence of Added ^{12}CO: The values of k_{obs} decrease with PMe_3 loading by a factor of ca. 2 over the range 0.5 to 2 PMe_3 molecule for each a-cage that contains an $Mo(^{12}CO)_6$ molecule. This decrease can be ascribed to PMe_3-induced reduction in the strength of binding to, and organization of, the $\{Mo(^{12}CO)_5....^{12}CO\}^{\ddagger}$ transition state with respect to cation anchoring. When the loading of PMe_3 is increased to an average of 6 PMe_3 per a-cage, i.e. when up to 4 physisorbed PMe_3 molecules are present in the windows of the a-cages containing the reacting $Mo(^{12}CO)_6$ molecules, the rates increase by about 50%. This enhanced reactivity in the presence of physisorbed PMe_3 molecules, in addition to the chemisorbed PMe_3, might be accounted for by the onset of an associative path, dependent on the population of physisorbed and, therefore, more reactive PMe_3 molecules in the vicinity of the $Mo(^{12}CO)_6$. Associative reactions of $Mo(^{12}CO)_6$ with more basic P-donor molecules are known to occur in homogeneous solution *(11)*. However, it would be expected that this bimolecular associative attack on an $Mo(^{12}CO)_6$ molecule would lead simply to $Mo(CO)_5PMe_3$ as product and this is not the case, the observed product still being entirely the *cis*-$Mo(^{12}CO)_4(PMe_3)_2$. This suggests that the enhanced reactivity in the presence of physisorbed PMe_3 might rather involve reduction of the effect transmitted by the chemisorbed PMe_3 and/or the introduction of an additional transmitted effect that acts in the opposite way to that of the chemisorbed PMe_3.

B) Mo(^{12}CO)$_6$ Loading Effects in the Absence of ^{12}CO: These experiments were carried out in the initial presence of 2 chemisorbed PMe_3 molecules. As the $Mo(^{12}CO)_6$ loading increases from 0.1-1.0 $Mo(^{12}CO)_6$ molecules per a-cage the rates decrease by a factor of ca. 4, but, when the loading is increased from 1 to ca. 2 $Mo(^{12}CO)_6$, the rates increase very substantially. (At 1.4 $Mo(^{12}CO)_6$/a-cage the rate is 27 times greater than that for 1.0 $Mo(^{12}CO)_6$/a-cage). The initial decrease in rates is evidence for an effect transmitted through the framework that is essentially analogous to that observed with PMe_3. Above a loading of 1 $Mo(^{12}CO)_6$ per a-cage the initially present chemisorbed PMe_3 must be displaced and changed into physisorbed PMe_3. However, the enhancement of the rates is much greater than the enhancement due to physisorbed PMe_3 observed in the PMe_3 loading experiments and the simple additional presence of physisorbed PMe_3 cannot account quantitatively for the enhanced reactivity

observed here.   We believe, therefore, that the spatial constraints
associated with the placement of a second $Mo(^{12}CO)_6$ molecule, orthogonal
to the first, in the a-cage leads to intermolecular Mo....OC interactions
which activate the $Mo(^{12}CO)_6$ molecules towards dissociative loss of ^{12}CO.

C) CO-Pressure Effects on the Rates: The rates of reactions with PMe_3
and with ^{13}CO were found to be progressively decreased as the applied
pressure of CO was increased and the data gave an excellent fit to eq. (5)
where a, b, and c are constants, the values of which are given in Table II.

$$k_{obs} = [a/(1 + bP_{CO}^2)] + c \qquad (5)$$

The data are all represented graphically by the plots in Figure 5. The most
striking feature of these data is the decrease in the rate of the ^{13}CO
exchange reaction with increasing pressures of ^{13}CO. This cannot be due
to reactions such as the reverse of (1) and (2) because the amount of ^{13}CO
vastly exceeds that of ^{12}CO and reaction of the coordinatively unsaturated
intermediates with ^{13}CO leads to product molecules.   Instead this
observation can be ascribed to a lowering of the anchoring ability of the
two Na^+ ions to which the $\{Mo(CO)_5....CO\}^{\ddagger}$ transition state becomes
attached, an effect similar to that found when the loading of $Mo(^{12}CO)_6$ or

Table II: Rate Parameters for Reactions of $Mo(^{12}CO)_6$ with PMe_3 and ^{13}CO
as a Function of CO Pressure

Reactant	$10^4 a/s^{-1}$	$10^5 b/Torr^{-2}$	$10^4 c/s^{-1}$
PMe_3[a]	2.9 ± 0.5	13.5 ± 4.0	1.42
PMe_3[b]	2.4 ± 0.2	2.4 ± 0.3	1.88
^{13}CO	4.9 ± 0.3	2.6 ± 0.3	0

[a] 2 molecules of chemisorbed PMe_3 in each a-cage reaction chamber
[b] 2 molecules of chemisorbed PMe_3 in each reaction chamber PLUS 4
molecules of physisorbed PMe_3, one in each of the 4 windows to the a-cage
reaction chamber

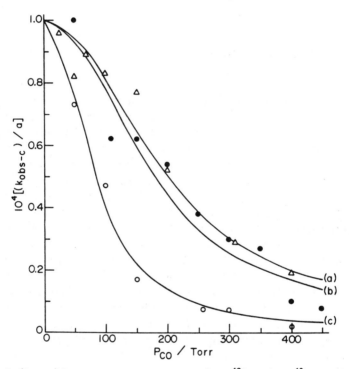

Figure 5: $\{k_{obs}-c\}/a$ versus P(CO) for (a) $2.8\{Mo(^{12}CO)_6\},m(^{13}CO)\}$-$Na_{56}Y$; (b) $1.2\{Mo(^{12}CO)_6\},48\{PMe_3\}$-$Na_{56}Y$, and (c) $1.2\{Mo(^{12}CO)_6\},24\{PMe_3\}$-$Na_{56}Y$, at 65.8°C.

chemisorbed PMe_3 is increased. In the presence of only chemisorbed PMe_3, the substitution reaction with PMe_3 is retarded more effectively by CO than is the exchange reaction as shown by the larger value of b in the former case. This indicates that reversal of reactions (1) and (2) is now significant, probably in addition to a process similar to that operating in the ^{13}CO exchange reaction.

When physisorbed PMe_3 is present, in addition to the chemisorbed PMe_3, the retardation of the rate by CO is greatly reduced. This can be ascribed to (i) a reduction of the activity and/or local concentration of CO in the reaction chamber, and (ii) the presence of an additional 4 PMe_3 molecules in the vicinity of the reaction chamber. These PMe_3 molecules, being physisorbed, would also have a greater reactivity than the chemisorbed PMe_3 molecules. The combined effect is that PMe_3 competes much more successfully than CO for the $Mo(^{12}CO)_4$ intermediate.

Another striking feature of the data is the existence of the term c for reactions with PMe_3 but not ^{13}CO. The term c does not appear to depend on the pressure of ^{12}CO but the range of ^{12}CO pressures available to us was limited and it may be that higher rates would be observed at higher pressure. The existence of the term does, however, represent a clear enhancement of the rate above what would have been expected from the deactivation caused by transmitted effects, as seen in the exchange reaction, and reversal of CO dissociation, as seen in addition in the substitution reactions. It is significant that this rate enhancement is observed even when only chemisorbed PMe_3 is present initially. This contrasts with the results found in the absence of CO when rate enhancement was only seen in the presence of physisorbed PMe_3. This suggests that the initially chemisorbed PMe_3 is activated in some way by the presence of CO.

Another distinctive feature of the reactions under CO is that significant yields of $Mo(CO)_5PMe_3$ are observed. When only chemisorbed PMe_3 is present initially, the yields of $Mo(CO)_5PMe_3$ remain essentially constant at ca. 30-40% over the whole range of CO pressures used. However, when physisorbed PMe_3 is present as well the yields of $Mo(CO)_5PMe_3$ rise monotonically to over 90%.

The reasons for these observations must be quite complex. The presence of CO will decrease the proportion of the reactions that proceed via successive dissociations of CO, as in reactions (1) and (2), and will increase the proportion of the reactions that proceed via the path governed by the constant c in eq. (5). The steady state concentrations of the coordinatively unsaturated intermediates, particularly $ZONa...(OC) Mo(CO)_3(CO) ... NaOZ$, will be changed, but so might the activities of the chemisorbed and physisorbed PMe_3 molecules that compete for the intermediates. At high pressures of CO, when virtually all the reactions proceed via the path governed by c, the effect of CO on the yields of $Mo(CO)_5(PMe_3)$ is less pronounced when only chemisorbed PMe_3 is present, than when physisorbed PMe_3 is present as well. This can be understood if the physisorbed PMe_3 is able to react with the $Mo(CO)_6$ by an essentially associative process, so leading to $Mo(CO)_5(PMe_3)$ in a single concerted step. This still leaves unexplained why the physisorbed PMe_3 does not lead to any appreciable yields of $Mo(CO)_5(PMe_3)$ in the absence of CO. This may, in part, be due to the fact that the latter reactions include a greater proportion that proceed via the dissociative path leading to *cis*-$Mo(CO)_4(PMe_3)_2$ but that does not seem to be a quantitatively adequate explanation. This aspect of the results calls for further investigation.

Finally, the effect of PMe_3 loading on the value of c was investigated by a study of the rates under 650 Torr CO when the dissociative reaction is almost completely inhibited (Figure 6). When there was only 1, chemisorbed, PMe_3 molecule in each reaction chamber the rate is quite low but, when the average loading of chemisorbed PMe_3 per a-cage was varied from 2 to 4 the values of k_{obs} were ca. 1.7 times higher and essentially constant. This is in accordance with the fact that only 2 chemisorbed PMe_3 molecules were present in the reaction chamber over the whole of this variation in total average loading. When the average loading increased to 5 and 6 PMe_3 molecules per a-cage, i.e. when the number of physisorbed PMe_3 molecules in the vicinity of the reaction chamber was increased from 0 to 2 to 4, the rates increased monotonically, again reflecting the greater number and greater activity of PMe_3 molecules available to cause rate enhancement.

D)   Summary of Intrazeolite Kinetic Behaviour: A number of pronounced differences, and some similarities, have been observed in comparison with reactions of $Mo(^{12}CO)_6$ in solution and in the gas phase. Thus:

(i)   Rate determining CO dissociation is a major contributor to the rates in the absence of relatively high applied pressures of CO.

(ii)   The activation parameters for such reactions reflect transition state stabilization through the extraframework alkali-metal cations of the a-cage lattice, an effect that substantially decreases the values of $\Delta H^{\ddagger}$ and $\Delta S^{\ddagger}$.

(iii)   The variation in the activation parameters with variation of M in the $M_{56}Y$ zeolite lattice is pronounced, and reflects the changes in the detailed spatial and electronic properties of the reaction chamber with changing M.

(iv)   Generalized "framework effects" of PMe_3, CO, and $Mo(^{12}CO)_6$ loading on the activating capacity of the reaction chamber towards $Mo(^{12}CO)_6$ are observed, the effect being to decrease the activation at higher loadings.

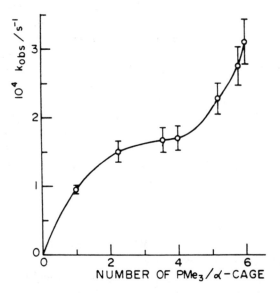

Figure 6: Effect of PMe$_3$ loading on reaction rates in the system 1.2-1.6{Mo(^{12}CO)$_6$},m{PMe$_3$}-Na$_{56}$Y, under 650 Torr of CO, at 65.8°C, showing the three observed regimes referred to as subsaturation, saturation and suprasaturation.

(v)   At even higher loadings of PMe_3 and $Mo(^{12}CO)_6$ a rate enhancement occurs and this is probably due to a combination of effects, one of which may be the inception of associative reactions under some conditions, another being intermolecular activation of $Mo(^{12}CO)_6$ towards dissociative loss of CO.

Concluding Remarks: This investigation has yielded the first quantitative measurements of the influence of the internal "perfect" surface of a zeolite host lattice on the chemical reactivity of encapsulated guests. In the case of zeolite Y the results of a comprehensive kinetic study of ^{12}CO substitution of $Mo(^{12}CO)_6$-$M_{56}Y$ by PMe_3 and ^{13}CO unveiled an appealing picture of the α-cage nanoreaction chamber as a rigid macrospheroidal multisite multidentate alkali metal cavitate (zeolate). A model of this kind aids one with the design of experiments and the interpretation of cation and framework anchoring and guest loading effects on activation parameters and reaction mechanism compared to gas, solution, surface or matrix phases. Quantitative experiments of this kind are expected to be of great value in understanding intimate details of size and shape selective catalytic reactions, the origin of host-guest inclusion and molecular recognition phenomena, and the parameters that control a range of intrazeolite synthetic and self-assembly processes that are basic to the preparation of new solid-state microporous materials of interest in chemoselective sensing, quantum electronics, nonlinear optics, information storage and artificial photosynthesis, to name but a few.

Acknowledgments: The generous financial assistance of the Natural Sciences and Engineering Research Council of Canada (GAO,AJP) is sincerely appreciated.   (Sö) is deeply indebted to the Chemistry Department, Middle East Technical University, Ankara for allowing him an extended leave of absence at the University of Toronto. (HOP and EJSV) express their gratitude to CNPq (Brasil) and FAPESP (Sao Paulo) for a graduate scholarship (HOP) and financial support of this project (EJSV).

Literature Cited:
1.   "Inclusion Phenomena and Molecular Recognition", Ed. J.L. Atwood, Plenum Press, New York, 1990 (and references cited therein).
2.   "Inclusion Compounds", Eds. J.L. Atwood, J.E. Davies and D.D. MacNicol, Academic Press, London, 1984, Vol. I,II,III (and references cited therein).

3. Ozin, G.A. and Gil, C., Chem. Rev., 1989, *89*, 1794; Ozin, G.A., Kuperman, A., and Stein, A., Angew. Chem. Int. Ed. Adv. Mat., 1989, *101*, 373; Stucky, G.D. and Macdougall, J.E., Science, 1990, *247*, 669 (and references cited therein).
4. June, R.L., Bell, A.T., and Theodorou, D.N., J. Phys. Chem., 1990, *94*, 8232; Vigné-Maeder, F., and Auroux, A., J. Phys. Chem., 1990, *94*, 316; Yamazaki, T., Watanuki, I., Ozawa, S., and Ogino, Y., Langmuir, 1988, *4*, 433 (and references cited therein).
5. Thomas, J.M., and Vaughan, D.E.W., J. Phys. Chem. Solids, 1989, *50*, 449.
6. Özkar, S., Ozin, G.A., Moller, K., and Bein, T., J. Am. Chem. Soc., 1990, *112*, 9575.
7. Fitch, A.N., Jobic, H., and Renouprez, A., J. Phys. Chem., 1986, *90*, 1311; Jobic, H., J. Chem. Soc. Chem. Commun., 1990, 1152.
8. Bein, T., Chase, D.B., Farlee, R.D., and Stucky, G.D., Studies in Surface Science and Catalysis, 1986, *28*, 311.
9. Ozin, G.A., Özkar, S., Pastore, H.O., Poë, A.J., and Vichi, E.J.S., (manuscript in preparation).
10. Lewis, K.E., Golden, D.M., and Smith, G.P., J. Am. Chem. Soc., 1984, *106*, 3905.
11. Howell, J.A.S. and Burkinshaw, P.M., Chem. Rev., 1983, *83*, 557.

RECEIVED February 4, 1992

Chapter 23

Self-Assembling Electron-Transport Chains in Zeolites

Molecular Bilayer Rectifiers and Photodiodes

Yeong Il Kim, Richard L. Riley, Edward H. Yonemoto, Daiting Rong, and Thomas E. Mallouk

Department of Chemistry and Biochemistry, University of Texas at Austin, Austin, TX 78712

Simple molecular diodes and photodiodes were prepared at the zeolite/aqueous solution interface. The rate of electron transfer between photoexcited $Ru(bpy)_3^{2+}$ and viologen electron acceptors co-exchanged onto/into zeolites L, Y, and mordenite was measured by transient spectroscopic techniques. In aqueous suspensions the largest pore zeolites (L, Y) show Stern-Volmer luminescence quenching, indicating that electron transfer involves diffusion of the acceptor. More complex systems containing covalently linked donor-acceptor diads at the zeolite surface also give interesting information about the relationship between molecular conformation, imposed by the zeolite, and intramolecular electron transfer rates.

Microporous inorganic solids, such as zeolites, clays, and layered oxide semiconductors offer several advantages as organizing media for molecular electron transport assemblies. Because these materials are microcrystalline, their internal pore spaces have well-defined size and shape. This property can be exploited to cause self-assembly, by virtue of size exclusion effects, ion exchange equilibria, and specific adsorption, of photosensitizers, electron donors, and electron acceptors at the solid/solution interface.

Molecular rectifiers. The effect of partitioning different redox molecules to opposite sides of an interface is to produce particles or devices whose function is mimetic of semiconductor-based devices. Figure 1 illustrates the analogy between a p-n junction device (a diode or photodiode) and a bimolecular "rectifier". At the p-n junction, there is a built-in electrical potential drop which determines which way current can flow. By applying a bias which partially cancels the built in potential, electrons flow readily from left to right. The opposite bias reinforces the built-in potential, and allows no current to flow. In the analogous molecular system, the direction of rapid electron transfer is from the molecule with the more negative redox potential to the more positive one. In this case, it is not a built-in electric field (or

0097–6156/92/0499–0333$06.00/0

electrical potential gradient), but an *electrochemical* potential gradient which is responsible for the rectification effect. Devices of this type were studied extensively by Murray and coworkers (1) and by Wrighton et al. (2), who developed techniques for preparing bilayers of different redox polymers on planar electrodes or microelectrode arrays.

We have demonstrated similar effects using a zeolite or pillared clay bound to an electrode surface (3). Molecules partition, according to their size, into sites on the external or internal surface of the zeolite or clay, and the externally bound monolayer mediates electron transfer between the electrode and molecules within the particle. An interesting example of such spontaneous partitioning of molecules occurs when zeolite Y is ion-exchanged with small redox-active cations such as FcR^+, and with large metal tris(2,2'-bipyridyl) complexes. The 13 Å diameter $Os(bpy)_3^{2+}$ ion is substitution inert, and is therefore blocked from entry into the internal

FcR^+ $Os(bpy)_3^{2+}$ Zeolite Y

pore network by the 7.4 Å diameter 12-ring windows. Figure 2 shows the cyclic voltammetry of zeolite Y-modified electrodes (prepared as in ref. 3a) exchanged with FcR^+ and $Os(bpy)_3^{2+}$ ions. In solution, these complexes show reversible anodic/cathodic waves centered at +.45 and +.60 V vs SCE, respectively. Only the $Os(bpy)_3^{2+}$ appears electroactive on zeolite Y, even though the ferrocene compound exchanges in readily; in this experiment its loading on the zeolite is about 5 times higher than that of $Os(bpy)_3^{2+}$. When these compounds are ion-exchanged *together* onto the zeolite, an enhanced anodic wave appears at the $Os^{2+/3+}$ potential. It appears that the Os complex can "communicate" with the previously silent $FcR^{+/2+}$ couple, mediating its oxidation, and the large anodic wave corresponds to oxidation of both Os^{2+} and FcR^+. However, the reverse process (reduction of the oxidized ferrocene complex) is too slow to be observable on the electrochemical timescale. Organized in space according to their site preferences on the zeolite internal/external surface, these two complexes form a molecular "rectifier" (see Figure 1) in the sense that only the oxidation half reaction is fast for the ferrocene.

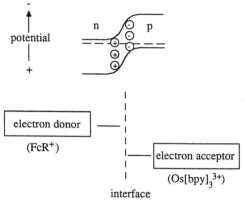

Figure 1. The analogy between semiconductor p-n junction and bilayer molecular rectifiers. In both cases a "forward bias" corresponds to electron flow from left to right.

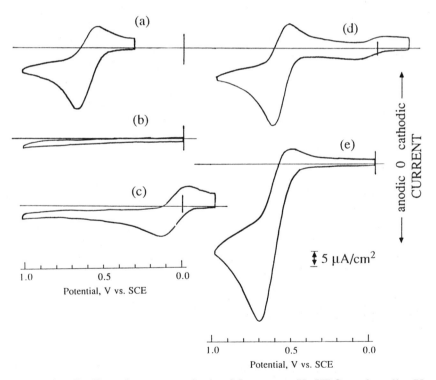

Figure 2. Cyclic voltammetry, in 1 mM aqueous K_2HPO_4, of zeolite Y modified electrodes exchanged with (a) $Os(bpy)_3^{2+}$, (b) FcR^+, (c) $Co(bpy)_3^{2+}$, (d) both $Os(bpy)_3^{2+}$ and $Co(bpy)_3^{2+}$, (e) both $Os(bpy)_3^{2+}$ and FcR^+. Scan rate 50 mV/sec.

Unlike $Os(bpy)_3^{2+}$, substution-labile $M(bpy)_3^{2+}$ complexes can be exchanged into zeolite Y. For example, the Co^{II} complex, which is nearly identical in size to the Os^{II} complex, can dissociate and then re-assemble inside the 13 Å diameter supercages. Consequently the maximum loading of $Co(bpy)_3^{2+}$ is about 2 x 10^{-4} moles/gram in zeolite Y, whereas the maximum for the Os complex (on the external surface of the particles) is 7 x 10^{-6} moles/gram. Figure 2 also shows the electrochemistry of $Os(bpy)_3^{2+}/Co(bpy)_3^{2+}/zeolite$ Y electrodes. In this case the surface Os couple appears to be unable to mediate electron transfer from the Co complex entrained in the zeolite. Since the size of the $Co^{2+/3+}$ and $Os^{2+/3+}$ waves are comparable in these experiments, it is likely that only the $Co(bpy)_3^{3+/2+}$ molecules on the zeolite external surface can be oxidized and reduced, and that the contribution of bulk charge transport diffusion to the current, on this timescale, is minimal.

An interesting question which relates to this system is the mechanism of charge transfer to the redox couple within the zeolite. Normally charge transport diffusion in chemically modified electrodes occurs by a combination of physical diffusion of molecules and electron exchange between them. We might imagine that physical diffusion of $FcR^{2+/+}$ could be reasonably rapid, but that $Co(bpy)_3^{3+/2+}$ would move very slowly within the zeolite. The latter complex must dissociate in order to move between supercages, and the oxidized form (a d^6 complex) does not dissociate readily. An additional complicating factor is the need to move charge compensating ions into and out of the zeolite as the entrained complex is oxidized and reduced. Several studies (4-6) have been directed at understanding the mechanism of intrazeolitic charge transport. In some cases electron transfer appears to occur only at the solid/electrolyte interface, whereas in others it must occur within the zeolite as well.

Light induced electron transfer. Just as a semiconductor p-n junction (Figure 1) can act as either a diode or photodiode, a rectifying molecular bilayer containing a photoactive component can transfer electrons in one direction under light excitation. This phenomenon has been reported for devices containing bilayer polymer films (7,8) as well as zeolites and clays (3d,9,10). $Ru(bpy)_3^{2+}$ is an extensively studied photosensitizer with an energetic metal-to-ligand charge transfer (MLCT) excited state that is quenched by electron transfer to either an electron donor or acceptor. By exchanging methyl-viologen (MV^{2+}), an electron acceptor, into zeolites and $Ru(bpy)_3^{2+}$ onto their external surface, bilayer "photodiodes" are created. We may evaluate the efficiency of electron transfer and separation of redox products by monitoring the MLCT state emission and transient absorbance of the oxidized ($Ru(bpy)_3^{3+}$) and reduced (MV^+) products.

At high loading of the photosensitizer (ca. 10^{-5} moles/g zeolite), quenching of the MLCT state by MV^{2+} is inefficient because of mutual electrostatic repulsions (11); however, at sub-monolayer coverage (below 2×10^{-6} moles/g), electron transfer quenching occurs readily and in fact obeys Stern-Volmer kinetics for hydrated large pore zeolites (Y and L) in which MV^{2+} can diffuse freely. Figure 3 shows plots of inverse luminescence intensity versus the effective concentration of quencher, M*. The latter is calculated from the loading of MV^{2+} and the zeolite free pore volume determined from tributylamine adsorption isotherms (12). The highest Stern-Volmer constants, k_{SV}, are found for aqueous suspensions of zeolites Y and L. A lower value (about half that of the aqueous suspension) is obtained with a powdered sample of zeolite Y air-dried at 40 °C, which still contains a about half the water of a fully hydrated zeolite in its pore network. Interestingly, mordenite, which has slightly smaller linear channels, does not show linear Stern-Volmer behavior. The motion of MV^{2+} is apparently severely restricted in the elliptical 12-ring channels in mordenite, and it is only at relatively high M* that the luminescence intensity of $Ru(bpy)_3^{2+}$ begins to be affected.

What is quite surprising in these plots is not the (solution-like) Stern-Volmer behavior that is found, but the diffusion coefficients which can be calculated for MV^{2+}. From the luminescence lifetime of $Ru(bpy)_3^{2+}$ in the absence of quencher, the k_{SV} values can be converted to bimolecular quenching rate constants k_q. The diffusion coefficient is derived from k_q by means of a modified Smoluchowsky equation (equation 1), where D_q is the diffusion coefficient of MV^{2+}, R_q is its radius (taken to be 6.7 Å), R is the average

$$k_q = p\pi R^2 D_q N_A / 1000 R_q \qquad (1)$$

radius of the zeolite channel (ca. 5 Å), p is the probability that encounter between the two molecules will result in quenching, and N_A is Avogadro's number. This calculation of D_q assumes that the diffusion coefficient of $Ru(bpy)_3^{2+}$, which is confined to the zeolite outer surface, is negligible, and that the effective encounter area of the two ions is πR^2, i.e., the cross sectional area of the zeolite channel. Taking p=1 (this gives a *lower limit* for the diffusion coefficient) we calculate $D_{MV2+} = 1.3 \times 10^{-7}$ cm^2 s^{-1} in zeolite L and 1.1×10^{-7} in zeolite Y.

While the presence of MV^{2+} on the zeolite external surface, or ion exchange of MV^{2+} with solution-phase cations such as H^+ might be invoked to explain these anomalously high quenching rates, we consider these explanations unlikely. The Stern-Volmer plots for zeolites L and Y are linear down to very low MV^{2+} loading, where we would expect little or no MV^{2+} to be present at external sites. Ion-exchange with solution-phase H^+ should give detectable concentrations of MV^{2+} and/or $Ru(bpy)_3^{2+}$ in solution, and none has been detected spectroscopically. Moreover, exchange with solution cations is not possible in the case of partially dehydrated zeolite Y powder, which still shows an unusually high quenching rate.

The diffusion coefficients of MV^{2+} measured by this technique are at least two orders of magnitude higher than charge transport diffusion coefficients measured electrochemically for the same zeolites, and indeed the k_q values $(9.1 \times 10^6$ and 7.9×10^6 M^{-1} s^{-1}, respectively) are only about 50 times smaller than the values of $(3-5) \times 10^8$ measured in aqueous solutions (13). Surprisingly, diffusion in the one-dimensional zeolite L channels is as fast or faster than in the three-dimensionally interconnected zeolite Y supercage network. The large difference between the photochemically and electrochemically determined diffusion coefficients may indicate that charge compensation by electrolyte cations is current-limiting in the latter case. In the electrochemical experiments, all the viologen ions within the diffusion layer are oxidized and reduced, and an equal number of electrolyte cations must enter the zeolite in order to preserve local electroneutrality. While cation motion is also needed to compensate charge in the photochemical experiment, only a small fraction of the ions are oxidized or reduced in a single laser flash.   Experiments by Shaw et al. indeed show that the electrochemical response at MV^{2+}/zeolite Y electrodes is greatly reduced by the use of (size-excluded) tetraheptylammonium bromide as the electrolyte, and that it is enhanced in the presence of mobile electrolyte cations such as Li^+(14).

At high values of M*, electron transfer quenching of $Ru(bpy)_3^{2+}$ is very efficient for both zeolites Y and L. However, the quantum yield for formation of charge separated products (measured by comparing the $Ru(bpy)_3^{2+}$ MLCT and $MV^+·$ transients at 360 and 400 nm, respectively) is in both cases only 6-7%, compared to a maximum value of 28% in aqueous solution (15). The charge separated ions, which are formed within a few tens of nanoseconds following photoexcitation, undergo back electron transfer on a timescale of tens of microseconds.   The decay kinetics of these pairs are non-exponential, and a significant fraction (30-40%) of the reduced viologen signal persists even after 100 μs. The kinetics of recombination depend not only on diffusion and electron self-exchange between viologens, but also, on these long timescales, on the presence of trace amounts of oxygen within the zeolite.

Covalently linked donor-acceptor molecules. One can gain greater control of the kinetics of electron transfer reactions in these systems by fixing the distance between donor and acceptor, and by introducing a third redox molecule, which serves as either a secondary donor or acceptor, into the chain. For example, we have shown that a $Ru(bpy)_3^{2+}$-diquat^{2+} donor-acceptor (**I**) exchanged onto the outer surface of zeolite L can rapidly transfer an electron to benzylviologen (**II**) exchanged into the bulk (10b).

What is quite interesting about this system is not only the self-assembly of a three-component redox chain, but also the fact that the back electron transfer rate within **I** itself is three orders of magnitude slower on the surface of the zeolite than it is in fluid solution (16). We proposed that rapid back transfer occurs in solution because of conformational flexibility, which can bring the donor and acceptor ends of the molecule into close contact. When this flexibility is removed, either by inserting a rigid spacer molecule between the donor and acceptor, or by immobilizing the molecule on the zeolite surface, back electron transfer is slow.

This point is illustrated clearly in two similar donor-acceptor systems. Figure 4 shows transient absorbance spectra taken 0.2 - 3.2 μsec after nanosecond flash photolysis of a solution containing **III**, a donor-acceptor molecule which is similar to **I**, except that the two ends are joined by an *inflexible* bridging group. The spectrum consists of a negative transient at

450 nm, corresponding to the bleaching of the $Ru(bpy)_3^{2+}$ chromophore, and positive transients at 360 and 480-580 which are assigned to the reduced

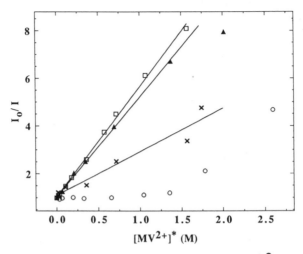

Figure 3. Stern-Volmer plots for quenching of Ru(bpy)$_3^{2+}$ luminescence by MV^{2+} in aqueous suspensions of zeolites L (□), Y(▲), and mordenite (o), and in air-dried zeolite Y powder (X).

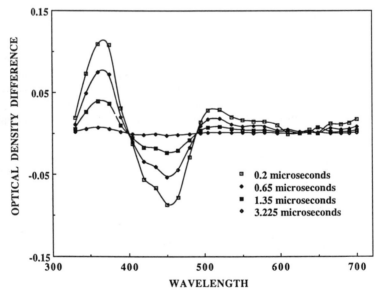

Figure 4. Transient UV-visible spectra of III in acetonitrile solution following 10 ns, 532 nm laser excitation.

viologen. The spectrum of the latter is shifted relative to dialkylviologen radical cations (which have absorbance maxima at 400 and 600 nm) because of electronic coupling to the unsaturated spacer. Upon photoexcitation, the MLCT excited state of the $Ru(bpy)_3^{2+}$ moiety is formed, and electron transfer quenching by the viologen group occurs in less than 5 ns. Back transfer, however, takes almost 1000 times as long; the lifetime of the first order decay of transients shown in Figure 4 is 1.4 µsec. Since the forward transfer actually proceeds via two short steps (metal to ligand, and then ligand to acceptor), and the back transfer must go in a single long step (acceptor to metal), the latter can be significantly slower than the former. Multi-step electron transfer mechanisms are found in nearly all molecular systems (for example, in natural and artificial photosynthetic systems (17)) in which long lived charge separated states are induced photochemically.

In stark contrast to the long-lived charge separated state found for **III**, molecules **IVa-c** show extremely fast back electron transfer rates in fluid solution. In all three cases, no transients are observable on a nanosecond timescale. Picosecond flash photolysis of **IVc** in acetonitrile shows only bleaching of the $Ru(bpy)_3^{2+}$ absorption at 450 nm, which decays with a 220 psec lifetime. There are no reduced viologen transients in these spectra, indicating that in this case the back electron transfer rate is in fact <u>faster</u> than the forward rate. Like **I**, these molecules can twist about the flexible alkyl spacers separating their donor and acceptor ends, and explore conformations which are favorable for back electron transfer on a picosecond timescale. CPK models show that in all three cases the viologen rings can be brought near the metal easily by twisting about C-C bonds.

IVa,b,c   R= H, x = 2,3,4

IVd   R= CH_2CH_3 , x = 4

The kinetics of electron transfer in compond **IVc** change dramatically when it is ion-exchanged onto the external surface of zeolites Y and L. Figure 5 shows transient diffuse reflectance spectra taken 12-130 µsec after nanosecond visible excitation of **IVc** on zeolite Y (similar spectra are obtained with zeolite L). Again, the observed transient corresponds to a charge-separated state which is formed in 10-15% quantum yield within a few nanoseconds of the laser flash. What is especially striking in this case is the extremely long lifetime of this transient. The decay is biphasic, with one

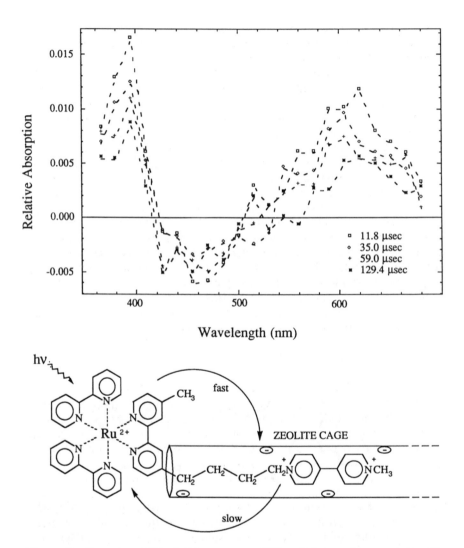

Figure 5. Transient UV-visible spectra of IVc exchanged onto zeolite Y
(2.2×10^{-6} mol/g) suspended in water. 10 ns excitation at 532 nm.

component (corresponding to about 60% of the total) decaying with a lifetime of 0.54 milliseconds. This decay rate is highly sensitive to small traces of molecular oxygen in the zeolite, and the intrinsic decay may in fact be even slower.

This dramatic change in electron transfer rates is consistent with loss of conformational flexibility on the zeolite surface. As indicated in Figure 5, the zeolite may enforce upon this molecule an extended conformation of the spacer chain, maximizing the distance between donor and acceptor. In order to probe this effect directly, compound **IVd** was synthesized with ^{13}C labels on both the viologen and ethyl terminal CH_3 groups, and was studied by CP-MAS NMR (18). Figure 6a shows the spectrum of this compound exchanged onto the surface of zeolite L. The peak centered at 13 ppm is assigned to the ^{13}C-labeled ethyl groups on the 2,2'-bipyridine rings, and the peak further downfield (at 45 ppm) to the viologen methyl group. Paramagnetic probes give an indication of the location of these groups, relative to the zeolite outer surface. In the solid state, signals from ^{13}C atoms near paramagnetic centers will be (inhomogeneously) broadened via dipolar interactions and hence should not be observable. Impregnating the zeolite powder with $Cr(acac)_3$, a size-excluded paramagnetic molecule, causes complete loss of signal from the ethyl carbons, but leaves 20-30% of the viologen signal. The small Gd^{3+} probe, which can be situated at both internal and external cation exchange sites, causes complete loss of the viologen signal, although some of the ethyl signal (presumably from terminal CH_3 groups on the side of the molecule which is remote from the viologen) is still observable. This experiment shows that an appreciable fraction of the **IVd** molecules are situated approximately as shown in Figure 5, with the viologen end in and the $Ru(bpy)_3^{2+}$ end out of the zeolite channels.

While forcing an extended conformation on **IVc** will increase its charge-separated state lifetime, the extreme increase observed is difficult to understand. Normally, one expects electron transfer rates fall off exponentially with increasing distance; most measurements made with saturated spacers indicate a factor of 3-4 in rate per CH_2 group. However, the back transfer rate in **IVc** on zeolites Y and L is approximately 1000 times slower than in **I**, which has a two-carbon spacer, on the same zeolites. Unfortunately, direct comparisons with **IVa** and **IVb** have not been possible, since both molecules appear to prefer a folded configuration on the zeolite surface. It is possible that lateral charge transfer diffusion, between $Ru(bpy)_3^{2+}$ or MV^{2+} (or both) units at the zeolite/solution interface, is sufficiently fast to compete with *intra*molecular back electron transfer in **IVc**, so that the unexpectedly long lifetime we observe actually corresponds to *inter*molecular charge separation. This possibility is currently being investigated.

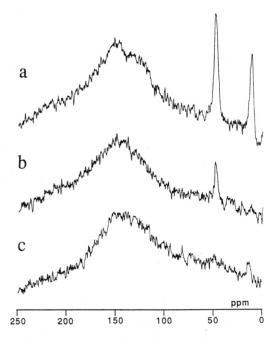

Figure 6. Solid state CP-MAS ^{13}C spectra of ^{13}C-labeled **IVd** (a) 6.0 x 10^{-6} mol/g on zeolite L, (b) sample $\underline{a}$ after impregnation with 1.1 x 10^{-4} mol/g Cr(acac)$_3$, and (c) sample $\underline{a}$ after exchange with 1.0 x 10^{-4} mol/g Gd^{3+}.

Conclusions. We have shown that ion exchange and size exclusion effects can be used to create molecular rectifiers at zeolite/aqueous solution interfaces. Photoactive bilayer assemblies containing $Ru(bpy)_3^{2+}$ derivatives transfer electrons upon photoexcitation, with reasonably good efficiency and unidirectionally into the zeolite. The zeolite pore network also controls the conformation of appropriately shaped donor-acceptor molecules, and in favorable cases leads to greatly enhanced lifetimes of charge separation within them.

Acknowledgments. This work was supported by the Division of Chemical Sciences, Office of Basic Energy Sciences, Department of Energy, under contract DE-FG05-87ER13789. We thank Dr. Stefan Hubig of the Center for Fast Kinetics Research, University of Texas at Austin, for carrying out the picosecond transient absorbance experiments, and Dr. Anthony Harriman for helpful suggestions. CFKR is supported jointly by the Biomedical Research Technology Program of the Division of Research Resources of NIH (RR00886) and by the University of Texas at Austin. T.E.M. thanks the Camille and Henry Dreyfus Foundation for support in the form of a Teacher-Scholar Award.

Literature Cited

(1) (a) Pickup, P.G.; Kutner, W.; Leidner, C.R.; Murray, R.W. *J. Am Chem. Soc.* **1984**, *106*, 1991; (b) Leidner, C.R.; Murray, R.W. *J. Am. Chem. Soc.* **1985**, *107*, 551.

(2) (a) Kittlesen, G.P.; White, H.S.; Wrighton, M.S. *J. Am. Chem. Soc.* **1985**, *107*, 7373;(b) Wrighton, M.S. *Comments Inorg. Chem.* **1985**, *4*, 269.

(3) (a) Li, Z.; Mallouk, T. E. J. Phys. Chem. 1987, 91, 643; (b) Li, Z.; Wang, C. M.; Persaud, L.; Mallouk, T. E. *J. Phys. Chem.* **1988**, *92*, 2592; (c) Li, Z.; Lai, C.; Mallouk, T. E. *Inorg. Chem.* **1989**, *28*, 178; (d) Rong, D.; Kim. Y. I.; Mallouk, T. E. *Inorg. Chem.* **1990**, *29*, 1531.

(4) Shaw, B. R.; Creasy, K. E.; Lanczycki, C. L.; Sargeant, J. A. *J. Electrochem. Soc.* **1988**, *135*, 869.

(5) Baker, M. D.; Zhang, J. *J. Phys. Chem.* **1990**, *94*, 8703.

(6) Rolison, D. R.; Hayes, E. A.; Rudzinski, W. E. *J. Phys. Chem.* **1989**, *93*, 5524.

(7) Oyama, N.; Yamaguchi, S.; Kaneko, M.; Yamada, A. *J. Electroanal. Chem.* **1982**, *139*, 215.

(8) Surridge, N; Hupp, J. T.; Danielson, E.; McClanahan, S.; Gould, S.; Meyer, T. J. *J. Phys. Chem.* **1989**, *93*, 304.

(9) Faulkner, L. R.; Suib, S. L.; Renschler, C. L.; Green, J. M.; Bross, P. R. in *Chemistry in Energy Production* (Wymer, R. G.; Keller, O. L., eds.), John Wiley and Sons, New York, 1982, pp. 99-114.

(10) (a) Persaud, L.; Bard, A. J.; Campion, A.; Fox, M. A.; Mallouk, T. E.; Webber, S. E.; White, J. M. *J. Am. Chem. Soc.* **1987**, *109*, 7309; (b)

Krueger, J. S.; Mayer, J. E.; Mallouk, T. E. *J. Am. Chem. Soc.* **1988**, *110*, 8232; (c) Krueger, J. S.; Lai, C.; Li, Z.; Mayer, J. E.; Mallouk, T. E., in *Inclusion Phenonomena and Molecular Recognition*; Atwood, J.L., Ed.; Plenum Press: New York, 1990; pp. 365-378.

(11) Krueger, J.S.; Mallouk, T.E., in *Kinetics and Catalysis in Microheterogeneous Media*; Grätzel, M.; Kalyanasundaram, K., Eds.; Marcel Dekker: New York, 1991; pp. 461-490.

(12) Breck, D. W. *Zeolite Molecular Sieves; Structure, Chemistry, and Use*; John Wiley and Sons: New York, 1974.

(13) (a) Grätzel, M.; Kiwi, J.; Kalyanasundaram, K. *Helv. Chim. Acta* **1978**, *61*, 2720; (b) Darwent, J.R.; Kalyanasundaram, K. *J. Chem. Soc. Faraday Trans. 2* **1981**, *77*, 373.

(14) Gemborys, H. A.; Shaw, B. R. *J. Electroanal. Chem.* **1986**, *208*, 95.

(15) Hoffman, M. Z., *J. Phys. Chem.* **1988**, *92*, 3458.

(16) Cooley, L. F.; Headford, C. E. L.; Elliott, C. M.; Kelley, D. F. *J. Am. Chem. Soc.* **1988**, *110*, 6673.

(17) For reviews see (a) Gust, D.; Moore, T. A. *Science* **1989**, *244*, 35; (b) Meyer, T. J. *Acc. Chem. Res.*, **1989**, *22*, 163;(c) Miller, J. R. *New. J. Chem.* **1987**, *11*, 83.

(18) Riley, R.L.; Yonemoto, E.H.; Hubig, S. M.; Richardson, B.R.; Haw, J.F.; Mallouk, T.E., in preparation.

RECEIVED January 16, 1992

Chapter 24

Synthesis of NaX Zeolites with Metallophthalocyanines

Kenneth J. Balkus, Jr., C. Douglas Hargis, and Stanislaw Kowalak[1]

Department of Chemistry, University of Texas at Dallas,
Richardson, TX 75083-0688

Aluminosilicate molecular sieves with the FAU structure have been crystallized in the presence of several metallophthalocyanines. A percentage of the complexes becomes included into the zeolites. The synthesis of NaX around the metal chelate represents a new method for encapsulating such complexes and modifying zeolite molecular sieves. The entrapped complexes were characterized by XRD, IR and UV-VIS spectroscopy. Preliminary results suggest the metal complexes may function as templates by modifying the gel chemistry.

The application of zeolite encapsulated metal chelate complexes in catalysis is a promising area of research. In particular shape selective oxidations catalyzed by metallophthalocyanines (MPc), shown in Figure 1, included in synthetic faujasite (FAU) type zeolites (2-10) appear to be competitive with other molecular sieve based catalysts that may have commercial potential. The restricted apertures (~7.4 Å) to the supercages (12Å) in FAU type zeolites precludes removal of the large MPc complex unless the zeolite lattice is destroyed. Such physically trapped complexes have been termed ship-in-a-bottle complexes as well as zeozymes (to reflect the biomimetic reactivity that is often associated with these catalysts).

The various strategies for preparation of zeolite encapsulated phthalocyanine complexes have largely involved the condensation of dicyanobenzene (DCB) around an intrazeolite metal ion to form the MPc complex. The efficiency of this template synthesis depends on the nature and location of the intrazeolite metal ion to be complexed. For example, metals have been introduced to the zeolite by ion exchange (7-13), metal carbonyls (14-19) and metallocene complexes (2-5, 19-21) prior to reaction with DCB. Some of the advantages and disadvantages of these methods have been detailed by Jacobs (2). However, there are several problems that are inherent to the template synthesis in general. Often there is incomplete

[1]Current address: A. Mickiewicz University, Faculty of Chemistry, ul Grunwaldzka 6, 60–780 Poznań, Poland

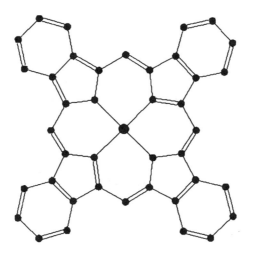

Figure 1. Metallophthalocyanine (MPc)

complexation of metal which can complicate reactivity studies or formation of free base ligand may be observed which may lead to blockage of diffusion pathways. The formation of the phthalocyanine ring requires two reducing equivalents. If this is derived from trace water then protons are formed that can be detrimental to the zeolite lattice. Additionally, the high temperatures and pressures involved in the template synthesis of MPc may be unsuitable for certain metal oxidation states. Furthermore, dicyanobenzene diffusing into the zeolite should first react with metal ions located at the outer portion of the crystal. Therefore, pores quickly become blocked and a heterogeneous distribution of MPc will result. This may explain typical MPc loadings on the order of ~10% occupation of the unit cells. Despite these limitations encouraging catalytical results have been reported.

We have recently reported an alternative strategy for the preparation of zeolite ship-in-bottle complexes which involves synthesis of the zeolite around a metal complex (22,23). This allows one to prepare well defined intrazeolite metal complexes without uncomplexed metal ions or free ligand complicating characterization or reactivity. This method also provides a mechanism for encapsulating metal complexes inside zeolites with restricted apertures that would not otherwise adsorb the metal complex or ligand precursors. The metal phthalocyanines are attractive for this application because of their size as well as chemical and thermal stability. In this paper we report further results for the synthesis of NaX zeolites in the presence of MPc complexes (M = Fe(II), Co(II), Ni(II)). XRD, IR, and electronic spectra provide evidence for the encapsulation of these complexes in NaX. Additionally, preliminary results suggest the MPc complexes may function as templates by modifying the gel chemistry during synthesis.

Experimental

Aluminum isopropoxide, silica and sodium hydroxide were obtained from Aldrich and used without further purification. Metallophthalocyanines Fe(II), Co(II), Ni(II)) were purchased from Strem Chemical. X-ray diffraction patterns were collected on a Scintag XDS 2000 using fluorite as an internal standard. IR spectra were obtained from KBr pellets using a MIDAC FT-IR. Electronic spectra were recorded on a Hitachi U2000. Solution ^{29}Si NMR spectra were obtained on a JOEL 200 using teflon lined 10mm NMR tubes and referenced to TMS.The silicate solution were scanned 500times (15 sec delay) and the aluminosilicate 10,000 times (15 sec delay). Elemental analyses were conducted by Galbraith Laboratories, Knoxville, TN.

The NaX molecular sieve containing metallophthalocyanines was prepared as follows. Freshly prepared silicate and aluminate solutions were combined with the MPc by adding the complex to the aluminosilicate gel, the aluminate or preferably the silicate solution. The highest levels of inclusion are obtained when the metal complex is mixed with the dry silica followed by NaOH and water. A typical mixture contains 2.0 g silica, 1.6 g NaOH, 0.04 g MPc and 4.0 mL H_2O. Addition of the aluminate solution (4.5 g Al(*i*OPr)$_3$, 1.6 g NaOH, 6.0 mL H_2O) results in a sticky gel of uniform color. An additional 18.0 mL of water are added and the gel transferred to a polypropylene bottle. The mixture is aged at room temperature with magnetic stirring for 24 hours, then heated at 90C for 6-10 hours. The resulting solid is washed with copious amounts of water, then dried at 90C for 18 hours. The molecular sieves containing metallophthalocyanines were placed in a vacuum sublimator at <1 torr and heated at 450C for 24 hours. Samples may be extracted with pyridine or DMF prior to sublimation to remove the bulk of the surface species. However, vacuum sublimation is more effective at removing all surface complexes. The sublimation of these samples was accompanied by a change in color from blue to bluegreen.

Results and Discussion

The zeolites containing metallophthalocyanines were synthesized by adding the metal complex to either the aluminosilicate gel or to the silicate solution. Heterogeneous dispersions of the complexes were obtained when the MPc was first added to aluminate solutions or the aluminosilicate gel regardless of stirring time. The resulting zeolites from this procedure have relatively low loadings of metal complex (Table I) such that a few percent of the supercages are occupied which is no more efficient than the template synthesis approach. Generally, attempts to increase the MPc concentration in the crystallization mixture produced different phases. However, if the MPc is first mixed with silica then a homogeneous distribution of the complex in the resulting gel is obtained. The MPc complexes are insoluble in water and hydroxide solutions but the dark blue silicate solutions appear to have dissolved the phthalocyanines. The zeolites formed from this procedure must be aged otherwise a mixture of phases is obtained. The highly crystalline NaX zeolites prepared after aging contain considerably more metal complex with up to ~50% of the unit cells being occupied. This is at least twice as much encapsulated MPc than has been observed for any template procedure.

Table I. Results for NaX Crystallization with MPc

$SiO_2:Na_2O:H_2O:MPc^a$	Hrs	Zeolite[b]	%M	Complex
3.2:4:155	4	X	--	H_2O Only
3.1:3.7:141:0.007	10	X	0.034	FePc
3.1:3.7:141:0.023	10	X, P	--	FePc
3.0:3.6:141:0.007	6,8,24	X, U[c]	--	FePc[d,e]
3.0:3.6:138:0.007	10	X	<0.014	CoPc
3.0:3.6:141:0.024	6	X	0.150	CoPc[d,e]
3.0:3.6:141:0.024	6	X	0.170	CoPc[d,e]
3.0:3.7:177:0.022	10	X, P	--	CoPc
3.1:3.7:141:0.005	10	X	0.022	NiPc
3.0:3.6:141:0.006	8	X	0.084	NiPc[d,e]
3.0:3.6:141:0.024	6	X	0.140	NiPc[d,e]
3.0:3.6:141:0.024	6	X	0.120	NiPc[d,e]
3.1:3.7:141:0.048	6	X, U[c]	--	NiPc[d,e]
3.0:3.6:141:0.006	10	X, U[c]	--	NiPc[d]
3.1:3.8:141:0.024	10	X, A	--	NiPc
3.1:3.7:177:0.022	10	X, P, A	--	NiPc
3.0:3.7:191:0.048	10	X, P, U[c], A	--	NiPc

a. per Al_2O_3 b. determined by XRD and FT-IR c. unknown phase d. MPc added to silica first e. aged 24 hours

The zeolites were characterized by XRD, IR and elemental analysis. The MPc complexes are best characterized by UV-Vis spectroscopy. For example Figure 2 shows the electronic spectra for FePc inside the zeolite and in H_2SO_4 after digestion of the zeolite. The characteristic $\pi^*\leftarrow\pi$ transition (Q band) associated with the phthalocyanine ring appears as a broad band (~630 nm) in Figure 2A. This spectrum is quite similar to other published spectra of intrazeolite FePc (2, 12, 18). Figure 2B indicates the FePc is still intact (Q ~790 nm) after partial digestion of this zeolite in concentrated H_2SO_4. This spectrum is typical for MPc complexes dissolved in H_2SO_4 which shows a bathochromic shift relative to the solid-state spectrum [24]. This shift results from protonation of the extracyclic nitrogens and axial coordination of HSO_4^-. Therefore, unlike the template synthesis [5] there is no evidence for inclusion of the free base Pc ligand which has a significantly different electronic spectrum Additionally, a recent Mössbauer study of intrazeolite FePc prepared by the template method seems to indicate the presence of two different types of FePc. In the case of CoPc/NaX prepared by the synthesis method we have shown there is only one electrochemically distinct CoPc complex encapsulated (25.) A more extensive comparison of the intrazeolite complexes prepared by the two techniques will be informative.

A further examination of the results in Table I indicate there is clearly a difference in reactivity between the different MPc complexes, especially if they are added to the silicate solution first. FePc is the least well behaved in terms of preparing a pure NaX. The most common impurity is zeolite P (GIS) which may arise from allowing the crystallization to proceed too long or from a decrease in

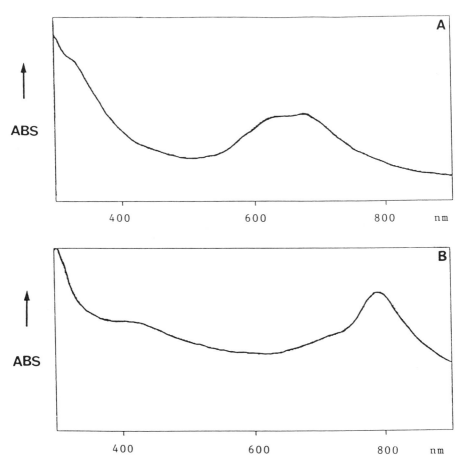

Figure 2. A) FePc/NaX (nujol) B) FePc/NaX dissolved in H_2SO_4.

NaOH concentration. If the metal complex is added to the gel then pure P can be prepared by simply crystallizing for several days. The MPc complexes are easily removed from this small pore zeolite. The occurrence of zeolite P increases as the MPc concentration increases such that at certain levels ($MPc/SiO_2 > 0.007$, MPc added to gel) the P phase appears at the onset of crystallization. When the metal complex is added to the silica first, the major impurity is an unknown phase. This phase will appear if the gel is not aged or with increasing MPc concentration. The dependence on MPc concentration in this case is much more sensitive to the type of metal. For example, when FePc is first added to the silica before crystallization, this unknown phase is invariably formed. The XRD pattern for this phase has the following d spacings: 7.13(100), 3.18(70), 4.11(64), 2.69(43), and 5.03(43). This aluminosilicate also appears to be stable to at least 550 C. In some cases the presence of zeolite A was detected which may result from a lower Si/Al ratio or a lower NaOH concentration. The addition of water further increases the occurrence of different phases even though the concentration of reagents is still outside the range where these impurities might normally form (*26*). It is possible that the metal complexes interact with hydroxide but the order of mixing would have little effect in this case. It is more likely that a MPc-silicate interaction exists that affects the free hydroxide concentration.

A template is not required for the preparation of X type zeolites, however, additives can affect the crystallization. We previously reported that NaX zeolites can be prepared in the presence of metal complexes (22,23) but phase purity and crystallization time were dependent on the concentration of MPc. If the properties of a structure directing agent include gel modification then the MPc complexes may be operating in that capacity. This is consistent with the variation in crystallization with changes in the type and concentration of MPc complex as well as a dependence on the order of mixing and aging. Additionally, there must be some driving force for inclusion of the complex since we observe loadings that are higher than for the template syntheses. The question is what is the nature of the MPc/gel interaction and ultimately can we exploit this to affect a new zeolite synthesis? The behavior of the gel after the complexes have been added to the silica seems to indicate some interaction between the MPc and the silicate. The ^{29}Si NMR spectra for the silicate solution with and without FePc are shown in Figure 3. A comparison of the spectra indicates a lower amount of silicate monomer (Q^0) when the metal complex is present. No additional resonances, that could be ascribed to an FePc-silicate adduct, were observed. The decrease in Q^0 signal intensity is consistent with a decrease in hydroxide concentration but a redistribution of the partially condensed species should also be observed. The ^{29}Si NMR spectrum of the gel formed after the addition of aluminate to the FePc/silicate solution is shown in Figure 4. The broad Q^4 resonance at -111 ppm is typical for the aluminosilicate gel, however, in this sample there is also silicate monomer (Q^0) present which is not observed in the

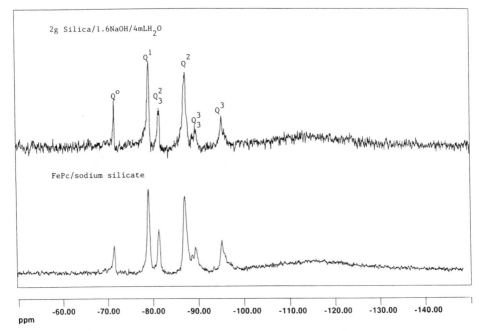

Figure 3. ^{29}Si NMR of sodium silicate solution with and without FePc.

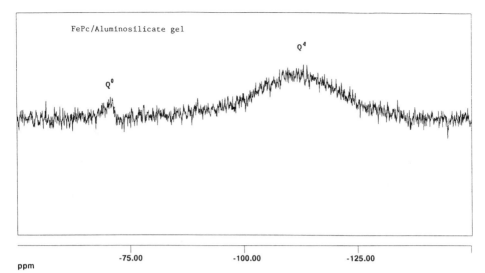

Figure 4. ^{29}Si NMR of FePc/aluminosilicate gel.

absence of iron phthalocyanine. This may be consistent with a weak interaction between FePc and the silicate monomers. The condensation to form the gel depends on the nucleophilicity of the monomer which would be reduced if associated with the metal complex. The result may be a local kinetic phenomenon such that the Si/Al ratio and the pH would increase slower than in the absence of metal complex. This would explain the need to age the gel in order to avoid the formation of phases with lower Si/Al ratios or that easily crystallize at lower hydroxide concentrations. Additionally, weak coordination of solution species may account for inclusion of the metal complexes. Further study of the gel chemistry and MPc-silicate interactions are currently in progress.

Conclusions

We have shown that NaX can be synthesized in the presence of metallophthalocyanines resulting in encapsulation of the complexes. Further work is required to better understand the gel chemistry, however, preliminary results suggest the metal complexes may interact with silicate species.

Acknowledgments

We thank the National Science Foundation (Grant No. CHE-9016705), the Robert A. Welch Foundation and the donors of the Petroleum Research Fund administered by the American Chemical Society.

Literature Cited

1. On leave from A. Mickiewicz University, Poland.
2. Parton,R.; De Vos,D.; Jacobs,P.A. *Proc. NATO Workshop, Portugal,* **1991,** In Press.

3. Huybrechts,D.R.C.; Parton,R.F.; Jacobs,P.A. *Stud. Surf. Sci. Catal.* **1991**, *60*, 225.
4. Parton,R.F.; Huybrechts,D.R.C.; Buskens,Ph.; Jacobs,P.A. *Stud. Surf. Sci. Catal.* **1991**, *65*, 110.
5. Parton,R.F.; Uytterhoven,L.; Jacobs,P.A. *Stud. Surf. Sci. Catal.* **1991**, *59*, 395.
6. Romanovsky,B.V. *Acta Phys. Chem.* **1985**, *31*, 215.
7. Diegruber,H.; Plath,P.J.; Schulz-Ekloff,G. *J. Mol. Catal.* **1984**, *24*, 115.
8. Tolman,C.H.; Herron,N. *Catal. Today* **1988**, *3*, 235.
9. Herron,N.; Stucky,G.D.; Tolman,A.T. *J.C.S., Chem. Comm.* **1986**, 1521.
10. Herron,N. *J. Coord. Chem.* **1988**, *19*, 25.
11. Meyer,G.; Wohrle,D.; Mohl,M.; Schulz-Ekloff,G. *Zeolites* **1984**, *4*, 30.
12. Kimura,T.;Fukuoka,A.; Ichikawa,M. *Shokubai* **1988**, *30*, 444.
13. Balkus,Jr.,K.J.; Welch,A.A,; Gnade,B.E. *J. Inclus. Phenom.* **1991**, *10*, 141.
14. Zakharov,A.N.; Romanovsky,B.V. *J. Inclus. Phenom.* **1985**, *3*, 389.
15. Zakharov,A.N.; Gabrielov,A.G.; Romanovsky,B.V.; Sokolov,V.I. *Vestn. Mosk. Univ., Ser. 2: Khim* **1989**, *30*, 234.
16. Tanaka,M.; Sakai,T.; Tominaga,T.; Fukuoka,A.; Kimura,T.; Ichikawa,M. *J. Radioanal. Nucl. Lett.* **1989**, *137*, 287.
17. Tanaka,M.; Minai,Y.; Watanabe,T.; Tominaga,T.; J. *Radioanal. Nucl. Lett.* **1991**, *154*, 197.
18. Kimura,T.; Fukuoka,A.; Ichikawa,M. *Catal. Lett.* **1990**, *4*, 279.
19. Romanovsky,B.V.; Gabrielov,A.G. *Mendeleev Commun.* **1991**, 14.
20. Gabrielov,A.G.; Zakharov,A.N.; Romanovsky,B.V.; Tkachenko,O.P.; Shpiro,E.S.; Minachev,Kh.M. *Koord. Khim.* **1989**, *14*, 821.
21. Zakharov,A.N.; Romanovsky,B.V.; Luca,D.; Sokolov,V.I. *Metalloorg. Khim.* **1989**, *1*, 119.
22. Kowalak,S.; Ly,K.T.; Hargis,C.D.; Balkus,Jr.K.J. In *Synthesis and Properties of New Catalysts: Utilization of Novel Materials Components and Synthetic Techniques*, Corcoran,Jr.,E.W.; Ledoux,M.J. Ed.; Materials Research Society, Pittsburgh, PA, 1990, 24, 37-.
23. Balkus,Jr.,K.J.; Kowalak,S.; Ly,K.T.; Hargis,C.D. *Stud. Surf. Sci. Catal.* **1991**, *69*, 93.
24. Berezin,B.D. *Coordination Compounds of Porphyrins and Phthalocynaines* J. Wiley, NY 1981.
25. Bedioui,F.; DeBoysson,E.; Devynck,J.; Balkus,Jr.,K.J *J. Electroanal. Chem., Interfac. Electrochem.* **1991**, *315*, 313.
26. Lechert,H.; Kacirek,H. *Zeolites* **1991**, *11*, 720.

RECEIVED February 18, 1992

Chapter 25

Synthesis via Superlattice Reactants

Low-Temperature Access to Metastable Amorphous Intermediates and Crystalline Products

Thomas Novet, Loreli Fister, Christopher A. Grant, and David C. Johnson

Department of Chemistry, University of Oregon, Eugene, OR 97403

A new synthetic approach in which thin (15-50Å), amorphous, elemental layers are sequentially deposited to create a uniquely tailorable initial "reactant" is described. This synthetic approach has two key steps. The initially layered composite is diffused at low temperatures to produce a homogeneous amorphous alloy. Nucleation of this amorphous alloy is the rate limiting step in the formation of a crystalline compound. This approach overcomes the limitations of traditional, diffusion limited, solid-state synthetic methods, which offer no control of the reaction pathway and therefore no selectivity over which intermediates are formed.

This synthetic approach has been used to prepare binary phases directly and selectively from homogeneous amorphous intermediates without the formation of intermediate compounds. The layered nature of the starting reactant permits the diffusion reaction to be followed in a quantitative manner using X-ray diffraction. The earliest stages of a reaction between two solids, before the nucleation of any crystalline phase, can be investigated. This ability to follow the course of a solid-state reaction permits the tailoring of the initial composite structure to obtain the desired intermediate state. This is illustrated with examples including iron silicides, iron aluminides, and molybdenum selenides. The importance of the ability to control reaction intermediates in the synthesis of ternary and higher order compounds is then addressed.

The ability to tailor the structure of solid-state materials rationally depends upon control of reaction intermediates. Low temperature synthetic methods for preparing extended solids, such as coprecipitation, sol-gel, and hydrothermal techniques rely on the atoms being intimately mixed in a fluid phase before they are condensed into the solid state (1-6). Indeed, the most used method of controlling the supramolecular architecture of a system involves the self-assembly of molecular precursors in solution onto a solid substrate.

Diffusional control has been used to limit and direct reaction pathways and control intermediates. In the last ten years, solid state chemists have developed a very clever "soft chemistry" approach to controlling the intermediates in solid-state syntheses (2,7,8). This approach takes advantage of large differences in diffusion rates often found in low dimensional solid-state compounds. For example, the oxide $K_2Ti_4O_9$ consists of two dimensional sheets of Ti octahedra sharing edges and faces. These sheets are separated by K^+ cations which are much more mobile than the Ti^{4+} cations. Washing $K_2Ti_4O_9$ in acid results in the formation of the cation exchanged materials $H_2Ti_4O_9$. This phase evolves H_2O upon heating, causing the sheets to fuse together via the elimination of a specific oxygen atom, the most acidic one, in the layered structure (9). The resulting solid via this "chemie douce" is a new metastable form of TiO_2. Its synthesis illustrates the importance of controlling the synthetic intermediates in preparing metastable solids and is an elegant example of using diffusional constraints to accomplish this goal.

The "chemie douce" approach, although extremely elegant, is limited by the availability of suitable precursor solids. Our research goal has been to develop a general, low temperature, controlled approach to preparing solids.

An amorphous alloy is an ideal general precursor or reaction intermediate. The amorphous state is metastable with respect to several different crystalline states but the crystalline state which is easiest to nucleate is the one which will form. Thus, the compound which crystallizes is not necessarily the most stable state. Controlling crystallization of amorphous alloys is a general route to both stable and metastable materials (10).

Several techniques have been developed to produce metastable amorphous alloys (11). Routes to amorphous alloys include the rapid cooling of molten alloys, referred to as splat cooling or melt spinning, the codeposition of the respective elements, and low temperature solid state amorphization reactions. All of these techniques are based upon limiting the opportunities for the system to nucleate. The important energies in this situation are that required nucleation, and that required for diffusion. Time is also important, as local rearrangements to form nuclei are limited by the diffusion rates. Each of these techniques has drawbacks for the general preparation of amorphous alloys.

We have developed the use of ultrathin-film multilayer composites, illustrated in Figure 1, as the starting point for the formation of amorphous intermediate states. Ultrathin-film multilayer composites offer several unique advantages over other methods of preparing amorphous solids. These include the ability to monitor the progress of the reaction by X-ray diffraction and scanning calorimetry, the ability to separate the processes of diffusion and phase nucleation, and the ability to obtain thermodynamic data as the reaction progresses.

If the modulation of the layers is coherent, then the multilayers act as artificial crystals in the direction perpendicular to the surface due to the regular repeat of electron density. This permits X-ray diffraction to be used to characterize the layering in the multilayer and determine the thickness of the interfacial regions between the layers. By monitoring the decay of the Bragg peaks in this diffraction pattern as a function of time and temperature, the interdiffusion reaction of the layered composite can be followed in a quantitative manner. Interdiffusion coefficients as low as 10^{-25} cm^2/sec can be measured. More importantly, the diffraction data give direct information as to how the interfaces evolve (*12*). This information allows us to tailor the structure of the initial composite to control the interdiffusion reaction and obtain the desired amorphous intermediate (*13*).

In addition to being a general synthetic method, our approach yields valuable insight to the reaction mechanism of solid-state reactions. Surprisingly little is known about the earliest stages of solid-state reactions, i.e. before the formation of a crystalline product at the interface. The interfaces of a bulk diffusion couple only make up a minute fraction of the sample and such buried interfaces are difficult to observe. Classic studies of bulk diffusion couples have shown that the growth of crystalline reaction products rapidly becomes limited by the diffusion of the reactants through the product layer. Every stable phase will form at the interface as shown in Figure 2, with the amount of each phase determined by the relative interdiffusion rates. Traditional solid-state techniques are likewise limited by the high activation energies for solid-state diffusion and the large diffusion pathlengths found in the initial reactants (*14*).

A working model for the course of solid-solid interface reactions before the crystallization of a product has resulted from studies of thin film diffusion couples (*15*). It has generally been assumed that an amorphous "reaction layer" is formed at the interface as the solid reactants begin to interdiffuse. As this amorphous interface becomes thicker, the composition gradients in this layer are reduced, and nucleation of an intermediate crystalline product occurs. As this crystalline product layer grows, the two resulting interfaces repeat this process to form additional crystalline products. This is illustrated in Figure 3.

Differential scanning calorimetry is an ideal companion measurement to the diffraction experiments discussed above. It is a rapid method of determining when interdiffusion begins (as a result of heat evolution not associated with crystallization) and when crystallization occurs. Because DSC measures heat flow directly, the heat of mixing of the elements and the crystallization enthalpy for the transition from the amorphous to crystalline state can be determined by integrating the areas under the peaks. However, difficulty in doing baseline corrections leads to large uncertainties in the integrated areas. The combination of differential scanning calorimetry and X-ray diffraction yields very fundamental information concerning the interfacial reaction of the studied composite (*16*).

Following a brief discussion of the experimental methods used, we present data on the evolution of several multilayer binary films in various phase diagram to illustrate the power of this approach in preparing stable and metastable materials. We begin this discussion by presenting data on a series of molybdenum-selenium composites which illustrate the importance of length scales in the initial modulated composite. This is followed by description of the preparation of the binary compounds FeSi$_2$ and Al$_5$Fe$_2$ from initially layered composites. The application of this approach to the rational preparation of ternary materials is then discussed.

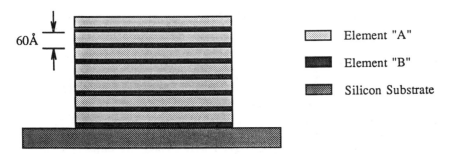

Figure 1. Idealized ultrathin-film multilayer composite, section view.

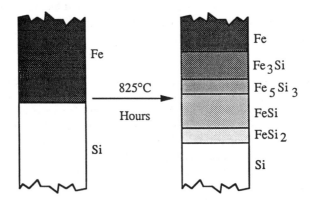

Figure 2. Schematic presentation of an iron–silicon bulk diffusion couple heated at 825 °C for several hours.

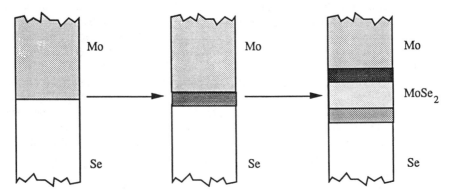

Figure 3. Schematic presentation showing concentration gradient formation at solid-solid interfaces during early stages of the evolution of a Mo/Se diffusion couple.

Experimental Section

Synthesis of Modulated Composites. The amorphous modulated multilayer composites were prepared in a custom-built ultra-high vacuum deposition system which is described in detail elsewhere (*17*). Iron, aluminum, molybdenum, and silicon were deposited using electron beam guns controlled via quartz crystal thickness monitors. The selenium was deposited using an effusion cell. The rate of selenium deposition was monitored via a quartz crystal thickness monitor and the deposition rate was controlled by controlling the temperature of the pressure of the cell. Deposition rates of 0.5 Å/sec were used for all of the elements and the background pressure of the chamber during deposition was approximately 5×10^{-8} torr. A set of samples was synthesized corresponding to the composition of the desired compound in each of the phase diagrams investigated (Fe-Si, Fe-Al, and Mo-Se). The thickness of the various layers was chosen to achieve the desired stoichiometry and to obtain a repeat unit thickness of approximately 60 Å.

Grazing and High Angle X-ray Diffraction. Both grazing and high angle X-ray diffraction data presented in this paper were collected on a Scintag XDS 2000 theta-theta diffractometer. The commercial sample stage was replaced with a sample mount containing optical flats against which the sample is held with spring tension. The vertical position of the sample stage can be varied using a vertical position stage with a 0.0001 inch micrometer movement. This movement is necessary to align the sample stage for the grazing angle studies. The alignment of the sample stage was confirmed before each diffraction session by reproducing a diffraction pattern from a reference sample. The high-temperature diffraction data were collected using a high-temperature diffraction attachment which is described in detail elsewhere (*12*).

Differential Scanning Calorimetry. Heat produced by the interdiffusion and crystallization of the multilayer reactants was quantified using differential scanning calorimetry. Approximately 1 mg of sample free of the substrate was used. These samples were obtained by first coating a three inch silicon wafer with polymethylmethacrylate (PMMA) using a 3% solution in chlorobenzene deposited by spin coating at 1000 rpm. The desired multilayer structure was then deposited upon the PMMA coated substrate. After the sample was removed from the chamber, it was immersed in acetone, which dissolved the PMMA, causing the multilayer film to float off of the substrate. Typically, the films broke up and rolled into many small pieces, which were collected via sedimentation into an aluminum DSC pan. The sample was dried under reduced pressure to remove residual acetone. Finally the pan was crimped closed.

The sample was then placed in a DuPont 910 DSC module, which was housed in an inert atmosphere chamber to prevent sample oxidation. The sample was heated at 10°C/min from room temperature to 600°C. After the sample had cooled to room temperature, it was reheated to obtain a baseline for the irreversible changes which occur during the initial heating. A second such background was collected to obtain a measure of the repeatability of the experiment. The net heat absorbed or released from the multilayer sample as it diffuses was obtained from the difference between the first DSC experiment and the subsequent runs. The two background runs were found to be within 0.05 mW/mg of one another.

Results

Our synthetic approach is based upon controlling the reaction pathway and making use of an amorphous alloy as the key synthetic intermediate. The desired reaction pathway contains two crucial steps: the diffusion of the initially layered reactant to a homogeneous amorphous alloy, and the nucleation and growth of the desired crystalline product from the amorphous alloy. The following discussion describes how these two steps can be rationally controlled and presents supporting data.

The first critical synthetic step is the formation of a homogeneous amorphous alloy. The modulation length of the initial multilayer ultrathin-film composites is used to attain this amorphous intermediate state in a controlled fashion. The concept on which this ability is based is that of competing timescales: a timescale for heterogeneous nucleation of a crystalline phase and a timescale for diffusion (18). If the sample diffuses quickly enough, the sample will become homogeneous before heterogeneous nucleation can occur. The diffusion time, based upon Fick's Law for diffusion, will be proportional to the square of the diffusion distance. To first approximation, however, the timescale for heterogeneous nucleation of a crystalline compound at an interface in a multilayer composite should be independent of the size of a repeat unit. At some small modulation length, the timescale for diffusion will be less than the timescale for nucleation (15).

To test this hypothesis, we decided to investigate the molybdenum-selenium system. This system was chosen because it contains a very stable crystalline solid, $MoSe_2$, which crystallizes with a very simple, two dimensional unit cell. We prepared a series of molybdenum-selenium composites of identical composition but varying layer spacings. Differential scanning calorimetry and complimentary X-ray diffraction experiments were then conducted upon these samples to determine if the mechanism of the reaction between molybdenum and selenium changes with the layering length scale.

Figure 4 illustrates the heat evolved from a layered composite with an initial layering of 80 Å and Figure 5 contains the X-ray diffraction patterns collected at the indicated times on Figure 4. These data are representative of a "bulk" reaction. The sample begins to interdiffuse causing the initial exotherm. While the sample is still interdiffusing, $MoSe_2$ nucleates and grows, causing the second exotherm. This "bulk" reaction mechanism exists in the molybdenum-selenium system until a layering lengthscale of 30 Å is reached.

Differential scanning calorimetry data representative of composites layered on a lengthscale less than 30 Å are shown in Figure 6 and the complimentary X-ray diffraction data are contained in Figure 7. The evolution of composites of this lengthscale is clearly broken into two reaction steps: the interdiffusion of the layers to form an amorphous intermediate and the subsequent crystallization of this amorphous alloy into the more thermodynamically stable crystalline compound.

The above experiment clearly demonstrates the ability of this synthetic approach to control the course of a solid-state reaction via lengthscales alone. The data demonstrate that nucleation of a crystalline compound from a modulated

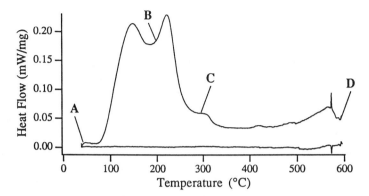

Figure 4. Heat flow rate as a function of temperature obtained for a layered Mo/Se composite with a repeat unit of 74Å. The upper curve was obtained by heating the sample at 10 °C/min and subtracting a subsequent run on the same sample obtained under identical conditions. The base line is the difference between the heat-flow rates of the second and third heating of the same sample. The points A through D show the temperatures at which the x-ray diffraction patterns shown in Figure 5 were obtained.

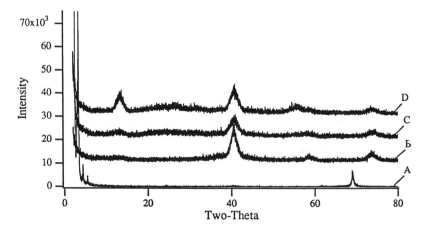

Figure 5. X-ray intensity as a function of 2θ for a layered Mo/Se composite with a repeat unit of 74Å showing sample evolution with annealing. Curves A through D correspond to points A through D in Figure 4. Peaks in curves B through D are [00l] reflections from highly oriented MoSe₂. Curves have been offset for clarity.

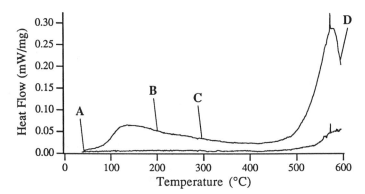

Figure 6. Heat flow rate as a function of temperature obtained for a layered Mo/Se composite with a repeat unit of 26Å. The upper curve was obtained by heating the sample at 10 °C/min and subtracting a subsequent run on the same sample obtained under identical conditions. The base line is the difference between the heat-flow rates of the second and third heating of the same sample. The points A through D show the temperatures at which the x-ray diffraction patterns shown in Figure 7 were obtained.

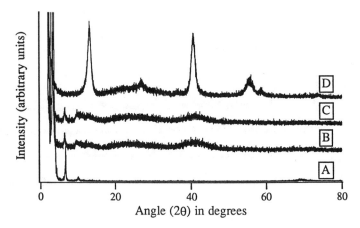

Figure 7. X-ray intensity as a function of 2θ for a layered Mo/Se composite with a repeat unit of 26Å showing sample evolution with annealing. Curves A through D correspond to points A through D in Figure 6. Peaks in curve D are [001] reflections from highly oriented $MoSe_2$. Curves have been offset for clarity.

composite (nucleation at 200°C in Figure 4) is much easier than nucleation of the same crystalline compound from a uniform amorphous alloy (nucleation at 375°C in Figure 6). This suggests a change in the nucleation mechanism in these two situations. Crystallization from a modulated composite is due to heterogeneous nucleation because the interfaces in such a composite provide composition gradients, stresses, and strains, all of which can lower the activation energy for nucleation. Crystallization from an amorphous alloy, which lacks these interfaces, is a much more "homogeneous" nucleation process. Thus, the mechanism of the solid-state reaction can be affected by the lengthscale of the modulation.

The control of the second critical step, crystallization of the amorphous alloy, is the focus of the following discussion. Iron-silicon and iron-aluminum systems are discussed. The critical lengthscales in the iron-aluminum and iron-silicon systems are much larger than that observed in the molybdenum-selenium systems. By critical lengthscale, we refer to the thickness of the repeat unit in the multilayer below which the multilayer evolves completely into an amorphous material without the nucleation of any crystalline phase. The samples discussed are all layered on a lengthscale which is less than this critical value. That is, they evolve from a layered initial state, through a distinct amorphous intermediate, to a crystalline compound.

Stoichiometry has long been used by solid-state chemists to control the final products of a reaction. However, traditional synthetic techniques do not have the ability to control reaction intermediates and all stable phases will form as illustrated in Figure 2. For example, in the iron-silicon system, thin film diffusion couples have been used to determine the sequence of phase formation (*19*). FeSi was always found to nucleate first, followed by the crystallization of $FeSi_2$ at the FeSi-Si interface and $FeSi_3$ at the FeSi-Fe interface. The following paragraphs provide evidence that stoichiometry of the amorphous intermediates can be used to control nucleation to obtain the desired crystalline compounds directly. Thus, we use stoichiometry to control the *mechanism* of the reaction.

Figure 8 contains the differential scanning calorimetry data for a layered iron-silicon sample with a stoichiometry of one iron to two silicons. This sample starts out as an initially layered composite with amorphous layering. It begins to interdiffuse at 80°C. Consistent with our first control objective, the interdiffusion reaction does not involve the crystallization of a compound. Through our ability to monitor the interdiffusion process, high temperature diffraction data obtained in the iron-silicon system suggest that Fick's Laws of diffusion are followed (*12*). The diffusion exotherm is clearly distinguishable up to approximately 350°C.

The complimentary X-ray diffraction patterns for this DSC trace are contained in Figure 9. Examination of the high angle diffraction data after heating to 350°C reveals that the sample is still amorphous, but the sample is no longer modulated. At 450°C, a very large exotherm occurs, corresponding to the direct nucleation of $FeSi_2$ from the amorphous state, consistent with our second control objective. The other crystalline iron silicides can also be directly synthesized by selecting their stoichiometry (*20*).

Figure 10 contains the differential scanning calorimetry data for a layered iron-aluminum sample with a stoichiometry of 5 aluminum to 2 iron. The DSC data contain two distinct exotherms. Diffraction data shown in Figure 11 indicate that the sample is initially modulated with poor coherence between layers. After the first

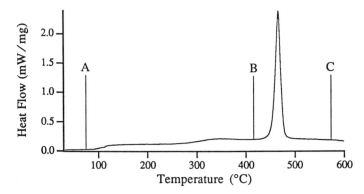

Figure 8. Heat flow rate as a function of temperature obtained for a layered composite of relative composition 1-iron:2-silicon. This curve was obtained by heating the sample at 10 °C/min and subtracting a subsequent run on the same sample obtained under identical conditions. The points A, B, and C show the temperatures at which the x-ray diffraction patterns shown in Figure 9 were obtained.

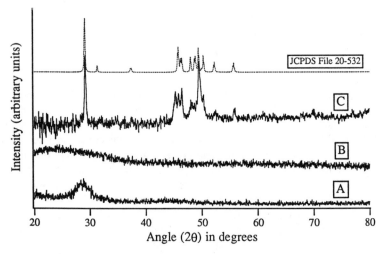

Figure 9. X-ray intensity as a function of 2θ for a layered Fe/Si composite of stoichiometry 1/2 iron/silicon showing sample evolution with annealing. Curves A through C correspond to points A through C in Figure 8. The upper curve is data synthesized from the JCPDS diffraction files for a sample of crystalline $FeSi_2$. The curves are offset for clarity.

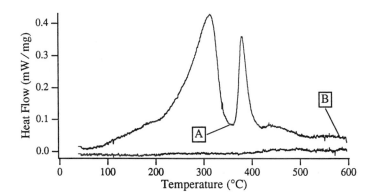

Figure 10. Heat flow rate as a function of temperature obtained for a layered composite of relative composition 2/5 iron/aluminum. The upper curve was obtained by heating the sample at 10 °C/min and subtracting a subsequent run on the same sample obtained under identical conditions. The base line is the difference between the heat-flow rates of the second and third heating of the sample. The points A and B show the temperatures at which the x-ray diffraction patterns shown in Figure 12 were obtained.

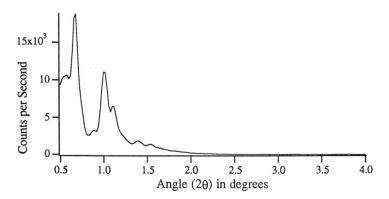

Figure 11. X-ray intensity as a function of 2θ at low angle for a layered Fe/Al composite of stoichiometry 2/5 iron/aluminum. These data show that the sample, as deposited, is layered, although with poor coherence beween layers.

exotherm, the sample appears to be amorphous; no crystalline phases can be detected. The second exotherm corresponds to the direct crystallization of the desired product. Figure 12 summarizes the evolution of the diffraction pattern as the sample is annealed.   These data along with the iron-silicon data presented above are representative of the ability to control the first phase which nucleates from an amorphous alloy by controlling stoichiometry.

The technique described here relies on control of the two crucial steps in the reaction pathway: the diffusion of the initially layered reactant to a homogeneous amorphous alloy, and the nucleation and growth of the desired crystalline product from the amorphous alloy. There is nothing fundamental which limits this approach to binary systems. If suitable lengthscales are selected, systems containing any number of elements may be diffused to a homogeneous amorphous state.   We anticipate that stoichiometry will also be an effective means of controlling the nucleation of ternary compounds from these homogeneous alloys. This has important implications for the synthesis of ternary and higher order compounds.   Current synthetic approaches to ternary compounds cannot avoid the formation of binary compounds as intermediates in these syntheses. Therefore, the resulting ternary compounds must be more stable than these intermediates. Our approach offers a general route to the synthesis of metastable ternary compounds by directing the course of the solid state reaction through rationally selected amorphous intermediates.

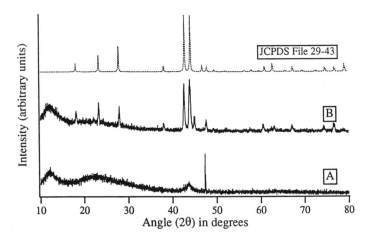

Figure 12. X-ray intensity as a function of 2θ for a layered Fe/Al composite of stoichiometry 2/5 iron/aluminum showing sample evolution with annealing. Curves A and B correspond to points A and B in Figure 10. The upper curve is data synthesized from the JCPDS diffraction files for a sample of crystalline Fe_2Al_5. The curves are offset for clarity.

Conclusions

The key to the success of the approach described herein is that the layered starting composite permits diffusional control of the reaction pathway. The diffusional control is a consequence of the ability to prepare modulated structures in which the lengthscales and layering sequence can be exquisitely controlled. The ability to direct the final crystalline compounds resulting from this synthetic approach depends upon the ability to control the nucleation of the desired phase from this amorphous intermediate. This can be accomplished by controlling stoichiometry of the amorphous intermediate.

Acknowledgments

We would like to acknowledge the assistant of J. McConnell in developing the experimental procedures used in this paper. This work was supported by a Young Investigator Award from the Office of Naval Research (#N00014-87-K-0543). Support by the National Science Foundation (DMR-8704652), the donors of the Petroleum Research Fund, and the University of Oregon are also gratefully acknowledged.

Literature Cited

(1)   Johnson, D. W., Jr. *Ceram. Bull.* **1981**, *60*, 221-224.

(2)   Horowitz, H. S. ; Longo, J. M. *Mat. Res. Bull.* **1978**, *13*, 1359-1369.

(3)   Gopalakrichnan, J. *Proc. Indian Acad. Sci., Chem. Sci.* **1984**, *93*, 421-432.

(4)   Rao, C. N. R.; Gopalakrishnan, J.; Vidyasgar, K.; Ganguli, A. K.; Ramanan, A.; Ganapathi, L. *J. Mater. Res.* **1986**, *1*, 280-294.

(5)   Matijevic, E. *Ann. Rev. Mater. Sci.* **1985**, *15*, 483-516.

(6)   Barrer, R. M. *Hydrothermal Synthesis of Zeolites;* Academic Press: New York, 1982.

(7)   Rouxel, J.; Meerschaut, A.; Gressier, P. *Synth. Met.* **1989**, *34*, 597-607.

(8)   Gerand, B.; Nowogrocki, G.; Guenot, J.; Figlarz, M. *J. Sol. State Chem.* **1979**, *29*, 429-434.

(9)   Marchand, R.; Brohan, L.; Tournoux, M. *Mat. Res. Bull.* **1980**, *15*, 1129-1133.

(10)   Coehoorn, R.; Van Der Kol, G. J.; Van Den Broek, J. J.; Minemura, T.; Miedema, A. R. *J. Less Common Met.* **1988**, *140*, 303-316.

(11)   Beck, H.; Güntherodt, H.-J. in *Introduction* in *Glassy Metals I: Ionic Structure, Electronic Transport, and Crystallization*; Beck, H.; Güntherodt, H.-J; Springer-Verlag, New York, 1981; pp. 1-17.

(12)   Novet, T.; McConnell, J. M.; Johnson, D. C. *Inorg. Chem.*, submitted.

(13)   Fister, L.; Johnson, D. C. *J. Am. Chem. Soc.*, **1992**, *114*, xxxx.

(14)   Brophy, J. H.; Rose, R. M.; Wulff, J. in *Thermodynamics of Structure;* Wulff, J.: John Wiley & Sons, New York, 1964; pp. 91-94.

(15)   Gösele, U.; Tu, K. N. *J. Appl. Phys.* **1989**, *66*, 2619-2626.

(16)   Johnson, D. C.; Fister, L.; Grant, C. A.; McConnell, J.; Novet, T. in *Using ultrathin multilayer composites as starting materials to produce homogeneous*

amorphous reaction intermediates: a technique for controlling and monitoring solid state reactions.; Mallouk, T. R.; JAI Press, Inc., Greenwich, CT, in preparation.

(17) Fister, L.; Li, X. M.; Novet, T.; McConnell, J.; Johnson, D.C., in preparation.

(18) Cotts, E. J.; Meng, W. J.; Johnson, W. L. *Phys. Rev. Lett.* **1986**, *57*, 2295-2298.

(19) Lau, S. S.; Feng, J. S. Y.; Olowolafe, J. O.; Nicolet, M. A. *Thin Solid Films* **1975**, *25*, 415-422.

(20) Novet, T.; Johnson, D. C. *J. Am. Chem. Soc.,* **1991**, *113*, 3398-3403.

RECEIVED January 16, 1992

Chapter 26

Solid-State Metathesis Routes to Layered Transition-Metal Dichalcogenides and Refractory Materials

John B. Wiley, Philippe R. Bonneau, Randolph E. Treece, Robert F. Jarvis, Jr., Edward G. Gillan, Lin Rao, and Richard B. Kaner

Department of Chemistry and Biochemistry and Solid State Science Center, University of California—Los Angeles, Los Angeles, CA 90024–1569

A large number of materials can be prepared via new metathetical (exchange) solid-state precursor reactions. This synthetic route is extremely rapid (often < 1 sec.), can be initiated at low temperature, and is potentially useful for controlling particle size and for preparing both cationic and anionic solid solutions. The often self-propagating and sometimes explosive behavior observed in these reactions is related to their exothermicity. Consequently, thermodynamic considerations can be employed to help select the best set of precursors as judged from reaction enthalpies. The control of particle size and product yield through manipulations of reaction temperature and the factors that influence reaction initiation in these systems are discussed.

Synthetic routes derived from molecular and non-molecular precursors have expedited the development of technologically important 2- and 3- dimensional materials. Such approaches have often proved superior to conventional ceramic techniques in that high purity bulk samples or thin films can be prepared at lower temperatures much more rapidly. Predominant among the precursor methods are those based on decomposition reactions. These either involve gaseous species, such as those used in chemical vapor deposition (CVD), or solids. Examples include the pyrolysis of the gas-phase precursor $[(CH_3)_2Al(NH_2)]_3$ to produce aluminum nitride (1) and the thermal decomposition of solid state carbonate precursors of calcium and manganese $(Ca_{1-x}Mn_xCO_3, 0 \leq x \leq 1)$ to produce several of the known ternary compounds in the Ca-Mn-O system (2). Single-displacement reactions are also common as precursor methods. These approaches usually involve gas-phase reactions and are also used in CVD techniques. Examples here include the formation of

0097–6156/92/0499–0369$06.00/0

elemental silicon (3) by hydrogen reduction of $SiCl_4$ and the preparation of gallium arsenide in hydrogen from $AsCl_3$ and gallium metal (4).

Metathetical reaction pathways are not as well studied as other precursor methods. This approach, though requiring the use of two precursors, can be performed in the liquid, gas or solid state. Chianelli and Dines (5) have carried out a series of investigations on the solution preparation of transition-metal dichalcogenides. Titanium tetrachloride, for example, was found to react with lithium sulfide in nonaqueous solvents to produce metal disulfide and lithium chloride (equation 1).

$$TiCl_4 + 2\,Li_2S \longrightarrow TiS_2 + 4\,LiCl \qquad (1)$$

The metal dichalcogenides formed are usually amorphous but can be crystallized on heating ($\geq 400°C$). Gas-phase exchange reactions are exemplified by the preparation of silicon nitride from silicon tetrachloride and ammonia (equation 2) (6).

$$3\,SiCl_4 + 4\,NH_3 \longrightarrow Si_3N_4 + 12\,HCl \qquad (2)$$

Very few examples of solid state metathesis reactions are known. In 1932 Hilpert and Wille (7) reported the preparation of mixed metal ferrites at moderate temperatures (400 - 500°C) by the reaction:

$$MCl_2 + Li_2Fe_2O_4 \longrightarrow MFe_2O_4 + 2\,LiCl \qquad (3).$$

The approach has not been fully explored as a synthetic technique (8).

The investigational void associated with metathesis reactions in the solid state suggested to us the need to explore this precursor method as a viable synthetic approach. We have found solid state metathesis reactions to be an extremely powerful route. Surprisingly, these reactions can often self-initiate at room temperature, internally produce enough heat to sustain themselves, and yield crystalline products in seconds. These reactions are also applicable to a very broad range of compounds. We have successfully synthesized transition-metal, main-group, and rare-earth chalcogenides (S, Se, and Te) and pnictides (N, P, As, and Sb), as well as, selected carbides, oxides, and silicides. This list of compounds includes 2- and 3- dimensional materials with important magnetic, electronic, catalytic, and refractory properties. Additional features of this synthetic technique, not typically seen in other precursor methods, are the facile control of crystallinity and the ability to prepare both cationic and anionic solid solutions.

Molybdenum Disulfide

The preparation of layered transition-metal dichalcogenides has been one of the areas of emphasis in this research (9). Molybdenum disulfide, an important lubricant, battery cathode and catalytic material, is readily prepared from the appropriate molybdenum halide and alkali-metal sulfide precursors. Though many different pairs of transition-metal halide alkali-metal sulfide precursors are available, the combination of $MoCl_5$ and Na_2S was found to be most effective (equation 4).

$$MoCl_5 + 5/2\ Na_2S \ ----> \ MoS_2 + 5\ NaCl + 1/2\ S \qquad (4)$$

Typically these metathesis reactions are initiated in one of three ways: 1) by self-initiation on light mixing or grinding, 2) local heating with a hot filament, or 3) low-temperature heating of the sample in a sealed evacuated glass tube. The particular method used depends both on the activation energy and the facility of the reaction. In the case of $MoCl_5$ and Na_2S, these precursors will self-initiate on light mixing at room temperature. The reaction, carried out in a helium-filled dry box, produces a bright white intense flash of light, a small mushroom cloud (due to the volatilization of the sulfur byproduct), and a dark product mixture. Pure highly crystalline molybdenum disulfide can be isolated simply by washing the product mixture with methanol to remove unreacted $MoCl_5$ and water to remove the halide salt byproduct and any unreacted Na_2S. (Excess sulfur can be removed with chloroform or carbon disulfide, though this is often not necessary because it essentially all boils away during the reaction.) The percent yield for the reaction is typically 80% of theoretical. The X-ray powder diffraction pattern of the MoS_2 (Figure 1) shows its high crystallinity. Thermogravimetric analysis (TGA) of this material heated in hydrogen (5% in nitrogen) to 1000°C (10°C/min) showed the sulfur content to be within 0.1% of the theoretical value.

WARNING: These metathesis reactions are often very exothermic and sometimes explosive. Some materials self-initiate and many can be activated with a drop of water. Reactions are typically carried out under inert atmosphere with the use of covered nonairtight sealed stainless steel bombs to contain fulminating mixtures. Before carrying out reactions in sealed glass tubes, gas law calculations should be done to determine the maximum pressure based on all the possible volatile reactants and products. Sample sizes should then be adjusted so as not to exceed the pressure limitations of the glass container (typically < 5atm.). Further caution should be exercised in the choice of the reaction container. When the reactions are quite exothermic, the heat produced can exceed the melting point of the glass and result in an implosion. In these instances, sealed quartz tubes (m.p. $\approx$ 1200°C) may be more appropriate than Pyrex (m.p. $\approx$ 550°C). However, metal or ceramic containers are preferred.

Many of the precursors and reaction byproducts, especially the pnictides, are toxic and potentially hazardous. On exposure to air or on washing with water, the product mixtures can evolve toxic gases from some of the unreacted starting materials e.g. H_2S and PH_3 from Na_2S and Na_3P, respectively. Workups should, at a minimum, be done in a fume hood.

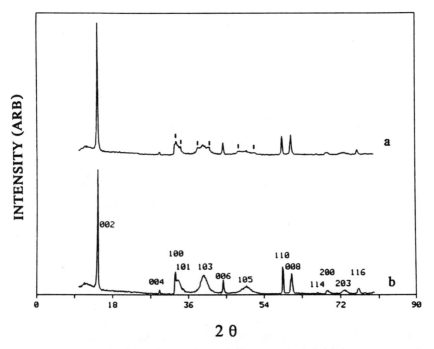

Figure 1. X-ray diffraction patterns of MoS_2 produced (a) in < 1 second from the self-initiated reaction of $MoCl_5$ and Na_2S and (b) from the elements heated to 900°C for 5 days. Miller indices for the major peaks of the 2H- polytype are given in (b) while only the noncoincident 3R-polytype lines are indicated in (a). (X-ray patterns were collected on a diffractometer equipped with Cu Kα radiation and a nickel filter.)

Additionally, precursors like Na_3P and byproducts like white phosphorus are pyrophoric and will often burst into flame when exposed to the air. Incomplete reactions can sometimes be a problem. If unreacted starting materials are present in the reaction mixture the addition of solvent can promote further reaction. This can be quite hazardous since the heat of the reaction is often enough to ignite flammable solvents such as methanol.

Thermodynamics and the Selection of Precursors

The success of these metathesis reactions can be primarily attributed to a conscious exploitation of thermodynamic factors. By designing reactions in which the formation of byproducts is extremely favorable, we are able to direct the formation of the desired product phase. Also, by maximizing the exothermicity of these processes, the large amount of heat released can often be significant enough to allow the reactions to be self-propagating. In this study the byproducts were chosen to be either alkali- or alkaline-earth halides. These salts are thermodynamically very favored and so their formation is a very effective driving force. The reactions, consequently, are often extremely exothermic. In fact, enough heat is produced to melt the halide salt byproduct. This is thought to contribute to the self-propagating and sometimes explosive nature of these reactions. The molten halide greatly enhances the mobility of the reactant species. More reactants, in turn, combine to produce more heat and more salt byproduct. If the diffusion and reaction rate of the precursors are very fast, a chain reaction can ensue and the overall process, short in duration (often < 1 sec.), can be quite vigorous.

The thermodynamics of these metathesis reactions are well illustrated by further consideration of the MoS_2 system. The large amount of heat produced in this reaction, emphasized by the accompanying pyrotechnic display (*10*), can be quantified from a Hess's law calculation where the heat of the reaction (ΔH_{rxn}) is -213 kcal/mole (*11*). (Bomb calorimetry experiments (*9*) resulted in a value of -174±5 kcal/mole, consistent with an 80% reaction yield.) This is comparable to heats of reaction produced in the combustion of low molecular weight fuels such as methane (ΔH_{rxn} = -210 kcal/mole) (*12*). Further consideration of heat capacities, heats of fusion (ΔH_{fus}) and heats of vaporization (ΔH_{vap}) of the products allows an estimate of the maximum tem-perature for this reaction. The rapidity of the reaction makes reasonable the assumption that the process is pseudo-adiabatic. The heat versus temperature plot shown in Figure 2 illustrates where heat in the reaction is consumed (*13*). Based on the heat consumption at every step, it is found that the maximum possible temperature is limited to 1686K, the temperature at which NaCl vaporizes. There is not enough heat produced in the reaction (-213 kcal/mole) to completely carry out this phase transition. Optical pyrometry (*9*) determined that the minimum temperature reached by this reaction is ca. 1400K, limiting the actual maximum temperature to a range of 1400-1700K.

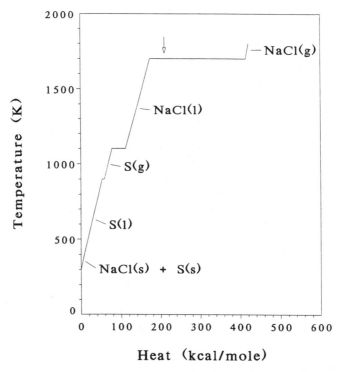

Figure 2. Plot of heat versus temperature as calculated for the products of the reaction of $MoCl_5$ and Na_2S. The value for the ΔH_{rxn} (-213 kcal/mole) is indicated by the vertical arrow.

In the above reaction, $MoCl_5$ and Na_2S were the chosen precursors. There are, however, many other possible molybdenum halide alkali-metal sulfide combinations. The selection of the appropriate set of precursors can greatly influence the success of these metathesis reactions. Consider the system with $MoCl_3$ as the halide precursor (equation 5). The effect of this

$$MoCl_3 + 3/2\ Na_2S \text{----} > 3/4\ MoS_2 + 3\ NaCl + 1/4\ Mo \qquad (5)$$

substitution can be seen in the X-ray diffraction pattern in Figure 3. The sample prepared from $MoCl_3$ is much less crystalline. The decreased crystallinity is directly related to the heat released in the reaction. The calculated ΔH values for the reaction between Na_2S and $MoCl_3$ or $MoCl_5$ are -121 and -213 kcal/mole, respectively. Consequently the temperature attained in the $MoCl_3$ reaction is considerably less, as is the crystallinity of the MoS_2 product. (An additional undesirable feature of the $MoCl_3$ precursor is the formation of molybdenum metal as a byproduct; this component cannot be easily removed without the destruction of MoS_2.)

Qualitatively, the differences in the reactions of the two precursors $MoCl_3$ and $MoCl_5$ can be attributed to several interdependent factors that all relate to the oxidation state of the metal. The greater the oxidation state, the more moles of halide salt produced in the metathesis reaction, and the more exothermic the reaction. It is therefore generally advantageous to choose the halide precursor (MX_m) where M has the highest possible oxidation state. In the case of late transition-metals, the higher oxidation states can be accessed through the preparation of alkali-transition-metal fluorides. Precursors like K_2NiF_6 have been found to be very useful for the synthesis of late-transition-metal dichalcogenides (e.g. NiS_2) (*14*).

In addition to the oxidation state of the metal in the halide precursor, the particular halogen used can also be important. A comparison of the heats of formation for different halide byproducts of a specific alkali-metal shows a drop of approximately one-half in absolute value going from the fluoride to the iodide (Table I). Similar trends occur for transition-metal, rare-earth, and main-group halides except that there is usually a significantly larger drop between the fluoride and the chloride than between any other adjacent halide pairs. As a consequence, use of the fluoride rather than the chloride precursor is sometimes accompanied by a significant decrease in the exothermicity of reaction. The difference in the ΔH_{rxn} for a particular reaction containing either ZrF_4 or $ZrCl_4$ is greater than 65 kcal/mole more exothermic in favor of the chloride when the salt byproduct is the sodium halide.

Variations in the heat of reaction as a function of alkali-metal appear to be minor compared to those which are a function of the halogen. Table I shows only a minor variation in ΔH_{rxn} going from the lithium to the sodium to the potassium halide. For the alkali-main-group precursors, it is difficult to ascertain any trend because there is not always an isotypic series and often the heats of formation for many of these materials are not known. Some exceptions include the oxides (A_2O) and the sulfides (A_2S). In the case of the oxides there is a increase in ΔH_f down the periodic table. For the sulfides, however, there is no significant variation and, as with the alkali-metal halides, the values are very close.

Table I. Heats of Formation (kcal/mole) at 298K for Selected Halides.

	F	Cl	Br	I
Li	-147.43	-97.58	-83.85	-64.55
Na	-137.52	-98.26	-86.38	-68.80
K	-135.90	-104.37	-94.12	-78.37
Mg	-268.70	-153.35	-119.70	-87.70

In spite of the fact that the ΔH_f's of the alkaline-earth halides are considerably more negative than those of alkali-metal halides (Table I), reactions are generally less exothermic with these precursors. In this case, only one mole of halide byproduct is formed relative to the two that would be formed with the alkali-metal precursor. A preference for an alkaline-earth precursor over an alkali-metal precursor may occur simply due to availability. One example is the readily available Mg_2Si precursor which is effective in the synthesis of FeSi (equation 6). (Iron metal can be removed

$$FeCl_3 + 3/4\ Mg_2Si ----> 3/4\ FeSi + 3/2\ MgCl_2 + 1/4\ Fe \qquad (6)$$

from the sample with an acidwash.) Other siliciding agents are known but are not as easily prepared.

So far, little has been said about the choice of the alkali-metal precursor with respect to the nonmetal main-group component. In many cases there is not a large selection of potential precursors and one must make use of the compounds that are available. When there is a choice, especially when the main-group byproduct is nonvolatile, it is usually better to choose the more metal-rich precursor. This is preferred both because it reduces the amount of main group byproduct and, as will be discussed in more detail below, the additional byproduct can act as a heat sink reducing the maximum temperature that the reaction mixture can reach.

The Initiation of Metathesis Reactions

Several of these metathesis reactions are known to self-initiate while others require some thermal input. It is expected that initiation is a function of the degree of initial surface reaction. If enough heat is generated to overcome the overall reaction barrier for the bulk sample, then reaction can occur. The apparent differences in activation between similar systems can be related to the structure and properties of the halide precursors. Compounds with molecular or low-dimensional structures and correspondingly low boiling points are more likely to self-initiate than comparable compounds with three-dimensional structures and high boiling points. This appears to occur with the precursors $ZrCl_4$, $NbCl_5$, $TaCl_5$, $MoCl_5$, WCl_6, and $ReCl_5$. These compounds all have molecular or low-dimensional structures and low boiling points and their reactions with alkali-metal precursors have been observed to self-initiate, sometimes with only light mixing. For precursors with three dimensional structures, initiation at elevated temperatures may be correlated to the decomposition of the halide precursor. Thermogravimetric analysis and X-ray powder diffraction have shown that K_2NiF_6 decomposes to K_2NiF_4 and F_2 gas at ca. 450°C in an inert atmosphere. Metathesis reactions with this precursor in evacuated sealed glass tubes are found to always begin at this same temperature. The increased reactivity of the precursor on decomposition is most likely due to the formation of more labile species, though it is conceivable that the decrease in the particle size of the precursor resulting on decomposition may also play a role. Effects due to particle size have been observed in other systems. When sintered, low-surface area Na_2S is used in the reaction of $MoCl_5$ and Na_2S (equation 4), the reaction will not self-initiate and requires exposure to a hot filament.

Effective Control of Temperature

As previously stated, the heat of these reactions is generally sufficient to melt the halide byproduct. ZrN, prepared by the reaction of $ZrCl_4$ and Li_3N (equation 7), was examined by scanning electron microscopy (SEM)

$$ZrCl_4 + 4/3\ Li_3N ----> ZrN + 4\ LiCl + 1/6\ N_2 \qquad (7)$$

before and after washing. Note the smooth continuous surfaces associated with the melt (Figure 4a). On washing the crystallites of ZrN are found to average less than 0.5μ in size (Figure 4b).

The melt itself is thought to aid the formation of product crystals in that it most likely acts as a medium where reactants can more easily diffuse and crystals can grow. This is certainly reasonable considering the number of materials that can be grown from halide fluxes (*15*). Because the halide requires a certain amount of heat to become molten and produce crystalline material, it is therefore possible to control the crystallinity of the product simply by adding a heat sink.

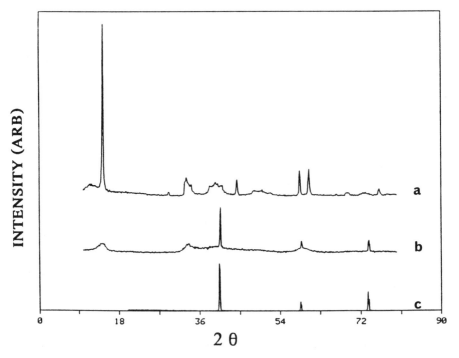

Figure 3. X-ray diffraction patterns of (a) MoS$_2$ product formed in the reaction of MoCl$_5$ and Na$_2$S, (b) product formed in the reaction of MoCl$_3$ and Na$_2$S at 300°C for four days, and (c) Mo metal.

(a)   (b)

Figure 4. Scanning electron micrographs of the ZrN reaction products (a) before and (b) after washing.

The heat sink may be introduced through one of the precursors. Consider the preparation of MoS_2 from $MoCl_5$ and the sulfiding agents Na_2S, Na_2S_2, and Na_2S_5. In each of these reactions there is excess sulfur byproduct. The amount of this simply increases as the sulfur content of the precursor increases (equations 4, 8 and 9).

$$MoCl_5 + 5/2 \ Na_2S_2 \ ----> \ MoS_2 + 5 \ NaCl + 3 \ S \qquad (8)$$

$$MoCl_5 + 5/2 \ Na_2S_5 \ ----> \ MoS_2 + 5 \ NaCl + 21/2 \ S \qquad (9)$$

The additional sulfur serves as a heat sink decreasing the maximum temperature attained in the system. This in turn decreases the crystallinity of the products. Figure 5 shows the X-ray powder diffraction patterns of products from each of these reactions. A significant decrease in crystallinity is observed as the amount of sulfur byproduct increases.

The heat sink need not necessarily be introduced as part of a precursor. The addition of any inert material could serve the same purpose. It is, however, most desirable to use a heat sink that can be easily removed after the reaction. We have found that the addition of a halide salt is very advantageous. Figure 6 shows the X-ray powder diffraction patterns of MoS_2 formed in the reaction of $MoCl_5$ and Na_2S in the presence of increasing amounts of excess NaCl. An obvious reduction in particle size is observable as the amount of salt increases. The reduction in particle size, as determined from the Scherrer and Warren equations (*16*), is presented in Table II.

Table II. Particle Size Control Through the Addition of NaCl.[a]

Reactants:NaCl[b]	Particle Size (Å)
1:0	450
1:0.25	310
1:0.50	180
1:1	120
1:2	80

[a] All values were obtained from the 002 reflection and were calculated relative to MoS_2 prepared from the elements at 900°C.
[b] Ratios are based on the mass of NaCl added to a 1.714g reactant mixture.

Heat sinks can be used to limit diffusion and lower crystallinity by lowering the reaction temperature. To increase diffusion, it helps to simply increase the amount of starting materials. The increase in sample size

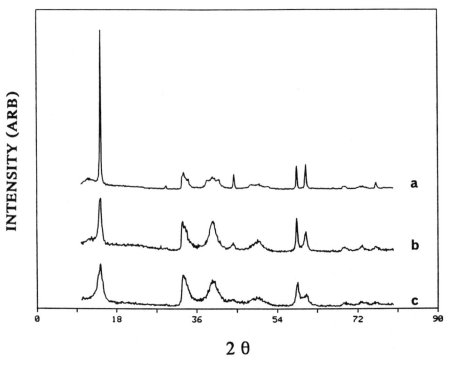

Figure 5. X-ray diffraction patterns of the MoS_2 product formed in the reaction of $MoCl_5$ with (a) Na_2S, (b) Na_2S_2, and (c) Na_2S_5.

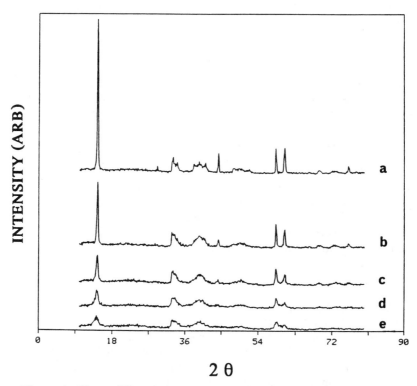

Figure 6. X-ray diffraction patterns showing the effect of varying amounts of a NaCl heat sink on the MoS_2 product formed from the reaction of $MoCl_5$ and Na_2S. NaCl was added to the 1.714g reactant mixture in the ratios (a) 0:1, (b) 0.25:1, (c) 0.5:1, (d) 1:1, and (e) 2:1.

essentially better insulates the bulk of the material from its surroundings, allowing more of the heat produced in the reaction to heat the sample. This in turn allows internal sample temperatures to be maintained for longer periods of time leading to increased overall diffusion of reactants and greater product yields. The yield of ZrN (equation 7) is increased significantly with an increase in sample size. When relatively small stoichiometric amounts of precursor are used (0.99g $ZrCl_4$ and 0.20g Li_3N), the percent theoretical yield is only $\approx 55\%$. If, however, the total amount of reagents is increased by about four times, the yield dramatically increases to $\approx 91\%$.

Solid Solutions

Metathetical precursor reactions are well suited to the synthesis of solid-solution compounds. This approach offers a diverse set of precursors and has the advantage over other methods in that both mixed cation and mixed anion reagents are readily obtained. Mixed metal chloride precursors of molybdenum and tungsten, for example, have been found to be efficacious in the preparation of solid-solution dichalcogenides such as $(Mo,W)S_2$. Similarly, anionic precursors like the mixed chalcogenide, $Na_2(S,Se)$, and the mixed pnictide, $Na_3(P,As)$, can be used to prepare transition-metal solid-solution dichalogenides $(Mo(S,Se)_2)$ and rare-earth solid-solution pnictides $(Sm(P,As))$. The composition of the product in these compounds does not always directly correlate with that of the precursor. The reasons for this result are currently being investigated.

Conclusions

Metathesis based precursor reactions offer a unique approach to solid state synthesis. They are rapid, can be initiated at low temperatures, and are easily manipulated for the control of particle size and the preparation of either cationic or anionic solid solutions. Their successful implementation can often be aided by a simple consideration of thermodynamic parameters. Our laboratory has employed these reactions to prepare a large number of compounds, but the full range of this precursor method is still under investigation.

Acknowledgments

This work was supported by the National Science Foundation (Grant No. CHE 86-57822) through the Presidential Young Investigator Award Program and by a David and Lucille Packard Foundation Fellowship in Science and Engineering.

Literature Cited

1. Interrante, L. V.; Carpenter, Jr., L.; Whitmarsh, C.; Lee, W.; Garbaukas, M.; Slack, G. A. *Mater. Res. Soc. Symp. Proc.* **1986**, *73*, 359.
2. Horowitz, H. S.; Longo, J. M. *Mat. Res. Bull.* **1978**, *13*, 1359.
3. Rao, C. N. R.; Gopalakrishnan, J. *New Directions in Solid State Chemistry*; Cambridge University Press: Cambridge, 1986.
4. Craford, M. G. *Prog. in Solid State Chem.* **1973**, *8*, 127.
5. Chianelli, R. R.; Dines, M. B. *Inorg. Chem.* **1978**, *17*, 2758.
6. Sato, K.; Terase, K.; Kijimuta, H. U. S. Patent No. 4,399,115, 1983.
7. Hilpert, S.; Wille, A. *Z. Phys. Chem.* **1932**, *18B*, 291.
8. Wold, A. *J. Chem. Ed.* **1980**, *57*, 531.
9. Bonneau, P. R.; Jarvis, Jr., R. F.; Kaner, R. B. *Nature* **1991**, *349*, 510.
10. See cover of journal from ref. 9.
11. *JANAF Thermochemical Tables*, 3rd ed.; Lide, Jr., D. R., Ed.; American Chemical Society and American Institute of Physics, Inc.: New York, 1985.
12. *CRC, Handbook of Chemistry and Physics* Weast, R. C., Ed.; CRC Press: Boca Raton, FL, 1983.
13. All thermochemical values and melting and boiling points were obtained from ref. 11 with the exception of the boiling point of NaCl obtained from ref 12. Melting points and boiling points were rounded off to the nearest 100K to simplify calculations. The melting point of MoS_2 was assumed to be $>>1700K$. Its melting point, reported to be 2023K under 1 atm $S_2(g)$, can be lowered as much as 350K at lower $S_2(g)$ pressures. Because in the reaction 0.5 moles of sulfur are produced for every mole of MoS_2, the assumption appears to be reasonable.
14. Bonneau, P. R.; Shibao, R. K.; Kaner, R. B. *Inorg. Chem.* **1990**, *29*, 2511.
15. Scheel, H. J.; Elwell, D. *Crystal Growth from High Temperature Solutions*; Academic Press: New York, 1975.
16. Cullity, B. O. *Elements of X-ray Diffraction*, 2nd Ed.; Addison-Wesley Pub. Co.: Massachusetts, 1978.

RECEIVED January 16, 1992

Chapter 27

Encapsulation of Organic Molecules and Enzymes in Sol–Gel Glasses

A Review of Novel Photoactive, Optical, Sensing, and Bioactive Materials

David Avnir[1], Sergei Braun[2], and Michael Ottolenghi[1]

[1]Institute of Chemistry and [2]Department of Biological Chemistry, Hebrew University of Jerusalem, Jerusalem 91904, Israel

The ability to trap organic and bioorganic molecules in inorganic oxides through the sol-gel process, first introduced in 1984, opened the road for a whole new class of materials and to intensive activity in many laboratories. The types of materials and their applications are reviewed. These include photocatalysts for redox reactions; photochromic materials and other information recording materials; filters and light-guides; fluorescent, phosphorescent and dye-laser materials; a variety of chemical sensors; and bioactive (enzymatic) glasses.

1. Background

1.1 The Sol-Gel Process.

One of the major revolutions in modern materials science has been the advent of a novel synthetic route for the preparation of glasses, known as the "sol-gel processes". Reference *1* is an excellent, comprehensive text on this topic, and no attempt can be made in this introductory section to even scratch the surface of this huge field. Consequently, following are only some of the very basic concepts needed for understanding this review, and the interested reader is referred to *(1)* for further details.

The basic idea of the sol-gel process is to split the glass preparation into two main stages: The first stage is a room temperature polymerization of a suitable monomer leading to a porous glass, usually an inorganic glass. The second stage is the closure of the pores at elevated temperatures (several hundreds degrees), forming the final glass. Two of the main advantages of this technique are: First, the overall energy-saving due to the lower temperatures utilized; and second, the ability to achieve unconventional glass compositions by a suitable choice of a mixture of monomers. Such compositions may be unobtainable through the classical methods due to phase-separation upon cooling.

0097–6156/92/0499–0384$06.25/0

Most of the glasses described in this review are porous silica glasses, (and their modifications) obtained by the general reaction (unbalanced):

$$Si(OR)_4 + H_2O \xrightarrow[H^+ \text{ or } OH^-]{-ROH} (SiO_2)_n$$

in which R = CH_3 or CH_3CH_2. We concentrate here on the porous materials obtained from the first stage of the sol-gel process. Other glasses reported in this review are alumina and titania, which are obtained by similar reactions. Reversed-phase porous glasses are also mentioned below. These are obtained by polymerization of monomers of the type $R^*Si(OR)_3$ in which R^* is a non-polymerizable substituent on the silicon, such as a long-chain alkyl. Finally, composite glasses in which organic polymeric molecules are incorporated in the inorganic glass, will also be mentioned.

Typical properties of the porous glasses are: a) high surface area (several hundreds of m^2/gr); b) low densities (around 1 gr/ml); c) narrow pores (around ten nm), and consequently; d) complete transparency to visible light and to the near u.v.; e) the ability to obtain any desired shape, including thin films; f) good chemical, photochemical and thermal stability. The brittleness of these glasses has been a major problem, but this is gradually being solved in various laboratories by modification of the synthetic procedures and by the use of proper additives *(1)*.

1.2 The Trapping of Organic Molecules in Inorganic Glasses: A Novel Class of Materials.

From the point of view of organic and bioorganic chemistry the most important application of the sol-gel process has been the ability, for the first time in the four-thousand years history of glass technology, to dope (oxide) glasses with organic molecules. Until recently doping was limited to salts and oxide-additives, which can withstand the high temperatures involved in the melting procedures. Since the vast majority of the $7 \cdot 10^6$ compounds known today are organic or bioorganic, the new horizons that have become open now, are limited only by imagination. The basic idea is simple: Adding a solution of the desired dopant to the starting polymerizing system, and stopping the glass preparation at the dry xerogel stage. Surprising as this may be, that simple idea is quite recent: It was first described in 1983 in a patent application *(2)* and a year later in *J. Phys. Chem. (3)*. It is, perhaps, less surprising that this glass-doping idea has since then expanded quite rapidly. Most of the applications developed at first concentrated on the preparation of materials for optical and photophysical uses, a trend which is still going on strongly. However, a new class of materials has been recently developed based on the doping idea, namely, chemically active sol-gel glasses. Catalysts, sensors and the exciting group of bioactive materials, are some examples to be discussed below.

Some of the properties and advantages of the doped sol-gel glasses which are of use for the applications reviewed here are the following:

1. The glasses are stable thermally, chemically and photochemically,

especially when compared to plastic matrices, which were the only materials used so far for encapsulation of organic molecules.

2. The encapsulated molecules are well protected again, much better than in plastics. For instance, their photodegradation rate is smaller, compared to solution (3).

3. The glasses are transparent well into the u.v., allowing a whole array of photochemical, photophysical and optical applications.

4. The trapped molecules are either non-leachable, or, in some cases, leachable over a very long time period.

5. A sub-population of the non-leachable molecules is trapped close enough to the pore-surface, to be able to interact with substrate molecules contained in the pore space. It is this portion of trapped-yet-exposed molecules which allows the development of chemically active sol-gel glasses.

We begin our review with the class of reactive glasses. We then proceed to bioactive glasses obtained by trapping of enzymes (Section 3). Section 4 describes the variety of applications that were developed in photophysics and optics. Spectroscopic properties of trapped molecules were used for the study of the properties of the glass cage and for follow-up of the structural changes that occur along the polymerization stages, from the sol to the xerogel. These studies are then outlined in Section 5.

2. Chemically Reactive Organically Doped Sol-Gel Glasses

2.1 Photoreactive Sol-Gel Glasses.

The ability to achieve photochemical interactions with organic substrates or catalysts trapped in sol-gel glasses was demonstrated by several research groups. First, we briefly review these reports and then proceed to a description, in some detail, of sol-gel systems exhibiting photoinduced charge separation.

Studies aiming at establishing the mechanism leading to and controlling the size of metal oxide and metal particles in sol-gel glasses were performed by Modes and Lianos (4). These authors examined the quenching of the luminescence of $Ru(bpy)_3^{2+}$ by methylviologen in SiO_2 gels prepared by adding both reagents to the initial polymerizing solution of tetramethoxysilane under a variety of (e.g., pH) conditions. It was shown that in the gel the lumophore and quencher are aggregated as molecular clusters and that glasses, in which $Ru(bpy)_3^{2+}$ is replaced by cadmium or lead salts, form CdS and PbS particles upon exposure to H_2S.

Quenching experiments were carried out with $Ru(bpy)_3^{2+}$ and with a pyrene quaternary ammonium salt doped within silica gels prepared from $Na_2SiO_3 \cdot H_2O$ by acid ion-exchange (5a). Quenching was effective for quenchers co-trapped with the fluorophore. However, exposure of the gels to atmospheric oxygen did not result in any change in excited state lifetimes indicating that the latter are inaccessible to atmospheric O_2. A different behaviour was observed in the case of pyrene-doped (crystalline or amorphous) TiO_2 powders prepared from titanium tetrapropoxide (5b). Pyrene was found to occupy sites which were readily accessible to water and ionic quenchers. However, photoproducts which are normally associated with excited pyrene were not observed in TiO_2 particles, a fact which was attributed to the close proximity of the products, which are formed under conditions of limited diffusion, thus enhancing back reaction.

Additional evidence showing that a substantial fraction of the trapped molecules do have access to the intra-pore volume was based on the observation that a trapped organic iridium complex $[Ir(bpy)_2(C_3,N')bpy]^{3+}$ (abbreviated as Ir(III)) is effectively quenched by dimethoxybenzene (DMB) dissolved in the liquid pore network of the gel matrix. Similarly, Pouxviel et al. reported *(6)* that pyranine (a hydroxyl pyrene derivative, ROH) molecules in a sol-gel matrix are sensitive to the composition of the intra-pore solution: The presence of water increases the polarity of the pyranine environment, enhancing dissociation of the excited state as monitored by the emission of the resulting RO^{-*} moiety.

The observation that the luminescence of a substantial fraction of a glass-entrapped Ir(III) complex undergoes quenching by DMB, not only opened the door to the use of doped sol-gel glasses as chemical sensors, but also suggested applications in photocatalysis *(7)*. The process was studied recently by Slama-Schwok et al. *(8)* for several donor-acceptor pairs: pyrene* $(Py^*)/MV^{2+}$, $Ir^*(III)/DMB$, $Ir(III)^*/S_2O_8^{2-}$, $Ru(II)^*/Fe^{3+}$, and $Ru(II)^*/S_2O_8^{2-}$, where the asterisk denotes the excited, glass-trapped sensitizer and the counterpart molecule or ion is present in the intra-pore liquid phase. (Ru(II) denotes the $[Ru(bpy)_3]^{2+}$ complex). By applying laser photolysis methods it was shown that photoinduced electron transfer between the donor and acceptor located in the two phases, appears to be a general phenomenon, independent of the nature of the specific pair. Thus, in all cases a substantial fraction of the photosensitizer population was quenched, leading to the corresponding charge-separated radical ion pairs:

$$^1Py^* + MV^{2+} \rightarrow Py^+ + MV^+ \tag{1}$$

$$Ir(III)^* + DMB \rightarrow Ir(II) + DMB^+ \tag{2}$$

$$Ir(III)^* + S_2O_8^{2-} \rightarrow Ir(IV) + SO_4^{-\bullet} + SO_4^{2-} \tag{3}$$

$$Ru(II)^* + Fe^{3+} \rightarrow Ru(III) + Fe^{2+} \tag{4}$$

$$Ru(II)^* + S_2O_8^{2-} \rightarrow Ru(III) + SO_4^{-\bullet} + SO_4^{2-} \tag{5}$$

By studying the concentration dependence of both quenching efficiency and quencher adsorption, it was concluded that the predominant mechanism involves long-range electron-transfer (LRET) from, or to, an adsorbed solute molecule and a photosensitizer embedded in the glass. The LRET mechanism is responsible for non-exponential quenching rates which reflect a distribution of distances between donor and acceptor.

The photoinduced electron-transfer process (2) may be generalized as:

$$A^*(t) + D_1(s) \rightarrow A^-(t) + D_1^+(s) \qquad (6)$$

where A and D_1 represent electron acceptors and donors, respectively, while (t) and (s) represent glass-trapped and intra-pore solute species respectively. Note that the situation is reversed in (1) where D_1^* is glass-entrapped and A is in the solution. A striking observation in the cases of (1) and (2) was the extremely long life of the radical ion-pair. In both cases the corresponding back reaction, e.g.,

$$A^-(t) + D_1^+(s) \rightarrow A(t) + D_1(s) \qquad (7)$$

was found to be slower, by 4-5 orders of magnitude with respect to the same (diffusion-controlled) process in homogeneous solutions. Such a retardation is among the highest achieved in any heterogeneous photochemical system *(9,10)*. The effect may be rationalized in terms of the same LRET mechanism and as suggested for the forward quenching reaction (6): A distribution of separations between $D_1(s)$ adsorbed at the interface and $A^*(t)$ trapped at different depths from the surface, will yield a distribution of separations between surface $D_1^+(s)$ and trapped $A^-(t)$. An important factor in retarding the back reaction (7) is associated with the adsorption of $D_1^+(s)$ at sites which are distant from a trapped $A^-(t)$ counterpart *(7)*.

These observations are relevant to the general problem of photochemical light-energy conversion. Thus, retardation of the back electron-transfer reaction is a major prerequisite in any artificial photosynthetic system *(9)*. In the specific Ir(II)*/DMB (sol-gel) system the absence of an effective back reaction allows (at acidic pH) the strong reductant Ir(II) to react with water yielding molecular H_2 according to *(7)*:

$$[Ir(II)](t) + H^+ \rightarrow [Ir(IV)H^-](t) \qquad (8)$$

$$[Ir(IV)H](t) + H^+ \rightarrow Ir[(IV)](t) + H_2 \qquad (9)$$

Since in the above (binary) system DMB$^+$ is incapable of completing the cycle by evolving O_2 from water, the net result is the generation of H_2 at the expense of DMB.

An ideal photosynthetic device is one in which the back reaction is totally inhibited by complete immobilization of both A^- and D_1^+. This will, in principle, allow one to carry out catalyzed reactions of the two species with, e.g., H_2O leading to the generation of both O_2 and H_2.

An important approach in reaching this goal is to overcome the difficulties encountered in two-component (A, D_1) systems by applying three (or more) molecules with appropriate redox properties, where, e.g., the primary donor D_1 acts as an intermediate charge-carrier ("shuttler") to a secondary donor D_2 *(10,11)*. This was achieved recently in a sol-gel glass by inhibiting the back reaction in a system where A and D_2 (and thus also A^- and D_2^+) are completely immobilized by trapping in the three dimensional network of the sol-gel glass matrix *(12)*. Accordingly, reaction (7) is inhibited by (10):

$$D_1^+(s) + D_2(t) \rightarrow D_1(s) + D_2^+(t) \tag{10}$$

in which the mobile charge carrier D_1^+ in the intra-pore phase transfers its positive charge to the immobilized secondary donor $D_2(t)$. The final products are $A^-(t)$ and $D_2^+(t)$, both immobilized in the glass matrix. The particular (ternary) redox system employed was: $A_1^*(t)$ = Ir(III), $D_1(s)$ = DMB, $D_2(t)$ = Ru(II). The final primary photoproducts were Ir(IV) and H_2 (generated via (8) and (9)) and Ru(III) generated via (10) (i.e., via $DMB^+ + Ru(II) \rightarrow DMB + Ru(III)$).

The ultimate goal is that of generating a long-lived radical pair where the reactions of both partners to yield useful fuels such as H_2 and O_2 would be controlled by means of appropriate catalysts. The prospective of applying doped sol-gel glasses to the photochemical conversion of light energy is especially appealing also in view of the inertness of the inorganic matrix. This offers considerable advances over microenvironments such as membranes, vesicles and organic polymers *(9,10)* where irreversible side reactions with the matrix decrease the efficiency and shorten the lifetime of feasible devices.

2.2 Chemical Sensors.

The observations described in section 2.1 suggested that suitable analytical reagents trapped in porous sol-gel glasses may be used for the preparation of a variety of chemical sensing materials. It was indeed shown that after condensation, drying and thoroughly washing with appropriate solvents, one obtains an almost non-leachable glass doped with an appropriate reagent which is sensitive to solutes in the adjacent liquid phase *(13a)*. Characteristic examples are classical reactions of cations and anions resulting in the coloration of the reagent-doped glass (rod or disc), e.g., Fe^{2+}/o-phenanthrolin, Ni^{2+}/dimethylglyoxime, Cu^{2+}/α-benzoinoxime, SO_4^{2-}/sodium rhodizonate + BaF_2, etc. Very high sensitivities, e.g., 100 ppt Fe^{+2} in water, were recorded *(13b)*. Due to the optical properties of the glass matrix the tests may be carried out visually as well as spectrophotometrically.

The exact mechanism of reagent trapping and exposure is still unclear. Trapping may be due to strong adsorption interactions at the glass-liquid interface, which are due to the morphology of the surface. It is also important to note that some of the color tests are associated with complex formation, with a reagent: metal ion ratio of 2:1 or even 3:1. This implies that at least some of the reagents are trapped as aggregates which are accessible to the metal analyte. Moreover, it is implied that tumbling of the reagent, sufficient to achieve appropriate intramolecular orientations in the complex, does take place. The latter hypothesis is indirectly supported by the e.s.r. experiments of Ikoma et al. *(14)*, showing that polyamine copper(II) complexes in wet alumina gels are almost free in tumbling motion. In spite of these open mechanistic questions, it is evident that doped sol-gel glasses may effectively function as sensors to inorganic and organic solutes. This also applies to their use as reversible pH indicators, based on the inclusion of appropriate indicators in the starting sol-gel reaction mixture *(13,15)*. Along the same lines, a glass-entrapped pH-sensitive dye was recently applied as sensor to atmospheric ammonia or acid *(16)*.

2.3 Trapping of Catalytic Metal and Oxide Particles.

Although the topic of this review is the trapping of organic molecules in sol-gel glasses, we believe it is important to include here also a brief comment on the application of the sol-gel method for the preparation of trapped inorganic catalytic particles, because of the major importance of this topic to industrial catalytic processes.

The use of oxides such as silica, alumina and titania as supports for dispersed metal and metal-oxide catalysts has been at the focus of catalytic chemistry for quite some time. Catalyst deposition methods such as metal evaporation *(17)*, mechanical admission *(18)*, ion exchange (followed by thermal decomposition of the surface-bound salt) *(19,20)*, or impregnation *(21)*, all involve the dispersion of catalyst particles on a predetermined surface of a porous solid or of colloidal particles. With the purpose of providing a novel technology, leading to a more convenient catalyst with better controlled dispersion, methods were devised, based on the incorporation of the catalyst particles in the bulk of porous silica or alumina matrices through sol-gel methods. The basic approach is to hydrolyze a mixed solution of the alkoxy compound in the presence of a soluble salt of the metal or oxide catalyst, leading to an active gel containing the catalyst particles *(18)*.

A typical example involves an iron particle catalyst prepared by drying and calcination of a gel obtained by hydrolysis of a mixed solution of ethylsilicate and iron (III) nitrate *(21)*. The Fe particle distribution (5-20 nm), determined by EXAFS and IR spectroscopy, was found to be better controlled than, e.g., by the alternative impregnation (i.e., adsorption) methods. Inorganic aerogels with large pore volumes and high surface areas prepared by the autoclave method, were later used for the incorporation of a variety of metal or metal oxide catalysts *(21)*. Systems investigated included NiO/alumina and NiO/alumina-silica (isobutylene oxidation), Ni/alumina (ethylbenzene hydrogenolysis), Cu/alumina (cyclopentadiene hydrogenation) and Ni/MoO_2. Relatively high selectivities were reported for the various catalysts and related reactions.

Studies of SiO_2-supported catalysts indicated advantages of high purity support which is free of possible poisons *(22)*. A comparison of the self-poisoning of such metal (e.g., Ru) catalyst, with one prepared by the classical impregnation method, was performed by Lopez et al. *(23)*. No systematic advantages or disadvantages of the sol-gel method (with respect to catalyst deactivation) were found for gas-phase hydrogenation reactions. It appears, however, that sol-gel glass-entrapped nickel-alumina catalysts exhibit a higher selectivity in the liquid phase hydrogenation of 1,3- and 1,5-cyclooctadiene and methyl linoleate than with Raney nickel or impregnated nickel-alumina catalysts *(24)*. A relatively lower activity of the sol-gel catalyst was attributed to the tendency of the nickel particles to be encapsulated within the alumina matrix.

The existence of the doping salts in the form of molecular aggregates was directly demonstrated using e.s.r. methods by Ikoma et al. *(20)* in the case of polyamine copper(II) complexes in silica (sol-gel) glasses. The complexes were thermally decomposed to copper(II) oxide and copper metal particles of the order of several hundreds Å. Similar experiments were carried out in alumina gels prepared by the sol-gel technique *(14)*. The important aspect of Ikoma's et al. work, from the point of view of this review, is the demonstration of the ability to obtain inorganic catalysts by doping with organic precursors, such as organic metal complexes and metal organic salts.

Very recently, Hardee et al. described the co-polymerization of n-triethoxysilyl-1,2-bis(diphenylphosphino)benzene with silicic acid, as a precursor for a rhodium catalyst for alkene hydroformylation and methanol homologation, but only low catalytic activity was observed *(24a)*.

3. Enzymes Immobilized in Sol-Gel Glass Matrices: Bioactive Glasses

In this section we describe what we believe to be the first successful trapping of cell-free enzymes in sol-gel glasses.

Attachment of biologically active molecules to an insoluble matrix is an essential, and in some instances an indispensable step in the development of biocatalysts. Immobilization allows reuse of enzymes, protects them from harsh external conditions, from microbial contamination, and prolongs their useful lifetime. A large number of existing immobilization techniques reflects the complexity of the biological material and the variety of its applications. Simple inexpensive general techniques, resulting in stable and active enzyme catalysts are yet in demand *(25)*.

A good carrier matrix should provide the enzyme with mechanical and chemical stability combined with large, accessible and hydrophilic surface. Porous glasses and hydrophillic silicas conform well with these requirements. However, the bonding of an enzyme to these matrices requires tedious chemical derivatization procedures.

The sol-gel glass technology provides the opportunity of bonding enzymes to glass using the most generally applicable immobilization procedure, namely, a simple entrapment. The main shortcoming of the entrapment technique as applied to a variety of natural or synthetic polymers is the loss of the enzyme by leakage through a nonuniform net of polymer molecule chains. In contrast, enzymes entrapped in sol-gel glasses are not removed from the catalyst particles even following extensive washing with salt or with slightly alkaline solutions. Only at pH above 10, some leaching of both the silicate and the protein was observed.

Diffusion of protein within the glass matrix is very limited. Thus, glasses were prepared containing brightly colored occlusions of hemoglobin aggregates in a clear glass. This glass was then immersed in water for as long as one month. No visible spreading of the red hemoglobin was detectable. The narrow uniform pore range of sol-gel glass (3-10 nm) may account for this phenomenon (30).

For an entrapment of an enzyme, the latter has to be added in water solution to a mixture of monomers at the onset of polymerization or later during the gel formation. When an enzyme is entrapped in a gel formed by a water-soluble polymer (gelatin, alginate, polyacrylamide) at low temperatures, denaturing conditions can be easily avoided. Use of organic solvents in the process of sol-gel formation may cause denaturation of the enzyme. Thus, Carturan et al. (26) have been able to demonstrate invertase activity in aggregates of whole yeast cells trapped in thin sol-gel films, whereas films with cell-free invertase were devoid of activity. This inactivation of the cell-free enzyme may be caused by denaturation. The enzyme in whole yeast cells was, probably, protected from the denaturing effect of alcohol by protein-rich cytoplasm. It seems, that a certain degree of protein aggregate formation is important for protection of enzymatic activity during immobilization (30). It is hardly accidental that for a successful immobilization of the cell-free alkaline phosphatase (27,28), it was essential to work at conditions near the precipitation of the enzyme. At the volume ratio of tetramethoxy orthosilicate (TMOS) methanolic solution to water 12:1 (compared to 6:1 used by Carturan et al. (26)) alkaline phosphatase actually precipitated during gel formation. Light scattering by protein aggregates (see Fig. 1 in ref. 27) was observed both before and after the hydration of the alkaline phosphatase glass. This aggregation or precipitation did not impair relatively high yield of enzymatic activity (about 30%).

Milder and less denaturing precipitating agents, such as polyethylene glycols were more effective in preservation of enzyme activity than lower alcohols, even in the absence of visible precipitation. Several enzymes have, thus, been successfully immobilized. The accumulated data do yet allow only preliminary conclusions concerning the influence of various additives and of gelation conditions. High yields of trypsin activity (50-60%) were obtained by mixing a slightly acidic enzyme solution in water containing PEG 6000 (3%) with an equal volume of TMOS at room temperature. The opaque mixture became homogeneous in 5-10 minutes, as TMOS was hydrolyzed, and congealed in about an hour. Clear monolithic glasses form after drying at 30-37°C for a week. These glasses sometimes crack during hydration (30).

Trypsin, aspartase, peroxidase and other enzymes have been entrapped in heterogeneous polymerization systems containing enzyme solution in water (0.2 ml/ml TMOS) and PEG 400 (0.6-0.8 ml/ml TMOS). Such mixtures separate into two phases: the upper (organic) phase contains TMOS saturated with PEG and traces of water, while the lower contains mostly PEG and water with little TMOS. Gel formation starts at the interface by initial hydrolysis of TMOS. Hydrophilic products of hydrolysis and polymerization undergo partition predominantly into the lower phase, where the gelation occurs. Protein aggregates or other relatively hydrophilic material (e.g. microbial cells) were always found only in the lower phase. Thus, during the formation of glass the proteins are not in contact with denaturing hydrophobic phase. After 3-4 h at

4°C, a solid glass-like material forms in the lower phase creating a diffusional barrier for further partitioning of precursors. The upper phase is then removed and the resulting homogeneous protein-containing sol-gel is dried. Cell free aspartase immobilized using this technique, resulted in highly active enzymatic catalyst, with a yield of enzyme activity similar to that of an industrial enzyme described by Chibata et al. *(29).*

In general, recovery of enzymes in polymerization mixtures containing methanol was poor. The volume change during the drying in the PEG containing polymerization systems was significantly lower than in methanol-based mixtures, and the latter, usually, shrank to about 10% of their initial volume. We believe that the shear generated during this shrinking may damage protein molecules.

Scanning electron microscope (SEM) pictures of protein-doped glasses show spongy uniform glass, sometimes dotted with protein aggregates. The degree of aggregation depends upon the protein and upon the conditions of precipitation. The recovery of enzyme activity in heavily doped glasses is usually low. We have noticed that the gelation is considerably speeded up by biological material such as proteins or microbial cells, probably leading to a denser, less porous glass.

The analysis of substrate dose-response curves for alkaline phosphatase (ALP), immobilized in methanol-containing polymerization mixture, revealed at least two kinetically different forms of the enzyme *(27).* The high affinity enzyme component had the Michaelis constant $K_m = 0.8$ mM, which was also measured for the soluble ALP. About 90% of the enzyme activity, however, had an average K_m of 7 mM. The pH maximum of immobilized ALP was about one pH unit higher than for the native enzyme. However, trypsin and acid phosphatase, prepared in the presence of PEG, demonstrated single form kinetics with K_m close to that of the soluble trypsin *(30).*

Sol-gel-immobilized trypsin activity towards small substrates was not inhibited either by soy bean trypsin inhibitor (20.1 kDa). Poly-L-lysine of molecular size above $M_r = 29.6$ kDa had no effect on immobilized trypsin activity, while in the range of $M_r = 1.5$-8.0 kDa this polymer inhibited the biological glass activity to the same extent, as that of the soluble enzyme. All this indicates effective pore sizes of about 2-4 nm *(30).*

We have compared the ability of the soluble and of the immobilized enzymes to form a complex with the dye Coomassie blue. This dye binds to the positive charges on protein molecules, and, thus, serves as a general protein surface probe. Coomassie blue binding was well correlated with the activity of the immobilized enzyme *(28).* This observation signifies equal accessibility of the protein to general and to active site-targeted probes. It could be interpreted as indicating that either enzyme molecules were partitioned in accessible and inaccessible phases, or that a part of a single protein molecule surface was interacting tightly and at random with the sol-gel matrix. The narrow effective pore size pointed out to the latter possibility. The existence of such tight interactions and the strong character of the binding forces could be demonstrated by diffusional limitations of the immobilized enzyme. Thus, loosely immobilized (e.g. entrapped in a polyacryl amide gel) trypsin can completely degrade itself when left at neutral or slightly basic pH for prolonged periods of time. Various sol-gel immobilized trypsin preparations have lost only 10-30% of their activity when incubated at 30°C overnight at pH 7.5 (28,30). This indicates that most of the trypsin molecules are not able to diffuse within the network of the sol-gel. Trypsin can be reversibly absorbed on sol-gel glass, while retaining its activity. As expected, such physisorption-immobilized trypsin was completely autodigested under the conditions described above (30).

Firm interaction between the immobilized protein molecule and the matrix is a characteristic of covalently bound enzymes, rather than of entrapped ones. In covalently immobilized enzymes a protective effect of the immobilization matrix against thermal inactivation, or the increase in the melting point of protein polypeptide chain, depends on the amount of binding sites per protein molecule (31). Multipoint (3 to 8 points) attachment of the immobilized enzyme results in a considerable increase of the denaturation temperature. Thus, one would expect significant stabilization of sol-gel immobilized enzymes. Indeed, it has been shown that sol-gel immobilized ALP is more stable at 70°C than its soluble form (27). An even more dramatic protective effect was observed in sol-gel immobilized preparations of acid phosphatase, which was an order of magnitude more stable than the soluble enzyme. We believe that during sol-gel formation the matrix and the glass form closely fitted surfaces generating strong binding forces. The high thermal stability of sol-gel glass immobilized enzymes allows to use them at elevated temperatures, thus increasing the reaction rates.

In conclusion of this section: We have demonstrated that the sol-gel method of enzyme immobilization by entrapment of the enzyme in a porous sol-gel glass produces a biocatalyst in a simple generally applicable procedure. A variety of enzymes representing different classes of catalysts have been already successfully immobilized. It has been shown that a large (up to 100%) population of trapped molecules can react with small external substrates through the pore network of the glass. The final product is an air-dried material with a shelf-life of more than six months. That makes it uniquely convenient for shipment and storage. In its hydrated active form, the catalyst is also stable for a very long time. This method of immobilization allows one to obtain highly active catalysts of the desired geometry: Beads, thin layer coating on various supports (such as glass, ceramics, metal), etc.

The sol-gel glass trapping of enzymes with a wide variety of applications are worldwide patent pending.

4. Sol-Gel Glasses for Optical and Spectroscopic Applications

4.1 Light Absorbing Materials

4.1.1. Photochromic glasses. The search for ever better information recording and information processing materials has never ceased in our information oriented society. A major family of molecules which has served for that purpose are the photochromic compounds *(32)*. These have been used for chemical computer-switches, as signal processors, as reusable information storage media, as microimaging materials, as protective materials against irradiation, as photomasking and photoresisting materials, to mention a few applications. The currently used photochromic glasses are based on a very limited selection of inorganic dopants. However, the ability to trap organic photochromic dyes in the sol-gel glasses, opens the possibility to use the thousands of existing photochromic molecules *(32)*, with the ability to tailor desired properties such as the nature of color change, the activating wavelength, the rates of response to light, the fading rate, etc.

The feasibility of this idea was first demonstrated briefly in reference *33* in which the trapping of the photochromic dye Aberchrome-670 in SiO_2 was described. Difficulties with long-lasting activity of that dye, have shifted the attention to the largest class of photochromic molecules, namely to the spiropyranes. These were used for both a mechanistic study of the sol-gel-xerogel process *((34)*, see Section 5) and for the preparation of good photochromic materials: Levy et al. have described two types of photochromic materials which are based on trapping of five different spiropyanes in modified SiO_2 matrices *(35)*. One type of matrix was obtained by polymerizing $CH_3CH_2Si(OCH_2CH_3)_2$. This results in a network which is less cross-linked than the usual SiO_2 matrix, and which has an a-polar cage comprised of $SiCH_2CH_3$ groups. The normal photochromic behaviour of the spiropyranes was obtained with these materials, namely colorless glass converted upon irradiation to colored glass. Interestingly, when the sol-gel process was employed with a 4/1 mixture of $Si(OCH_3)_4$/polydimethyl siloxanes (of various molecular weights) materials with reverse photochromism were obtained, namely colored in the dark and fading through the action of light. The difference between the normal and reversed photochromisms in these two types of materials was explained in terms of the differences in the cage properties within which the dye was trapped, namely, low polarity in the former and higher polarity in the latter.

Reversed photochromism of spiropyranes was observed in sol-gel glass obtained from the polymerization of $Si(OCH_3)_4$ *(34)* and also, as reported by Matsui et al. *(36)* from $Si(OCH_2CH_3)_4$. However, in these unmodified glasses, the photodynamics slow down significantly in the aged glass, due to the high rigidity of the final cage. However, the two organically modified glasses described above provide long lasting materials. The added flexibility of those matrices is the direct cause of that improvement.

Stable photochromic materials were also obtained by encapsulating a number of spiropyranes in alumino-silicate sol-gel glasses *(37)*. These were obtained by the polymerization of $(CH_3CH_2CH_2O)_2Al-O-Si(OCH_2CH_2CH_3)_3$. The photochromic properties were used for the investigation of this polymerization (Section 5).

4.1.2 Glasses for Hole-Burning and Nonlinear Optics. Another photochemical method with great potential for information recording purposes is the process of hole-burning *(38)*. It is a high-resolution photobleaching technique which can, in principle, store data at a density which is a 1000 fold higher than in present optical disc systems *(38)*. Sol-gel glasses have been studied for these purposes by Tani et al. *(39,40)* by Locher et al. *(41)*, by Kobayashi et al. *(42)* and by Tanaka et al. *(43)*. In a typical study *(40)*, 1,4-dihydroxyanthraquinone was used as a dopant for SiO_2 glasses prepared by acidic polymerization of $Si(OCH_2CH_3)_4$. The observed burning yield was $1.2 \cdot 10^{-4}$ which is comparable to yields obtained in organic glasses. An intrinsic holewidth of 0.9 cm^{-1} was obtained at zero burning time. The annealing of the burned hole indicated two types of mechanisms which dominate the temperature dependence of the holewidth: A reversible one which peaks below 27°K and an irreversible one above that temperature.

Tanaka et al. *(43)* observed efficient hole-burning activity in sol-gel alumina thin films, having in mind wiring applications of these materials.

On a very basic level, it has been demonstrated that the sol-gel technique can be used for the preparation of filters: A large variety of organic molecules can be used for that purpose and this has been demonstrated for SiO_2 films *(44)* and blocks *(39)*, for SiO_2-TiO_2 films *(44)*, for Al_2O_3 films *(42)* and for SiO_2-poly-methyl methacrylate composite *(45)*. On a more advanced level, since all of the dyes trapped in these studies are asymmetric π-conjugated molecules, many of these glasses can be used in principle for nonlinear optic purposes. These dyes include, e.g., acridines *(39)*, 2-methyl-nitroaniline *(45)*, fluorescein *(44,46)* and coumarines *(46a)*. Third order nonlinear effects were observed by Knobbe et al. *(47)* in polyaniline doped SiO_2-sol/gel films and in coumarins and rhodamines, trapped in originally modified silica sol-gels *(48)*.

4.1.3 Electro-optical Liquid Crystals Sol-Gel Glasses. A most interesting application of sol-gel glasses was suggested recently by Levy, Serna and Oton *(49)*. These authors prepared thin films of reversed-phase silica (obtained from $(CH_3CH_2)Si(OCH_2CH_3)_3$), in which microdomains of liquid crystals of 4-methoxybenzylidine-4'-n-butylaniline were encapsulated. The drive for that study has been the current major obstacle in the application of liquid crystals for electrooptical devices, namely the lack of suitable materials which have both the required electrooptical properties and can also be prepared in suitable forms such as thin films and monolithic blocks. Microscopy studies showed that the microdomains vary in size from 0.5 to 3 microns. The orientation of these domains was found to be affected by the ethyl groups on the walls of the pores. Transitions between nematic phase and liquid phase were observed by following laser diffraction patterns. Most importantly, it was found that the trapping process does not alter the electrooptical and thermal behaviour of the liquid crystals.

4.2 Light Emitting Materials

4.2.1 Fluorescent and Dye-Laser Materials.

Chronologically speaking, fluorescent dyes were both the first organic molecules which have been incorporated in sol-gel glasses *(2,3,50)* and the most numerous as a class. Examples include rhodamine 6G *(3,39,42,44,51,52)*, naphthazin *(39)*, acridines *(39)* coumarines *(39,42)*, rhodamine B *(39,42,44,45,53,54)* ruthenium complexes *(44,55,56)*, fluorescein *(44,46,57)*, crystal violet *(42,44)*, malachite green *(44)*, oxazines *(16,42,44)*, 4-methylembelliferone *(45)*, nile-blue *(42)*, porphyrines *(42)*, resorufin *(42)*, cresyl-violet *(42)* and phthalocyanine *(42)*. Additional ones are mentioned in Sections 2 and 5.

Many of these glass preparations were aimed at studying the properties of the glass cage to which we return in Section 5. Some were aimed at the development of thin-film light-guides *(44)* and luminescent solar concentrators *(58)*, but perhaps the most fascinating has been the potential use of these materials in dye-lasers. That was the very reason for our choice of rhodamine 6G as the first trapped organic molecule *(3)*. However, it was not until 1988 that this concept was successfully demonstrated simultaneously by research groups in Japan and in the USA. Kobayashi et al. *(59)* have doped rhodamine 6G and B and oxazine-4 in thin films (10-100 = μm) of alumina, prepared by hydrolysis of $AlCl_3$. One of the main drawbacks of dye lasers is their tendency to aggregate in aqueous solutions. The sol-gel method overcomes this difficulty *(3,44)* and indeed Kobayashi et al. were able to reach dye concentrations as high as $10^{-2}M$ without aggregation. Irradiation of the doped film with a nitrogen-laser produced laser emission with a beam divergence of 0.1 rad. It was found that the width of the dye laser emission was 10 nm, which is typical of dye lasers. The calculated conversion efficiency was 2.1% and it was found that the power decreases linearly with the number of shots, which is indicative of a single-photon process. Later it was found *(60)* that increasing the concentration of the dye increased the pumping efficiency. This dependence indicates that the doped films operate as compact dye-laser films. The behaviour of laser-dye pairs was reported in the ref. 60, as well.

Dunn, Zink, Knobbe and their colleagues achieved successful lasing activity from rhodamine 6G in SiO_2 sol-gel glass *(61)*. Samples were pumped with the 308 nm output of a XeCl excimer laser with pulses of 10-50 mJ and the output was a blue-shifted (585.8 nm) lasing emission with FWHM of 8 nm. The output power from the gel was comparable to that obtained from a control ethanol solution. Most importantly, the pump power from the excimer laser was at levels which would bleach the dye when incorporated in an organic polymer matrix. The gain of the glass was 4.6 cm^{-1}, compared to 6.9 cm^{-1} in ethanol. In subsequent studies *(48,62)*, this group successfully extended their studies to other laser dyes (rhodamines and coumarines) trapped in organically modified silicas where improved photostability (compared to, e.g., polymethyl methacrylate) was observed *(62)*.

Lasing activity from sulforhodamine-640-doped silica sol-gel was recently reported by Salin et al. *(63)*. The doped glass was pumped with a frequency-doubled radiation from a Q-switched Nd:YAG laser, and a 20%

conversion efficiency was obtained. The important aspect of that report is the 40 nm tunability of that material.

One should also mention a different approach in which laser-dyes and optically nonlinear dyes were not trapped by the sol-gel method, but adsorbed on the pores of the glass via an organic polymer solution. This approach which is outside the scope of the present review, is described in *(64)*.

Finally, an interesting application of fluorescent sol-gel glasses is for radiation scintillating detectors. This has been described recently by Nogues et al. *(65)*.

The incorporation of fluorescent and lasing dyes in sol-gel glasses, is perhaps the fastest growing application. After completion of this review, several additional papers on this subject were brought to our attention; these are collected in reference *65a*.

4.2.2 Phosphorescent Sol-Gel Glasses (33,66). One of the most remarkable manifestations of the special properties of the silica sol-gel cage is the observation that many trapped organic molecules exhibit efficient phosphorescence at room temperature when exposed to air, in many instances without the aid of a heavy atom, and in several cases even in the wet gel. The reader probably recalls that phosphorescence is a delicate process which is usually quenched at temperatures which are not cryogenic and by exposure to oxygen. The generality of the phenomenon was demonstrated with the following molecules: Phenanthrene, naphthalene, quinine, 4-biphenyl carboxylic acid, 1-naphthoic acid, eosin-y and pyrene. It was observed that under various gelation conditions, most dyes exhibited not only phosphorescence but also delayed fluorescence. Specific glasses were needed to observe phosphorescence from the various compounds. For example, neutral gelation conditions were sufficient to observe phosphorescence from phenanthrene, naphthalene and quinine, but basic conditions were needed for the two carboxylic acids and for eosin. A heavy atom (Br^-) was needed to observe phosphorescence from pyrene, and so an SiO_2 glass doped with NaBr was prepared. Very long emission lifetimes were observed. For instance, when biphenylcarboxylic acid was trapped in an SiO_2 glass prepared under basic conditions, the lifetime reached the order of several seconds. Lifetimes of the order of milliseconds were obtained even from the wet gels. A detailed mechanistic study of the sol-gel transition was carried out with that carboxylic acid.

5. The Use of Trapped Molecules for Mechanistic Studies of the Sol-Gel-Xerogel Transitions and of the Cage Properties

5.1 Pyrene and its Derivatives as Probe Molecules.

The relatively long singlet lifetime of excited pyrene (≈ 100 nsec), the clear vibronic features of its fluorescence emission, its ability to form an excimer and the sensitivity of all these parameters to environmental conditions, have popularized the use of this molecule as a dopant probe for the study of sol-gel glasses *(5b,33,67-71a)*. A brief description of some of these studies follows:

Kaufman et al. *(67)* performed a detailed study of the structural changes at the molecular scale that a polymerizing $Si(OCH_3)_4$ system undergoes along the sol-gel-xerogel transition. Changes in the emission spectra of pyrene and of its excimer revealed complex structural changes which proceed well beyond the gel point. The vibronic fingerprint of pyrene was used for a follow-up of polarity changes around the probe molecule, along the polymerization. Tested polymerization parameters were pH, water/silane ratio and the nature of the alkoxy group. It was found that polymerization-gelation occurs at low water/silane ratio, whereas colloidal gelation occurs at the higher water/silane ratios. A remarkable observation has been that at the final xerogel stage, no excimer emission was observed, demonstrating again the efficient trapping and isolation of doped molecules in these glasses. The excimer/monomer ratio which changes along the polymerization from ≈ 0.4 to $\approx$ zero was consequently used for the identification of isostructural gels, obtained under different reaction conditions. Matsui et al. *(70)* have confirmed most of these conclusions by performing a comparative study, using pyrene and pyrene-3-carboxyaldehyde for the acidic polymerization of $Si(OCH_2CH_3)_4$.

Evidence for some ground state pyrene association was provided by Yamanaka et al. *(69,71)* in doped SiO_2 thin films, based on time-resolved fluorescence spectroscopy.

An interesting observation, made by Kaufman et al. (described in detail in *72* and briefly in *33*), has been the following: When the polymerization of $Si(OCH_3)_4$ is carried out with pyrene as a dopant and in the presence of a surface active agent, prolonged (over 1,000 hours) structural oscillations occur at the xerogel stage. The oscillations, as revealed from changes in the monomer/excimer ratio are of large amplitude, they are slow (several hrs/period) and they are chaotic in nature. The origin of this phenomenon is not clear. It has been suggested tentatively that the driving force for the oscillations is the structural relaxation of the secondary polymeric gel structure and the dispersion of the adsorbed pyrene molecule to thermodynamically favored cage sites; further exploration of this phenomenon is needed.

A pyrene derivative, 8-hydroxy-1,3,6-pyrenesulfonic acid (pyranine) was successfully used *(73)* for the detection of water consumption during the early stages of the polymerization of $Si(OCH_3)_4$. The idea of using this probe molecule is based on the fact that its excited state is a short-lived strong acid, capable of efficient proton transfer to neighboring water molecules, leaving behind the anion of pyranine in its excited state. Since the anion and the undissociated acid fluoresce at distinctly different wavelengths (510 and 430 nm, respectively) the ratio of intensities of the fluorescence at these two wavelengths can be used as a water-probe. Indeed, this probe showed high sensitivity to the variations in the dynamical changes of water contents (water consumption during the hydrolysis stage and water release during the condensation) as a function of variations in the polymerization conditions, i.e., changes in water/silane ratio and pH. A similar study was carried out by Pouxviel et al. on aluminosilicate sol-gels *(6)*. Here too, pyranine proved to be a highly sensitive probe to changes in water contents of this polymerizing system.

5.2 Various Other Probe Molecules.

The rhodamines were also used intensively to study cage properties of the glasses *(3,53,74,75)*. Thus, the cage polarity of SiO_2 sol-gel was determined from spectral shifts of rhodamine 6G *(3)* and it was found that the cage is hydroxylic and highly polar, although its polarity is somewhat less than water (but higher than methanol). Fujii et al. used rhodamine B for a follow-up of the polymerization of $Si(OCH_2CH_3)_4$ *(74)* and found a remarkable thermal stability of that trapped dye *(53)*. This stabilization effect of the cage seems to be general: It was found for fluorescein *(46)*, and for trapped enzymes *(27,28)*. A detailed study by Pouxviel et al. demonstrated that the cage electrical charge influences the trapped molecules, which in their study were rhodamine 6G and rhodamine B.

Various aromatic polycyclic molecules were employed for structural studies. Matsui et al. *(76)* have used 7-azaindole to study the polymerization of $Si(OCH_2CH_3)_4$ by following spectral shifts in the 380-430 nm range. It was found that 7-azaindole forms a strong complex with the silanols in the cage, stabilizing both the ground state and the excited state. Fujii et al. used 1- and 2-naphthols *(77)* and reported that they find these two molecules to be very convenient photoprobes. Interestingly, these authors observed an inversion of the fluorescent levels 1La and 1Lb after the gel point. Yamanaka et al. observed very short events that trapped naphthalene undergoes in SiO_2 glass *(78)*. They used picosecond time-correlated single-photon counting techniques and observed excimer formation on the ps time scale. Finally, Levy et al. studied the room temperature phosphorescence of three aromatic polycyclic hydrocarbons: naphthalene, phenanthrene and pyrene *(66)* trapped in SiO_2 glasses and found that the phosphorescence efficiency increases with decrease in the number of benzene rings; this was attributed to the parallel decrease in the T_1-S_0 gap. A mechanistic study with these and other trapped molecules led to the conclusion that phosphorescence is observable in these glasses due to strong adsorption interactions with the cage walls; due to an almost complete elimination of solvent molecules from the cage; and due to the exceptional rigidity of the final cage *(66)*.

Indeed, the rigidity of the gelating system was the topic of a study by McKiernan et al. *(79)*. They successfully employed the emission maximum of $ReCl(CO_3)$bipyridine blue shifts as a probe for the increasing rigidity of polymerizing aluminosilicate and silica systems. It was found that the major spectral shifts in the SiO_2 system occurred at the drying stage, while in the Al_2O_3-SiO_2 system it occurred at much earlier stages. This was attributed to an early cage formation in the latter system.

Fluorescence polarization of several organic dopants in a polymerizing SiO_2 sol-gel system was recently described by Winter et al. *(80)*. Their main conclusion has been that the mobility of small and medium sized probe molecules is hardly affected by the gelation and that there is no rigidity at the microscopic level; however, this conclusion seems to be in variance with other previously cited studies, including McKiernan's *(79)*.

Photochromic compounds, which were already mentioned in Section 4 in the context of information recording materials, were also used for the investigation of the sol-gel-xerogel transition in $Si(OCH_3)_4$ *(34)*. The continuous environmental changes along that transition were reflected by gradual change in the photochromic behaviour of the trapped spiropyranes. In particular, the photochromism changed at a certain point into reversed photochromism. The same approach was recently applied by Preston et al. *(37)* for the study of the abovementioned alumino-silicate system: Aided with spiropyrane probes, these authors were able to identify four distinct stages of the gelation by following photochromism rates and spectral changes.

6. Copolymerizations

The concept of co-polymerization (instead of doping) was developed by Schmidt et al. *(81)*. For instance, organometallic complexes of cobalt and rhodium were incorporated in SiO_2 by co-polymerization of tetraethoxy silane with tri-ethoxy silane complex ligands of these metals *(82)*. Co-polymerizations were employed also in a recent Kodak patent *(83)* describing the preparation of dye-polymer/sol-gel composites. Copolymerizations aimed at achieving optical non-linearity were already mentioned in Section 4.1.2. Two additional examples are given in references *84,85*.

Acknowledgments: We are deeply indebted to our students and co-workers in this project: S. Druckman, O. Lev, D. Levy, S. Rappoport, C. Rottmann, A. Slama-Schwok, S. Shtelzer and R. Zusman. These studies are currently supported by the Materials Division of the USA RDSG-UK, the Harry Kay Foundation, the Krupp Foundation, the German BMFT Foundation and by the Israel NRCD. D.A. and M.O. are members of the Farkas Center for Light Energy Conversion and of the Fritz Haber Research Center for Molecular Dynamics.

Literature Cited

1. Scherer, G.; Brinker, J. *Sol-Gel Science; Academic Press:* San-Diego, 1990.
2. Avnir, D.; Reisfeld, R.; Levy, D. Israel Patent Application 69724, Sept. 23, 1983.
3. Avnir, D.; Levy, D.; Reisfeld, R. *J. Phys. Chem.* **1984,** *88,* 5956.
4. Modes, S.; Lianos, P. *Chem. Phys. Lett.* **1988,** *153,* 351.
5. (a) Thomas, J.K.; Wheeler, J. *J. Photochem.* **1985,** *28,* 285. (b) Chandrasekaran, K.; Thomas, J.K. *J. Coll. Interface Sci.* **1985,** *106,* 532.
6. Pouxviel, J.C.; Dunn, B.; Zink, J.I. *J. Phys. Chem.* **1989,** *93,* 2134.
7. Slama-Schwok, A.; Avnir, D.; Ottolenghi, M. *J. Phys. Chem.* **1989,** *93,* 7544.
8. idem, *Photochem. Photobiol.* **1991,** *54,* 525.
9. e.g., (a) Gratzel, M. *Acc. Chem. Res.* **1981,** *14,* 376. (b) *Photochemical Energy Conversion;* Fox, M.A.; Chanon, M., Eds. Elsevier: Amsterdam, 1988.
10. e.g., Rabani, J.; Sasson, R.E. *J. Photochem.* **1985,** *29,* 1.
11. Rabani, J., in ref. 9b, pp. 642-696.
12. Slama-Schwok, A.; Avnir, D.; Ottolenghi, M. *J. Am. Chem. Soc.* **1991,** *133,* 3984.

13. (a) Zusman, R.; Rottman, C.; Ottolenghi, M.; Avnir, D. *J. Non-Cryst. Solids* **1990**, *122*, 107. (b) Lev, O.; Iosefson-Kuyavskaya, B.; Gigozin, I.; Ottolenghi, M.; Avnir, D. *Fresenius J. Anal. Chem.*, **1991**, submitted.

14. Ikoma, S.; Kawakita,, K.; Yokoi, H. *ibid*, **1990**, *122*, 183. Ikoma, S.; Takano, S.; Nomoto, E.; Yokoi, H. *ibid* **1989**, *113*, 130.

15. Rottman, C.; Ottolenghi, M.; Lev, O.; Avnir, D. **1991**, manuscript in preparation.

16. Chernyak, V.; Reisfeld, R.; Gvishi, R.; Venezky. D. *Sens. Mat.* **1990**, *2*, 117. Eyal, M.; Gvishi, R.; Reisfeld, R. *J. Phys.* **1987**, *C7*, 471.

17. Tatarchuk, B.J.; Dumesic, J.A. *J. Catal.* **1981**, *70*, 308.

18. Bianchi, D.; Lacross, M.; Pajonk, J.M.; Teichner, S.J. *J. Catal.* **1981**, *68*, 411.

19. Tominaya, H.; Ono, Y.; Keji, T. *J. Catal.* **1975**, *40*, 197.

20. Ikoma, S.; Takano, S.; Nomoto, E.; Yokoi, H. *J. Non-Cryst. Solids* **1989**, *113*, 130.

21. Gardes, G.E.E. ; Pajonk, G.; Teichner, S. *J. Bull. Soc. Chim. (Fr.)* **1976**, 1321.

22. Carturan, G.; Fachin, G.; Gottardi, V.; Gugliemi, M.; Movazio, G.J. *Non-Cryst. Solids* **1982**, *48*, 219.

23. Lopez, T.; Lopez-Gaona, A.; Gomez, R. *Langmuir* **1990**, *6*, 1343. Lopez, T.; Villa, M.; Gomez, R. *J. Phys. Chem.* **1991**, *95*, 1690.

24. Ishiyama, J.I.; Kurokawa, Y.; Nakayama, T.; Imaizumi, S. *Appl. Catal.* **1988**, *40*, 139.

24a. Hardee, J.R.; Tunney, S.E.; Frye, J.; Stille, J.K. *J. Polym. Sci.*, **1990**, *28A*, 3669.

25. Kennedy, J.F.; White, C.A. in: Handbook of Enzyme Biotechnology Wiseman, A. ed.; Ellis Horwood Ltd: Chichester, 1985, pp. 147-207.

26. Carturan, G.; Campostrini, R.; Dire, S.; Scardi V.; de Alteriis, E. *J. Mol. Cat.* **1989**, *57*, L13-L16.

27. Braun, S.; Rappoport, S.; Zusman, R.; Avnir, D.; Ottolenghi, M. *Mat. Lett.* **1990**, *10*, 1.

28. Braun, S.; Rappoport, S.; Zusman, R.; Shtelzer, S.; Druckman, S.; Avnir, D.; Ottolenghi, M. in: Biotechnology: Bridging Research and Applications Kamely, D.; Chakrabarty, A.; Kornguth, S.E. Eds., Kluwer Acad. Publ.: Amsterdam, 1991, p. 205.

29. Chibata, I.; Tosa, T.; Sato, T. *Meth. Enzymol.* **1976**, *44*, 739.

30. Shtelzer, S.; Rappoport, S.; Avnir, D.; Ottolenghi, M.; Braun, S., *Biotech. Appl. Biochem.*, submitted 1991.

31. Martinek, K.; Mozhaev, V. V. *Adv. Enzymol.* **1985**, *57*, 179.

32. *Photochromism;* Brown, G. H., Ed.; Wiley: New York, 1971

33. Kaufman, V.R.; Levy, D.; Avnir, D.J. *Non-Cryst. Solids* **1986**, *82*, 103.

34. Levy, D.; Avnir, D. *J. Phys. Chem.* **1988**, *92*, 4734.

35. Levy, D.; Einhorn, S.; Avnir, D. *J. Non-Cryst. Solids* **1989**, *113*, 137.

36. (a) Matsui, K.; Morohoshi, T.; Yoshida, S. *Proc. Int. Meet. Adv. Mat.*, Tokyo 1988. (b) Morohoshi, T.; Matsui, K. *Kenkyiu Hokoku* **1989**, *32*, 229 (Chem. Abstr. 1989, 111:47962r).

37. Preston, D.; Pouxviel, J.-C.; Novinson, T.; Kaska, W.C.; Dunn, B.; Zink, J.I. *J. Phys. Chem.* **1990**, *94*, 4167.

38. Gutierez, A.R.; Friedrich, J.; Haarer, D.; Wolfrum, H. *IBM J. Res. Div.* **1982**, *26,* 198.
39. Makishima, A.; Tani, T. *J. Am. Ceram. Soc.* **1986**, *69,* C-72. Makishima, A.; Morita, K.; Inoue, H.; Tani, T. *Proc. SPIE,* **1990**, *1328,* 264.
40. Tani, T.; Namikawa, A.; Arai, K.; Makishima, A. *J. Appl. Phys.* **1985**, *58,* 3559.
41. Locher, R.; Renn, A.; Wild, V.P. *Chem. Phys. Lett.* **1987**, *138,* 405.
42. Kobayashi, Y.; Imai, Y.; Kurokawa, Y. *J. Mat. Sci. Lett.* **1988**, *7,* 1148.
43. Tanaka, H.; Takahsashi, J. Tsuchiya, J.; Kobayashi, Y.; Kurokawa, Y.K. *J. Non-Cryst. Solids* **1989**, *109,* 164.
44. Avnir, D.; Kaufman, V. R.; Reisfeld, R. *ibid,* **1985**, *74,* 395.
45. Pope, E.J.A.; MacKenzie, J.D. *MRS Bull.* **1987**, March 17, 29.
46. Fujii, T.; Kawauchi, O.; Kurikawa, Y.; Ishii, A.; Negishi, N.; Anpo, M. *Chem. Express* **1990**, *5,* 917.
46a. Shen, P.; Li, M.; Najafi, I.; Currie, J.F.; Leonelli, R. *Proc. SPIE* **1990**, *1328,* 338.
47. Knobbe, E.T.; Dunn, B.; Fuqua, P.D.; Nishida, F.; Kaner, R.B.; Pierce, B.M. *Ceram. Trans.* **1990**, *14,* 137.
48. Knobbe, E.T.; Dunn, B.; Fuqua, P.D.; Nishida, F.; Zink, J.I. in: *Proc. 4th Int. Conf. Ultrastructure of Ceramics.* Wiley: New York, 1989. Zink, J.I.; Dunn, B.; Kaner, R.B.; Knobbe, E.T.; McKiernan, J. in: *"Materials for Non-Linear Optics-Chemical Perspectives",* Marder, S.R. et al., Eds. Am. Chem. Soc., Washington, 1991, p. 541.
49. Levy, D.; Serna, C.J.; Oton, J.M. *Mat. Lett.,* **1991**, *10,* 470. Oton, J.M.; Serrano, A.; Serna, C.J.; Levy, D. *Liquid Crystals,* **1991**, in press.
50. Levy, D. M.Sc. Thesis, The Hebrew University, Jerusalem, 1984.
51. Reisfeld, R.; Eyal, M.; Brusilowski, D. *Chem. Phys. Lett.* **1988**, *153,* 210.
52. Y. Haruvy et al., this symposium.
53. Fujii, T.; Ishii, A.; Anpo, M. *J. Photochem. Photobiol.* **1990**, *A54,* 231.
54. Santos, D.I.; Aegerter, M.A.; Brio-Cruz, C.H.; Scarparo, M.; Zarzycki, J. *J. Non-Chryst. Solids* **1986**, *82,* 231.
55. Eyal, M.; Reisfeld, R.; Chernyak, V.; Kaczmarek, L.; Grabowska, A. *Chem. Phys. Lett.* **1991**, *176,* 531.
56. Reisfeld, R.; Brusilovski, D.; Eyal, M.; Jorgensen, K. *Chimia* **1989**, *43,* 385.
57. Reisfeld, R.; Gvishi, R. *Chem. Phys. Lett.* **1987**, *138,* 377.
58. Reisfeld, R. in: Sol-Gel Science and Technology; Aegerter et al., Eds.; World Scientific: Singapore, 1989; pp. 323-345.
59. Kobayashi, Y; Kurokawa, Y. Imai, Y; Muto, S. *J. Non-Cryst. Solids* **1988**, *105,* 198.
60. Sasaki, H.; Kobayashi, Y.; Muto, S.; Kurokawa, Y. *J. Am. Ceram. Soc.* **1990**, *73,* 453.
61. Dunn, B.; Knobbe, E.T., McKiernan, J.M.; Pouxviel, J.C.; Zink, J.C. *Mat. Res. Soc. Symp. Proc.* **1988**, *121,* 331.
62. Knobbe, E. T.; Dunn, B.; Fuqua, P. D.; Nishida, F. *Appl. Opt.* **1990**, *29,* 2729.
63. Salin, F.; Le Saux, G.; Geoges, P.; Brun, A.; Bangall, C.; Zarzycki, J. *ibid,* **1989**, *14,* 785.

64. Reisfeld, R.; Seybold, G. *Chimia* **1990**, *44*, 295. Reisfeld, R. et al. *Chem. Phys. Lett.* **1989**, *160*, 43.

65. Nogues, J.-L. et al. *J. Am. Ceram. Soc.* **1988**, *71*, 1159.

65a. Fujii, T.; Ishii, A.; Anpo, M. *J. Photochem. Photobiol.* **1990**, *54A*, 231. Fittermann, J.; Doeuff, S.; Sanchez, C. *Ann. Chim. Fr.* **1990**, *15*, 421. Lecomte, M.; Viana, B.; Sanchez, C. *J. Chim. Phys.* **1991**, *88*, 39. McKiernan, J.M.; Yamanaka, A.S.; Dunn, B.; Zink, J.I. *J. Phys. Chem.* **1990**, *94*, 5652. Haruvy, Y.; Webber, S.E. *Chem. Mater.* **1991**, *3*, 501. Hinsch, A.; Zastrow, A.; Wittner, V. *Sol. Energy Mater.* **1990**, *21*, 151. Kobayashi, Y.; Sasaki, H.; Muto, S.; Yamazaki, S.; Kurokawa, Y. *Thin Solid Films* **1991**, *200*, 321. Gvishi, R.; Reisfeld, R. *J. Non-Cryst. Solids* **1991**, *128*, 69. Chernyak, V.; Reisfeld, R. *Chem. Phys. Lett.* **1991**, *181*, 39. Several papers in *Proc-SPIE* **1990**, *1328* Session 3.

66. Levy, D.; Avnir, D. *J. Photochem. Photobiol. A: Chem.* **1991**, *57*, 41.

67. Kaufman, V.R.; Avnir, D. *Langmuir* **1986**, *2*, 717.

68. Brusilovsky, D.; Reisfeld, R. *Chem. Phys. Lett.* **1987**, *141*, 119.

69. Takahashi, Y; Kitamura, T.; Uchida, K.; Yamanaka, T. *Jap. J. Appl. Phys.* **1989**, *28*, L1609. See also, *idem., J. Luminesc.* **1991**, *48/49*, 373.

70. Matsui, K.; Nakazawa, T. Bull. Chem. Soc. Jap. **1990**, *63*, 11. Matsui, K.; Usuki, N. *Bull. Chem. Soc. Jap.* **1990**, *63*, 3516.

71 Yamanaka, T.; Takahashi, Y.; Kitamura, T.; Uchida, K. *Chem. Phys. Lett.* **1990**, *172*, 29. See also, *idem., J. Luminesc.* **1991**, *48/49*, 265.

71a. Guizard, C. et al. *Proc-SPIE* **1990**, *1328*, 208. Matsui, K.; Nakaza, T.; Morisaki, H. *J. Phys. Chem.* **1991**, *95*, 976.

72. Kaufman, V. R.; Avnir, D. *Mat. Res. Soc. Symp. Proc.* **1986**, *73*, 145.

73. Kaufman, V.R.; Avnir, D.; Pines-Rojanski, D.; Huppert, D. *J. Non-Cryst. Solids* **1988**, *99*, 379.

74. Fujii, T.; Ishii, A.; Nagai, H.; Niwano, M.; Negishi, N.; Anpo, M. *Chem. Express* **1989**, *4*, 1.

75. Pouxviel, J.C.; Parvaneh, S.; Knobbe, E.T.; Dunn, B. *Solid State Ionics*, **1989**, *32/33*, 646.

76. Matsui, K.; Matsuzuka, T.; Fujita, H. *J. Phys. Chem.* **1989**, *93*, 4991.

77. Fujii, T.; Mabuchi, T.; Mitsui, I. *Chem. Phys. Lett.* **1990**, *68*, 5.

78. Yamanaka, T.; Takahashi, Y.; Uchida, K. *ibid*, **1990**, *172*, 405.

79. McKiernan, J.; Pouxviel, J.C.; Dunn, B.; Zink, J.I. *J. Phys. Chem.* **1989**, *93*, 2129.

80. Winter, R.; Hua, D.W.; Song, X.; Mantulin, W.; Jonas, J. *ibid*, **1990**, *94*, 2706.

81. Schmidt, H.; Popall, M. *Proc.-SPIE* **1990**, *1328*, 249.

82. Schubert, U.; Rose, K.; Schmidt, H. *J. Non-Chryst. Solids* **1988**, *105*, 165.

83. Robert, M.R.; Coltrain, B.; Melpolder, S.M. *U.S. Patent 4*, 948, 843, Aug. 14, 1990. See also, *Proc. 2nd Int. Ceramic Sci. Tech. Congress*, Orlando, Nov. 1990.

84. Nishida, F.; Dunn, B.; Knobbe, E.T.; Fuqua, P.D.; Kaner, R.B.; Mattes, B.R. *Mater. Res. Soc. Symp. Proc.* **1990**, *180*, 747.

85. Wung, C.J.; Pang, Y.; Prasad, P.N.; Karasz, F.E. *Polymer* **1991**, *32*, 605.

RECEIVED January 16, 1992

Chapter 28

Sol–Gel Preparation of Optically Clear Supported Thin-Film Glasses Embodying Laser Dyes

Novel Fast Method

Y. Haruvy[1,4], A. Heller[1,2], and S. E. Webber[1,3]

[1]Department of Chemistry and Biochemistry, [2]Department of Chemical Engineering, and [3]Center for Polymer Research, University of Texas at Austin, Austin, TX 78712

Fast sol-gel processes are facilitated by the direct reaction of alkoxysilane monomers with water at 60-80°C, using near-stoichiometric water-to-siloxane ratios. Following partial polymerization and distilling-out of the produced alcohol within several minutes, the viscous polymer is spin-cast onto a support. The facile polymerization and crosslinking processes of the glass-film, typically 10-25μm thick, are completed within a few hours at 60-70°C, enabling preparation and measurements on the same day. No additive or surfactant is required to eliminate the typical fracturing of the glasses during the fast curing step. The process is carried out with single layer, as well as with multi-layered assemblies of varying indices of refraction and total thickness up to 50μm. Chromophore-encaging glasses are attained by introduction of laser-dye into the reaction mixture. Minor tuning of the process is needed for different dyes. The absorbance and fluorescence spectra of the laser-dyes embodied in the glasses are typical of monomeric chromophores, only slightly shifted to the red, even at high concentrations of the dyes ($>10^{-2}$ M). The convenience of this fast and facile method provides a promising route for non-linear optics applications such as a two-dimensional dye laser system.

The sol-gel techniques extensively investigated for almost three decades have been comprehensively reviewed in several symposia (1-3) and books (4-5). The sol-gel synthetic routes have been used to prepare glasses and ceramics for use in a wide variety of applications, employing various precursors, catalysts, additives and procedures, which have been reviewed in detail by several authors (6-8). The most investigated precursors were the alkoxysilanes (9-14), while titania (15), alumina (16-18) and mixed glasses (19) have been investigated to a lesser extent. Numerous chromophores have been incorporated into sol-gel produced glassy materials, laser-dyes in particular. The latter exhibited promising characteristics for use in non linear optics (NLO), especially for laser systems (9-20). Dye embodying supported glass thin films were accomplished by the sol-gel technique, aiming at surface laser systems (9,11,16,21), yet prolonged and complex processes were required to facilitate crack-free glasses of the desired properties.

[4]On Sabbatical leave from Soreq NRC, Yavne 70600, Israel

0097–6156/92/0499–0405$06.00/0

Tetraethoxysilane is the favored sol-gel precursor due to its moderate reaction rate. The commonly employed water-to-siloxane molar ratios (MR) are 5:1 to 10:1. A co-solvent (e.g. ethanol) is regularly added to maintain a one-phase reaction solution (11), although in the reaction of tetramethoxysilane the methanol produced by the hydrolysis becomes sufficient to maintain a single-phase at a very early stage (22):

$$Si(OMe)_4 \ + \ H_2O \ \longrightarrow \ Si(OMe)_3OH \ + \ MeOH \quad (1)$$

Following hydrolysis of the alkoxide groups polycondensation takes place via Si-O-Si bonds. The glassy matrix formed by this polymerization is capable of encaging large molecules (e.g. chromophores, enzymes) which have been introduced into the reaction mixture. During this stage of the sol-gel synthesis severe cracking and fragmentation of the formed glass are the common obstacles that impede the fabrication of articles and films in general, and supported films in particular. This is due to the extensive volume-contraction which accompanies the condensation reactions and the corresponding expulsion of the solvent and the condensation products (23). Several investigators overcame this obstacle by using substantial quantities of additives (24, 25) or surfactants (11, 21) but slow and very cautious drying of the sol-gel glass was still necessary to attain a fracture-free glass, making this synthetic route more of an art.

The cracking problem is aggravated for sol-gel glasses cast onto a rigid support since the gelling matrix is no longer free to contract. Crack-free supported glass-films can be maintained by a surfactant, yet a high concentration (up to 3% in the final glass) is required (21) and its durability towards decomposition under conditions of high energy-density (e.g. solid-state laser) is questionable.

In the present work we aim at a fast and convenient synthetic route for the sol-gel preparation of glass thin-films attached to a rigid support. We discuss the rationale for the different phase-separation phenomena observed during the reaction as well as the rationale for the cracking syndrome observed in the latter stage, aiming at better understanding of these features. We show that the rationale for these phenomena provides the methodology to maintain a single-phase reaction throughout the various stages as well as the synthetic route to overcome the cracking and detachment obstacles.

The effect of incorporation of various guest-molecules on the sol-gel process is further discussed as we load the glasses with high concentrations of NLO active molecules. Finally, the facile features of this fast route to crack-free supported films which are up to 50 μm thick and capable of encaging discrete chromophore molecules at high concentrations are described and their research applications are discussed.

Experimental Methods

Materials. Monomers: tetraethoxysilane (TEOS), tetramethoxysilane (TMOS), methyltrimethoxy-silane (MTMS) dimethyldimethoxy-silane (DMDMS) and trimeth-oxysilane (HTMS) were purchased from Aldrich. Chromophores: Polyphenyl-1 (PP), Coumarin-153 (COU), Rhodamine-6G (RH) and Pyridine-1 (PY) were purchased from Lambdachrome, p-nitroaniline (PNA) and 4,4' diaminodiphenylsulfone (DDS) were purchased from Aldrich. Catalysts: hydrochloric acid (Baker, AR) and dimethylamine (Kodak, AR). All the materials were used without further purification.

Preparation of the Support Glasses. Glass slides (Pyrex, 1"x1" and 3"x1", and Quartz, φ = 1") were precleaned with detergent, washed with deionized water , dried and immersed 24 h in a 6N NaOH solution, or in a mixture of 1:10 H_2O_2 (30%) and H_2SO_4 (98%). Then, they were rinsed thoroughly with deionized water, dried and kept dry until the casting. FTIR samples' supports made of Aluminum foil or slab were wiped with Kimwipe paper soaked with acetone and then dried.

Set-up. The experimental set-up is schematically described in Figure 1 (adapted from (26) copyright ACS). The reactions were carried out in disposable vials equipped with a screw-cap, immersed in a thermostated bath and magnetically stirred. As a safety precaution necessitated by splashing during the vigorous reaction of the some alkoxy monomers at elevated temperatures, the reactions were started with tightened screw-cap which was gradually released after ca. 30 s to allow evaporation of volatile products.

1. THERMOSTATED BATH
2. HOLLOW BEADS
3. STYROFOAM BAR
4. MAGNETIC BAR
5. THERMOMETER
6. "REACTOR" (VIAL OF 1-4 DRAMS)
7. "PRESSURE REGULATOR" (CAP)
8. "VISCOMETER" (BUBBLES)

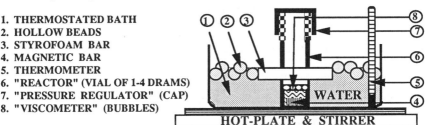

Figure 1. Experimental set-up

Monitoring the Progress of Polymerization and Casting. The progress of the hydrolysis reaction was monitored gravimetrically by recording the weight-loss following the distillation-out of the alcohol produced therefrom. The propagation of the condensation was also watched by the size and the duration of the bubbles produced by the alcohol boiling out from the viscous polymerizing solution.

Films were spin-cast on the support using a bench-top spin-coater (Headway Research, Model 2-EC101-R485). The pretreated support was attached to the spinner chuck prior to the polymerization. The polysiloxane liquid was poured onto the support at the appropriate viscosity and spun-cast for 240 s at 1000 RPM. The samples were left to cure until their surface was no longer sticky. Typical curing duration was a few hours (at 60-70°C) or a few days (at ambient temperature and humidity).

Determination of film Thickness. The thickness of the dry films was calculated from the measured absorbance of the dye in the glass, the coefficient of extinction, the dye concentration and the density of the glass film (measured to be ca. 1.3 g/cm^3). Direct measurements were carried out with an Inverted-Stage Epifluorescent Video Microscope (Leitz Fluovert; Rockleigh, NJ).

Observations, Results and Discussion

Phase Separation. The phase separation phenomena observed can be classified into two types: liquid - liquid separation, and liquid - gel (glass) separation, commonly referred to as precipitation or sedimentation (27). These phase-separations depend to a large extent on both the type and concentration of the catalyst employed for the reaction. The following discussion refers to sol-gel reactions catalyzed with HCl at a typical concentration of 10^{-2} M.

Liquid - liquid phase separation typical of the initial stage of the reaction occurs due to the immiscibility of water in the hydrophobic siloxane precursor(s). Phase merging follows the hydrolysis reaction (1) due to formation of alcohol and hydroxy siloxane species as well as consumption of water in the reaction (especially true in water-lean recipes). Elevation of the temperature to 60°C or higher accelerates these processes and phase merging occurs within a few seconds.

The second type of phase separation is typically observed in the reaction of TEOS carried out at water-to siloxane molar ratios (MR) smaller than 5:1. It results in precipitation of glass-gel-particles from the sol syrup at an early stage of the hydrolysis:

$$Si(OEt)_4 + H_2O \longrightarrow Si(OEt)_3OH + EtOH \rightarrow\rightarrow\rightarrow Si(OH)_4 \qquad (2)$$

This sedimentation reflects fast aggregation of small sol particles to larger ones which can no longer be held by the solution. To eliminate this aggregation, one needs to shift the condensation reaction (3) to the left during the hydrolysis stage:

$$\equiv\!Si\text{-}OH \quad + \quad HO\text{-}Si\!\equiv \quad \rightleftarrows \quad \equiv\!Si\text{-}O\text{-}Si\!\equiv \quad + \quad H_2O \qquad (3)$$

A surplus of water is required, therefore, to accomplish a precipitate-free sol-gel process of TEOS. It is noteworthy that impeding the condensation also may be attained by employing high concentrations of HCl (23,27). Entirely different behavior is observed when TMOS is employed: the MR can be set as low as 1:1 without any precipitation, even at ambient temperature when a relatively slow reaction takes place. This may be attributed to the higher miscibility of the polymethoxysiloxane intermediates in the methanol produced in the hydrolysis. Nevertheless, all our preparations using this monomer failed to yield sols which could be cast and cured on a rigid support without undergoing severe cracking shortly after being spin-cast.

The immediate questions are whether the phase separation can be eliminated by manipulating the speed of the reaction, e.g. by elevating the temperatures or by using different monomers, and can these changes also eliminate the cracking? These questions drive us to the need to choose between two possible routes of the sol-gel progress: one, via a reaction-controlled mechanism, typical of diluted systems, in which the glass is attained through the formation of small sol particles and their subsequent aggregation to the gel (7,8 [c.f. Figure 1]) and the alternative, via a diffusion-controlled mechanism, typical of concentrated systems, in which fast polymerization of the monomers yields primarily linear chains which undergo slow crosslinking via the remaining active groups (7, 8, 28). A combination of the two is also plausible.

It seems reasonable that the faster the reaction is conducted, the less phase separation (of both types) should take place. Our observations during all the experiments indicated that this is indeed the case, and strongly suggested that the fast polymerization route should be preferred. This issue seemed to be related not only to the phase separation problem, but also tightly connected with the contraction-fracture syndrome, as described in the following subsection.

Condensation, Contraction and Fracturing of the Sol-Gel. The salient observation during numerous sol-gel experiments performed at room temperature, using a wide variety of tetraalkoxysilane-recipes and various hydrolysis routes was the contraction-fracture syndrome. This syndrome occurred upon casting the viscous sol onto the support (with or without spinning) within a period of a day or two. It was manifested not only in fracturing of the cast-glass layer, but also in subsequent fragmentation and detachment from the support. This syndrome is an inevitable outcome of the drastic compaction of the gel, which occurs as the products of the hydrolysis and the condensation reactions, namely, alcohol and water, are expelled from the glass while it crosslinks (6,7,8 [c.f. Figure 1], 28). Calculated values of this contraction for several glass compositions are displayed in Table I.

From this data it becomes clear that one of the first steps towards a crack-free supported glass should be to minimize the volume of the reactants. A few experiments lent support to this rationale: the cracking was reduced, yet not eliminated. It was obvious that we still need to acquire more explicit understanding why this syndrome takes place and to suggest better synthetic routes to overcome it. Hence, the inter-relation between the recipe, the procedure, the contraction and the fragmentation and detachment were further analyzed.

From the data in Table I it is obvious that the commonly employed super-stoichiometric molar ratio (MR) of 5:1 substantially enhances the severity of the problem. This MR is required for sol-gel reactions of TEOS to eliminate precipitation, but since it is unnecessary for TMOS, the latter monomer was preferred. Thus, we could employ a much lower MR and also benefit from the lower molar volume of this monomer, 149 cm^3 as compared to 223 cm^3 for TEOS, reducing the volume contraction from 5 to 3.1 (c.f. Table I). It is also advantageous that the much higher reaction rate of TMOS drives us further towards the polymerization route (**28**).

Table I. Calculated Contraction for SiO$_2$ Glass Prepared by the Sol-Gel Technique

Reactants	SiO$_2$ Weight Fraction	Volume (a) Contraction	Longitudinal Contraction
One-phase Si(OEt)$_4$/H$_2$O/MeOH 1 : 5 : 5 (m/m)	0.13	7.6	2.0
Two-phase Si(OEt)$_4$/H$_2$O 1 : 5 (m/m)	0.20	5.0	1.7
Si(OMe)$_4$/H$_2$O 1 : 5 (m/m) (**b**)	0.25	4.0	1.6
Si(OMe)$_4$/H$_2$O 1 : 1 (m/m) (**b**)	0.32	3.1	1.5
MeSi(OMe)$_3$/H$_2$O 1 : 1.5 (m/m) (**b**)	0.41	2.4	1.34
"Distilled" recipes: Si(OMe)$_4$/H$_2$O 1 : 1.5 (m/m) (**b**) -2MeOH	0.52	1.9	1.24
MeSi(OMe)$_3$/H$_2$O 1 : 1.5 (m/m) (**b**) -2MeOH	0.68	1.5	1.14

(a) all densities were taken as **1**. Actual density of the sol-gel glass may vary from 0.9 to 1.4 (**1-3**), according to the recipe and the procedure employed.
(b) single-phase is formed within 10-30 sec.

It is interesting to note that drying of <u>non-supported</u> glasses from TEOS or TMOS may involve contraction of 50-90% in each dimension, dictating very cautious, complex and tricky drying procedures to obtain crack-free glass (**11,21,24,25,28, 29**). The primary precaution is a very slow rate of drying and periods of a few weeks are typically cited. Slow drying alone is effective only for the preparation of <u>self-supporting</u> matrices. The problem is drastically aggravated when the gel is attached to a rigid support, the contraction in two dimensions is restricted, and the gel has to rearrange itself to maintain the full volume contraction <u>(200-600%)</u> in one dimension only or else, undergo extensive cracking. Hence, additives had to be employed in order to maintain crack-free drying of supported thin-films (**11,21,24,25**).

In this respect, an important observation was the effect on the detachment of the two procedures for the preparation of the support glass (c.f. experimental section): much less detachment occurred on the base-washed supports as than on the acid-washed ones. These findings can be attributed to some etching of the glass-support effected by the concentrated NaOH, thus increasing the surface area and thereby the adhesion of the cast-gel to the support.

Careful observations of gradual contraction and fragmentation of the gels attached to the rigid supports reveal that there occur simultaneously at least two major processes: cracking and fragmentation of the film, and detachment of the film from the support. Most probably, there also take place some stress-relaxation processes, via bond-reformation and segment-relocation processes. A very cautious drying results in a film of which part is intact but detached from the support, part is fractured and (small) parts are neither fractured nor detached, manifesting the occurrence of all three processes.

It is obvious, therefore, that in order to eliminate the occurrence of both the cracking and the detachment phenomena, we ought to **enable the gel-reformation and stress relaxation processes to occur predominantly!** It is most reasonable to analyze these processes from the point of view of polymer science. Doing so we can distinguish between three principle processes which take part in the course of this relaxation, as displayed in Figure 2 (from (26) copyright ACS).

Si - ATOM RELOCATION	POSSIBLE ONLY WHEN TEMPORARILY TWO OF THE FOUR Si- BONDS ARE "FREE"
I I -O-Si-O-Si-O RELOCATION I I OF SEGMENT	POSSIBLE ONLY FOR A 1-D POLYMER-SEGMENT (SLIGHTLY CROSSLINKED POLYMER)
RELOCATION OF A GEL PARTICLE	POSSIBLE ONLY IF THE COLLOIDAL-PARTICLES OF THE GEL ARE SEGREGATED (e.g.BY A SURFACTANT)

Figure 2: the main processes of gel reformation and relaxation

In light of the this rationale, three possible strategies can be employed to overcome the contraction-fragmentation obstacle. The first one, suggested by Avnir, Reisfeld and collaborators, is the use of a surfactant (**11, 21**) which most probably maintains some segregation of the gel-particles. Thus, a certain amount of mobility is imparted and relaxation processes can take place, most probably by some type of gliding of the particles along surfactant-coated surfaces. In such a case, however, there arises an interesting question: will the dye molecules preferably be trapped inside the particles, or onto their surfaces? This question may be of special importance when dye-containing glasses are used in environments where dye may be leached out by contact with a solution. An alternative route, suggested by Hench (**25**) and Sakka (**24**), is to use drying-control chemical additives such as DMF (**24**), formamide or organic acids (**25**) to maintain a very slow drying rate and thus, to allow sufficient time for evaporation of solvent and condensation products to take place without cracking the matrix. However, this method results in a highly porous glass, and a high temperature post-curing vitrification process is required to endow the desired optical quality to glass articles prepared by this technique.

Because an additive-free glass was our goal in this sol-gel research, an alternative to the earlier approaches to overcome the fracture problem was necessary. The preferred approach seemed to be to directly maintain the processes of the gel-reformation and relaxation, by imparting substantial <u>inherent internal mobility</u> to the formed polymeric siloxane chains/particles and thus, to enable their rearrangement. The principle precondition for such an inherent mobility is that <u>a substantial part of the Si atoms, at least part of the time, will be bound to the polymer by only two substituents</u>, while the other two must remain free to move. This principle is illustrated in Figure 3.

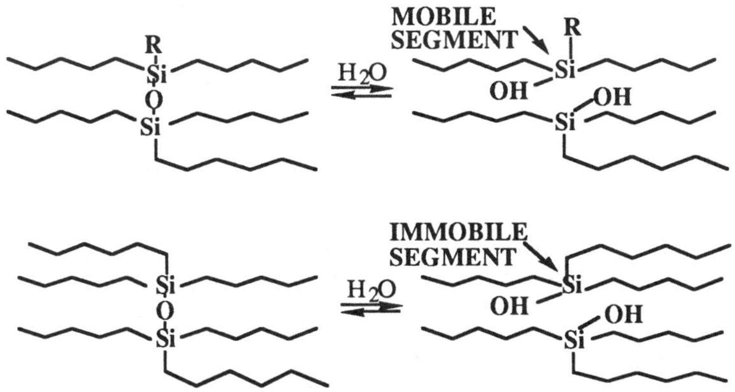

Figure 3: mobile and immobile Si-segments formed via temporary hydrolysis of one Si-O-Si bond of an atom which has three and four polymeric substituents, respectively.

TEOS and TMOS can form up to four bonds to the polysiloxane matrix. In this case, as crosslinking proceeds the possibility of any relaxation processes vanishes (28) since the probability that two of the Si-O-(GEL) bonds will open at the same time is extremely low. To eliminate cracking we need to allow relaxation to proceed. This implies that the maximum number of bonds between an Si atom and the gel must be 3. Thus, upon a temporary rehydrolysis of any Si-O-(GEL) bond, its number of bonds with the gel is reduced to 2, which allows mobility of that Si atom, as illustrated in Figure 3. It is noteworthy that by applying this precondition, we completely shift the problem from a sol-gel aging-drying process towards a polymer-crosslinking one.

From the above discussion, it becomes obvious that what is needed to impart mobility to backbone-atoms during the gelling of the polymer is to block one substituent of the silicon. Blocking by a less-reactive bulky alkoxy-substituent is undesired since it implies a penalty of an increased contraction due to the increased volume, and yet cracking can start eventually when this group will be hydrolyzed. The better alternative is to select a monomer with a single blocked substituent, which is non-reactive under the mild conditions of the sol-gel reaction. Such monomers are the alkyl-trialkoxy-silanes: $R'-Si(OR)_3$. This type of monomer is analogous to $Al(OR)_3$, monomers for which successful formation of thin-films by the sol-gel technique has been recently reported (16-18).

The smallest stable single blocked siloxane monomer is MTMS [MeSi(OMe)$_3$]. The sol-gel processes of this monomer were investigated in the past for the preparation of crack-free cladding of laser rods (30-32). The effectiveness of this application was enhanced because of the very low index of refraction of the polymethylsiloxane thus formed. However, the method employed then was "too slow" for industrial uses (32) and hence abandoned. Later, it was investigated in relation to the formation of optical fibers (33) and this direction was also discontinued. Nonetheless, the special stability of the Si-C bonds in this monomer and the absence of C-C bonds ensure that the polymer derived from this monomer is most unlikely to undergo any chemical changes when exposed to intense illumination conditions (it can withstand even the very high doses of ionizing-radiation in space environment (34)).

It should be stressed that our objective is to block one Si-bond, and the small methyl group is sufficient for this purpose, yet it does not introduce an organic-polymer modification of the silicate (14), which is not necessary for our purposes. Crack-free supported thin films were attained right from the first sol-gel preparations with this monomer, and these films were stable throughout the drying stage. We focused our efforts in sol-gel synthesis at the lower MR range, i.e. between 2:1 to 1:1, to achieve formation of mono- and dihydroxy monomers in order to drive the process towards linear polymerization, while keeping minimal volume of the reactants:

$$\text{n } RO\text{-}\underset{\underset{OR}{|}}{\overset{\overset{R'}{|}}{Si}}\text{-}OR + \text{n } H_2O \quad ---\!\!> \quad [\text{-}\underset{\underset{OR}{|}}{\overset{\overset{R'}{|}}{Si}}\text{-}O\text{-}]_n + 2n\ ROH \qquad (4)$$

With MTMS as the monomer, phase-merging of the reactants occurred within a few seconds even at room temperature, glass-sedimentation was no longer observed at MR as low as 1:1, and the reaction remained homogeneous throughout the hydrolysis and condensation steps. Several representative experiments and observations are summarized in Table II (adapted from (26) copyright ACS). The most important advantage of this monomer was that the syrup produced therefrom could be cast or spin-cast onto appropriately prepared supports (c.f. experimental section) and dried at ambient conditions while remaining crack-free and shock-insensitive. This facile drying is effective for films up to 25μm thick, while for thicker films and multilayered assemblies a little more cautious drying is required, as discussed in a later subsection.

Table II. Sol-Gel Experiments and Observations Under Acid Catalysis [a]

REACTION CONDITIONS			OBSERVATIONS			
Monomer	Water Ratio (m/m)	Temp (°C)	Ph.M (b) (min)	Cracking (hours) [c]	Clarity	Remarks
Si(OEt)$_4$	5.7	25	~10	~6 ; M	OK	Ph.S after casting (d)
-"-	5.7	60	< 1	24 ; M	OK	--
-"-	1.9	60	< 1(e)	--	--	--
Si(OMe)$_4$	5.0	25	< 0.1	24 ; M	OK	detachment from the glass support
-"-	2.0	25	-"-	-"-	OK	-"-
-"-	1.5	60	-"-	-"-	OK	-"-
-"-	1.0	60	-"-	-"-	OK	-"-
MeSi(OMe)$_3$	1.5	25	< 0.1	24 ; F	Milky	Ph.S after casting (ca. 24 h)
-"-	1.5	60 / 25 (f)	< 0.1	24 ; F	-"-	-"-
-"-	1.5	65-70	< 0.1	none	-"-	Ph.S after casting (v. slight)
-"-	1.5	70	< 0.1	none	Clear	Regular Dye Abs. & Fluores. Spectra
-"-	1.5	70-80	< 0.1	none	Clear	-"-

(a) Catalyzed with HCl, 10^{-2} M.
(b) Ph.M - Phase Merging: the time until reactants merge into single phase is indicated.
(c) Time of earliest observation of cracks.   F - few;   M - many.
(d) Ph.S - Phase Separation occurring after the film is cast.
(e) Glass precipitation occurs within 90 min. at this water-to-monomer ratio.
(f) Performed in two steps, 10 min. each, at 60°C and then 25°C.

It is noteworthy that the supported glass films prepared so far by the sol-gel technique not only needed more complex and prolonged preparation but also were limited to thickness of 1 μm or less (5 [p. 506]). Although the use of organic modification of the monomers and the formation of composite siloxane-organic-polymers to attain thick supported glass films have been widely investigated (35), the principle of blocking one substituent has not been explicitly suggested.

Phase Separation in Cast Films and the Homogeneity Principle. The sol-gel study was first done in the temperature range of 25-60°C. In most of the experiments with MTMS carried out in this range of temperatures the cast gel-film dried crack-free, yet a short time after casting all these films developed a milky opaque surface which ruined any prospective use for optical-related devices.

Careful observation of the cast sol-gel films during their drying steps revealed that the appearance of these milky surfaces was the result of a <u>new phase-separation</u>: a new liquid-phase was expelled from the gel-phase shortly after it has been cast and was floating on top of it, as discrete droplets or as continuous layers. In experiments in which this liquid-layer was wiped off the remaining gel dried to fine clear glass. These glasses were also prepared with low concentrations of dye ($\sim 10^{-3}$ M). Thus, it could be clearly observed that the more hydrophobic dye (e.g. Coumarin-153) tended to concentrate in the liquid upper-phase, while the more hydrophilic one (Rhodamine 6G) tended to remain in the gel and the expelled liquid layer was dye-depleted. It was further observed that upon aging at ambient conditions, solidification of this upper layer led to the formation of the opaque ("milky") coating. The latter often developed cracks which progressed into the clear lower layer. When the upper liquid layer was wiped-off before solidifying, the remaining clear gel cured to a crack-free glass.

These observations clearly imply that at the low MR, two different (and immiscible) populations of siloxane intermediates are present simultaneously in the apparently homogeneous reaction medium. The occurrence of phase-separation only some time after the casting can be attributed to rapid evaporation of the methanol (acting as a co-solvent) following the great increase in the surface area of the gel. This implies that our assumption that stoichiometric hydrolysis of MTMS results in a uniform population of semi-hydrolyzed intermediates (5) may be erroneous at low temperatures:

$$2\,MeSi(OMe)_3 \;+\; 3\,H_2O \;\xrightarrow{\;2\;}\; 2\,MeSi(OMe)_{1.5}(OH)_{1.5} + 3\,MeOH \quad (5)$$

$$[MeSiOMe(OH)_2 \;+\; MeSi(OMe)_2OH]$$

In other words, these observations should alert us towards the classical question about **"where would the second reacting molecule go"** (Markovnikov's Law), as illustrated in Figure 4 (from (26) copyright ACS).

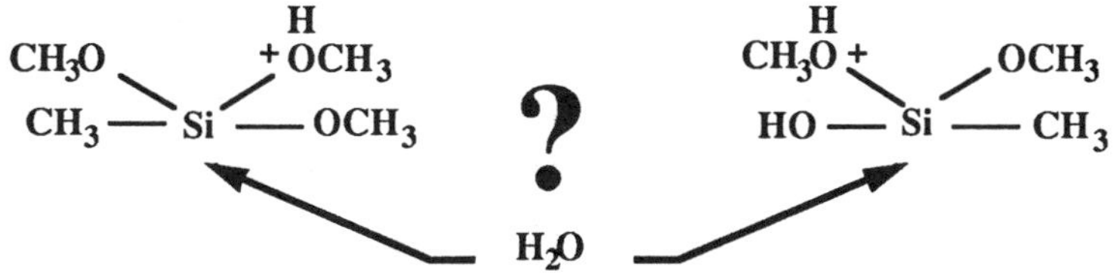

Figure 4.   Where would the second molecule go?

The answer for this basic question is that the second molecule of water will react, via a nucleophilic attack, with the Si nucleus which has the less electron-donating substituents. Therefore, the subsequent hydrolysis will take place at the monomer which has already been hydrolyzed once and acquires a hydroxyl-group instead of a methoxyl one. Thence, the further these molecules are hydrolyzed, the greater their reactivity becomes. Thus, following subsequent hydrolysis reactions, the majority of the molecules will become either completely hydrolyzed or remain intact, while only a small fraction of the molecules will be partially hydrolyzed. A ^{29}Si-NMR study is currently being conducted to verify this point.

The consequent question is how to drive the molecules to undergo uniform partial hydrolysis while "overruling" their different reactivities. In other words, we require the hydrolysis steps to occur stochastically, regardless of the reactivity of the molecules. This leads us to the key to the answer: since the reactivity reflects the energy of activation for the reaction, to overrule it we should simply raise the temperature! In more general terms, we follow the basic rule that whenever undesired products are formed by a thermodynamically driven reaction, it is reasonable to try to obtain the desired products by allowing the kinetics to take over, by means such as elevated temperature, more reactive alkoxy-groups, or enhanced catalysis. Thus, while at lower temperature the reaction of MTMS with water at a molar-ratio of $1:1.5$ resulted in a mixture composed primarily of $MeSi(OH)_3$ and $MeSi(OMe)_3$, at higher temperatures the desired semi-hydrolyzed $MeSi(OMe)_2OH$ and $MeSiOMe(OH)_2$ species can prevail.

This hypothesis was strongly supported by the observation that the post-hydrolysis phase separation occurred to a lesser extent in the experiments performed at 60°C than those performed at 25°C. Further elevation of the temperature to 70-80°C resulted in a viscous liquid which, after being spin-cast, dried without any significant phase separation and yielded the desired thin-film glass.

The accumulated experimental evidence of glass sedimentation in TEOS recipes complies well with our hypothesis and lends further support to this mechanism: at a molar ratio of water to TEOS well above the stoichiometry of complete hydrolysis $(4:1)$, a uniform population of fully hydrolyzed species is formed and hence, no post-casting phase-separation is observed. Conversely, if a lower ratio is employed at ambient temperature, according to our hypothesis part of the TEOS molecules quickly undergo quadruple hydrolysis (due to their increasing reactivity) and separate from the remaining hydrophobic monomer as a second phase, most probably accompanied by the remaining water. Thence, this water- and catalyst-rich phase undergoes fast condensation to gel particles which manifests in the early glass-sedimentation typical of TEOS. The much later sedimentation in TMOS recipes should be attributed to the higher reactivity of its alkoxy-groups and more even hydrolysis as compared to TEOS.

The importance of the water-to-monomer ratio in sol-gel processes is well recognized (11,33,35-39), as well as the question of "where would the second molecule go", inherent at low MR to the multistage hydrolysis reactions (Avnir, D., Personal Communication, 1990, Univ. Texas at Austin). Sakka et al. (36,37) in their research on the preparation of silica fibers and films by the reaction of TEOS with sub-stoichiometric quantities of water had solved this very problem by two alternative methods. In one case, elevated temperature (80°C, MR of 1.5 to 2) was employed (36) while in the other case, a dropwise addition of water-ethanol mixture to a vigorously stirred TEOS-ethanol solution was employed to maintain a clear sol (37). However, the principle of temperature-elevation during the hydrolysis stage as a means to maintain homogeneity throughout the curing stage by the formation of evenly-hydrolized siloxane monomers has not yet been explicitly suggested.

Finally, it is widely accepted that sol-gel systems which employ sub-stoichiometric MR and/or enhanced reaction rates proceed via the polymerization mechanism, rather than via the sol aggregation to a gel (for example: Brinker (40), Sakka (36,37,39), Yoldas (41,42,43) Schmidt (6)). Our synthetic route employs both low MR and elevated temperature. Hence, most probably in our synthesis polymeric chains rather than particles have been formed and contribute to the mechanical stability as well as to the flexibility of the resultant glasses (c.f. Figure 5 presented later). The gel formed from MTMS is most probably *entirely different* from that of TMOS since its dimers can form long segments of linear ladder-like polymer, as already suggested by Sakka (33), segments that have only methyl side-groups and hence do not undergo crosslinking:

$$CH_3O - \underset{\underset{OCH_3}{|}}{\overset{\overset{CH_3}{|}}{Si}} - OCH_3 \quad \xrightarrow{H_2O} \quad \begin{bmatrix} CH_3 \\ | \\ -Si-O- \\ O \\ -Si-O- \\ | \\ CH_3 \end{bmatrix}_n \quad (6)$$

Although additional support for such a hypothesis is required, e.g. from ^{29}Si-NMR studies for the MTMS sol-gel system (currently in progress), it is interesting to note the early appearence of *four-membered-ring Si species* Q^2(4c) and Q^3(4c) in the TMOS system (c.f. (5): p. 164) which hints at the preferred coupling of dimers and lends some support to our model for the MTMS polymerization. This model explains why gels made of PMSO could be shaped into supported films and cladding layers as well as non-supported fibers (30-33). In this context it is noteworthy that in our experiments the casting of the viscous syrup onto the support was often accompanied by the formation of fibers which could sometimes be drawn to length of a few feet.

The Molar-Ratio Limiting-Value. A molar-ratio ≥ 1 appears to be a limiting value for attaining a phase-separation-free sol-gel reaction, even at temperatures as high as 80°C. This observation, which is common to TEOS, TMOS and MTMS can be rationalized as follows: we should keep in mind that all the hydrolysis and condensation processes (1) to (4) occur, as a matter of fact, via reversible reactions. Therefore, when we manage by elevated temperature to drive the hydrolysis to form partially-hydrolyzed siloxane species, there is always the possibility that given sufficient mobility and time these species will undergo disproportionation reactions:

$$MeSi(OMe)_2OH + MeSiOMe(OH)_2 \rightarrow MeSi(OMe)_3 + MeSi(OH)_3 \quad (7)$$

The key factor is mobility. At MR ≥ 1, the siloxane monomers can polymerize at an early stage to yield a syrup in which the mobility of all species is restricted by viscosity. Conversely, only shorter oligomers can be formed at MR < 1 and hence, disproportionation processes can easily take place. Consequently, the phase separation and the glass-sedimentation processes discussed earlier are inevitable.

The typical recipe which is fast yet easily controlled is at MR = 1.5: 1 g MTMS is reacted with 0.2 g aqueous HCl (10^{-2} M) at 80-82°C. A typical loading of most guest molecules is 2 mg, which corresponds to ca. 10^{-2} M (at M.W. $\cong 500$) in the dry glass. The typical reaction time before casting is 5 min. or more accurately, until a weight-loss of 0.5 g methanol is measured, as discussed in the next subsection.

Distillation-Out of the Methanol. By elevating the temperature of the sol-gel reaction to nearly 80°C, fast distillation-out of methanol takes place, which provides a series of fringe benefits such as:
– further reduction of sol-volume thus decreasing the contraction during the gel-curing and impeding cracking (c.f. bottom of Table I).
– increase of the viscosity at an early stage of the condensation, allowing spin-casting while the molecular weight is still low so that rearrangement and stress-relaxation processes can still take place.
– the ability to monitor the reaction progress gravimetrically and improve the timing of the spin-casting, as exemplified in Table III (adapted from (26) copyright ACS)

– the ability to monitor by the grouth of vapor-bubbles the increase in the polymer viscosity and predict the point at which the spin-casting will be successful.
Similar effects can be attained with TEOS and similar monomers, by elevating the temperature to c.a. 90°C to distill out the ethanol produced.

Table III.   Monitoring Hydrolysis and Condensation by Methanol Distilling-Out

1g MeSi(OMe)$_3$ (MTMS) REACTING WITH 0.2g WATER (1.5 m/m)			
MeOH wt. loss (mg)	$\dfrac{m\ MeOH}{m\ MTMS}$	No. Hydrolyzed -Si-OMe groups (average) [a]	No. Condensed -Si-OH groups [b] (average) [a]
235	1.0	1.0	—
353	1.5	1.5	—
470	2.0	2.0	1.0 (dimers)
529	2.25	2.25	1.5 (tetramers)
588	2.5	2.5	2.0 (polymer)
> 588			GEL
705	3.0	3.0	3.0 - GEL (entirely crosslinked)

(a)   a minimum value (evaporation lags behind methanol formation), assuming the hydrolysis is much faster than the condensation.
(b)   see text.

We should note in Table III that the primary water can account only for the hydrolysis of 1.5 moles of methanol per each mole of siloxane. Additional methanol must result from the condensation reactions, one molecule per two siloxane mer-units:

$$\equiv\!Si\text{-}OH \quad + \quad Me\text{-}O\text{-}Si\!\equiv \quad \text{---}> \quad \equiv\!Si\text{-}O\text{-}Si\!\equiv \quad + \quad MeOH \qquad (8)$$

$$\equiv\!Si\text{-}OH \quad + \quad HO\text{-}Si\!\equiv \quad \text{---}> \quad \equiv\!Si\text{-}O\text{-}Si\!\equiv \quad + \quad H_2O \qquad (9)$$

(and further hydrolysis of $\equiv\!Si\text{-}OMe$ to $\equiv\!Si\text{-}OH$ + MeOH)

Two additional interesting features of the progress of this reaction should be noted. The first one is the drastic increase in the concentration of the catalyst at a very early stage of the reaction due to the consumption of most of the water. This results in an acceleration of the hydrolysis rate, but also in slowing the condensation. The second is the rapidly increasing viscosity of the polymer and the consequent caging of the reacting molecules. This results in an acceleration of reaction rates, an effect which resembles the gel effect in reactions of addition polymerization (**44**), although the restricted diffusion eventually brings the processes to a stop. Both effects induce a steep rise of the viscosity at a certain stage of the reaction which impose some difficulty in casting glass films of reproducible thickness.

Curing and Final Stabilization of the Supported Glass. Immediately after being cast, the gel attached to the support is still soft and partly acetone-miscible. It solidifies slowly and becomes non-sticky after half a day at room temperature. Condensation at such a slow rate is typical of tetraalkoxysilane gels prepared at low pH (<2.5) and a substoichiometric water-to-siloxane ratio (<2) (**28**). Their slow condensation was attributed to the formation of linear polymeric chains which could undergo rearrangement and compaction while the gel was weakly crosslinked. These features are believed to allow stress relaxation at the early stages of the alcohol evaporation. However, at a later stage the polymeric skeleton comprising the gel becomes substantially crosslinked and additional evaporation of alcohol leads to the formation of pores (28). This mechanism results in a stable non-cracking film as long as its thickness does not exceed 1μm (5 p.506). In this respect, the PMSO gel is entirely different since it is capable of growing linear ladder-like polymer segments enveloped in methyl side-groups which do not undergo crosslinking.

The crosslinking of PMSO takes place, therefore, at a much slower rate than the silicate gels: it becomes acetone-immiscible within a few hours but still remains soft and tacky. Rearrangement and stress-relaxation processes most likely continue in this type of glass long after it is seemingly solidified, since its polymeric structure complies with the second condition set in Figure 2. Further hardening is attained by additional aging for a few days at RT or a few hours at 60-65°C. Nonetheless, the overall duration of the process permits preparation and use of the glasses on the same day!

When the PMSO glass is apparently cured it may still contain a substantial amount of ≡Si-OH groups which can not undergo condensation due to steric hindrance and/or restricted mobility. Further, this glass contains permanent moieties of ≡Si-CH$_3$ hydrophobic groups. Being prepared from water-lean recipes, some of the ≡Si-O-CH$_3$ groups in these glasses most probably become trapped inside the hydrophobic regions and remain inaccessible to hydrolysis, even if at a later stage the glass is treated with boiling water. The existance of residual ≡Si-O-CH$_3$ groups was verified from the C^{13} NMR spectrum of the MTMS sol-gel glass powder, which had been exposed to hot water for a prolonged period of time and then dried under vacuum. For this sample, the single ≡Si-CH$_3$ peak at 0 ppm was accompanied by a smaller peak of ≡Si-O-CH$_3$, at 50 ppm (Shoulders, B., Univ. Texas at Austin NMR Lab, unpublished data, 1990). The main Si^{29} peak was accompanied, similarly, with a smaller neighboring peak, a few ppm higher. These findings should alert us to various effects which may emerge from the existence of such moieties inside the glass, as well as to the possible binding of reactive guest molecules to the matrix through residual reactive groups.

Fast Sol-Gel Reactions with Monomers other than MTMS. An intriguing question derived from the MTMS fast sol-gel process is whether the principle of "three hydrolyzable substituents" applies only to a pure alkyl-trialkoxy monomer or perhaps it can be applied also to a mixture of tetra- and di- alkoxysilanes and may be even to multi-componet mixtures. A series of experiments was carried out with binary mixtures of tetramethoxysilane (TMOS) and dimethyldimethoxysilane (DMDMS), employing exactly the MTMS reaction procedure and casting process. The molar ratio of water to siloxane was kept at 1.5 (substoichiometric, in most cases). The results of these experiments are collected in Table IV.

Table IV. Cracking of Sol-Gel Glasses Prepared from TMOS-DMDMS Mixtures [a]

TMOS (mol.%)	DMDMS (mol.%)	No. -OMe (average)	Time before cracking (h) [b]
100	0	4.0	<1
75	25	3.5	12
67	33	3.3	18
60	40	3.2	36
55	45	3.1	100
50	50	3.0	∞
<50	>50	>3	∞

(a) all the experiments were carried out at 80°C using HCl 10^{-2} M. as catalyst. Water to siloxane molar ratio was 1.5 : 1.

(b) approximate duration until more than 50% of the area is cracked.

These findings suggest that the 1:1 molar mixture of TMOS and DMDMS can form linear segments of non-crosslinked polymer which is similar to that of MTMS. The assumed structures of these linear polymers are illustrated in Figure 5. Additional support to the existence of such flexible polymeric features is lent by the behavior of sol-gel multilayered assemblies as discussed in a following subsection.

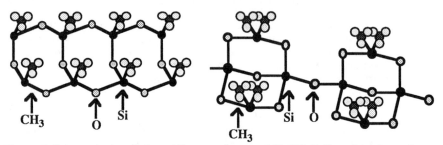

Figure 5. Schematic description of linear segments of PMSO (left) and a 1:1 copolymer of orthosilicate and dimethylsiloxane (right).

Further extension of the principle of three hydrolysible substituents led us to apply the fast sol-gel method with an even smaller monomer: hydrogentrimethoxysilane (HTMS): $H-Si(OCH_3)_3$. This monomer has no organic constituent and should result in a glass of an improved optical quality. Following its hydrolysis and polymerization (without any catalyst) the Si-H bond oxidizes to Si-OH which undergoes further condensation, which results in some cracking. Elimination of this cracking is the subject of the presently on-going research towards the formation of silica films and fibers through this approach.

The Scope of Guest Molecules Embodied in Fast Sol-Gel Matrices. Embodying of guest molecules in the sol-gel matrix is accomplished simply by the incorporation of these guest molecules (e.g. laser dye) in the reaction mixture and proceeding according the regular fast sol-gel procedure (see a typical recipe in a former subsection). Generally, this results in dye-embodying glass-films, usually with a marginal effect on the fast-sol-gel process itself. For laser dyes a typical loading of 2 mg per 1 g MTMS corresponds to $1.1 \cdot 10^{-2}$ M and $1.7 \cdot 10^{-2}$ M in the PMSO glass ($d \sim 1.3$ g/cm^3) for Rhodamine and Coumarin, respectively, which are fairly high concentrations. Their absorbance spectra are typical of non-aggregated dye molecules (see later subsection). Higher loadings of these laser dyes, as well as embodying of laser dyes of larger molecular weight (>600) could be attained with the same fast sol-gel synthesis by means of an increased acid concentration, yet the optimal degree of increased acid concentration is currently being investigated. Embodying of discrete Pyridine species can be attained with the present recipe at dye loading of 0.5-1 mg/g monomer. Loadings smaller by two orders of magnitude can be attained presently with Polyphenyl-1 due to its low solubility in the reactants. The route to increase the loading of Pyridine-1 beyond 10^{-2} M while ensuring molecular dispersion in the glass is also being investigated at present and apparently increasing the acid concentration is the key to achieve this goal. Nevertheless, when proper dissolution of the guest molecules in the polymerizing siloxane is maintained, the resultant glass exhibits absorbance and fluorescence spectra typical to the discrete embodied species, as described in a later subsection.

Donor-Acceptor type molecules such as p-nitroaniline (PNA) and diamino-diphenylsulfone (DDS) are embodied in the PMSO glasses by the same fast sol-gel process as laser dyes. Higher loadings (10-15%) of these chromophores could be prepared with the same fast sol-gel synthesis and again, an increased concentration of acid was required to retain the fast rate and the good optical quality of the glass.

Metal ions (e.g. Cu, Tl, Ce, Ag) could also be easily incorporated into the fast sol-gel recipe. However, most metal ions tended to induce aggregation of PMSO particles, resulting in a grainy appearance. Therefore, their incorporation (an aqueous solution of their salt, at a typical loading of 2 mg/g MTMS) was most successful when the sol-gel process was almost completed, and additional polymerization was allowed for a few seconds only. The spin-casting was then immediately carried out before in-vial gelation or aggregation could take place. The resulting glass had the typical coloration and absorbance of the metal salt as well as the optical clarity of the PMSO glass.

Multilayered Glasses and Waveguides. Successive spin casting of several layers of PMSO on a support results in a multilayered assembly. The sol-gel and drying processes of a single-layer can be straight-forward applied to the preparation of a multilayered assembly (1"x 1") of total thickness up to 50μm. For films which are thicker or of larger dimensions, an extended period of drying is required.

In order to convert such an assembly into an efficient wave-guide, we prepared a supported three-layer assembly in which the middle layer has a higher index of refraction than the other two, which thus serve as cladding layers for the middle one. This increase in the index of refraction was attained either by maintaining a high concentration of dye in the middle-layer, or by incorporating in it monomers which carry aromatic groups (e.g. phenyltriethoxysilane). The thus achieved wave-guide property was exemplified in measuring front-face vs. right-angle fluorescence of three-layer assembly comprising a laser-dye in its middle layer: in spite of the fact that the right-angle dye-area was three orders of magnitude smaller than that of the front-surface (~10 μm vs. ~10 mm), the fluorescence intensities differ only by a factor of 10. This proves that in such an assembly, most of the fluorescence intensity is indeed guided through the dye-layer and only a small portion comes out through its surfaces. This property may favor lasing in the dye-layer if sufficient pumping power is applied and appropriate cavity conditions are accomplished.

It is interesting to note that a slight increase in the water-to-siloxane molar ratio was required to ensure proper adherence between the layers when multilayered films larger than 1" x 1" were prepared. These observations comply with the interesting capability of the surface of cast PMSO films to rearrange rapidly to match the environment it is facing: the surface of a PMSO gel cast onto Parafilm™ (American Can Company) and separated from this support after curing became hydrophobic (contact angle >86°) while that cast onto cellophane was hydrophilic (contact angle <37°). The impressive rearrangement capability complies well with the existence of non-crosslinked segments of substantial length as suggested for PMSO (c.f. Figure 5). Hence, it is most probable that when the gel surface is exposed to air for sufficient time it will become hydrophobic and allow gliding of a successive gel-layer on its methyl-rich surface. Such phenomena manifesting in inhomogeneous appearance of the second and third layers were observed in a few cases (on large supports). These phenomena were successfully eliminated by tuning of the molar-ratio.

Fast Sol-Gel Process Under Basic Catalysis. A general problem for many NLO molecules which carry primary and substituted amino-groups, pyridine groups, etc. is that their inclusion in sol-gel glass prepared under acid catalysis may face some difficulties if the guest molecules undergo protonation. Among the laser dyes, Pyridin 1 is an example which requires additional measures if we wish to embody it in the sol-gel glass at high concentrations. It seemed desirable, therefore, to try and modify the new fast synthetic route and adapt it for basic catalysis.

The main difficulty observed was the occurrence of the two types of phase-separation processes, when strong bases such as NaOH or KOH were used, even at low concentration (i.e. 10^{-2} M). Firstly, the basic aqueous phase and the siloxane phase remained immiscible even in the presence of the hydrolysis-generated methanol. Secondly, the hydrolyzed siloxane molecules formed at the interface of the phases undergo rapid base-catalyzed condensation and separate immediately as a third phase. Therefore, it was plausible to look for basic catalysts which may exhibit higher miscibility in organic media.

Basic catalysis with ammonium hydroxide overcame the phase separation obstacle, but at the penalty of a very low reaction rate, since ammonium hydroxide is a relatively weak base ($K_b \sim 1.9 \; 10^{-5}$ at 50°C). When the catalyst concentration was increased to maintain the fast reaction rate, phase separation occurred once again.

When dimethylamine ($K_b \sim 3.3 \; 10^{-3}$ at 25°C) was used, fast and facile sol-gel reactions could be maintained using concentrations between 10^{-2} to 4.4 M. These observations reflect first of all the higher compatibility of the organic base with both the aqueous and oleophilic phases, thus assisting their miscibility during the early stages of the sol-gel process. In this respect, the organic base acts as a micro-surfactant. Another advantage of this catalyst is that it is volatile. Upon the consumption of most of the water, the base-protonation reaction shifts to the left:

$$\text{Me}_2\text{NH} + \text{H}_2\text{O} \; \rightleftharpoons \; \text{Me}_2\text{NH}_2^+ + \text{OH}^- \qquad (10)$$

and the volatile free amine is expelled from the gel, thus slowing down the condensation. The latter self-regulation phenomenon is of special importance for the sol-gel curing process: the more crosslinked the glass becomes, the more time is needed for stress relaxation. This relaxation period is extended by the catalyst evaporation and the consequent decrease of the condensation rate. Hence, the glasses formed following casting onto the support behaved similarly to those prepared by the acid catalysis and all the dyes investigated could thus be embodied in the sol-gel glass. It should be noted, however, that the residual amines in the glass may quench the fluorescence of the glass-embodied dyes (see next subsection) and therefore, have to be removed from the glass films by either washing with water or by means of vacuum drying.

The facile basic-catalysed route equips us with a second method of fast sol-gel preparation of glass-embodied dyes and other NLO compounds. Both types of catalysis are complementary and actually enable us to embody a wide variety of dyes, regardless of their preferential miscibility in acid or base.

Absorbance and Fluorescence Characteristics of Laser Dyes Embodied in the Sol-Gel Glasses. Absorbance and fluorescence spectra of several laser-dyes embodied in supported PMSO glass films prepared under acid catalysis are shown in Figure 6 (from (26) copyright ACS). The spectra are typical of discrete dye molecules and are comparable to data in the literature (11,45). In Table V the absorbance and fluorescence maxima in solution and in PMSO glasses prepared under acid and base catalysis are compared for the four laser dyes studied. The absorbance maxima in the glass are very close to those in solution, and the slight shifts exhibit a mixed trend.

The slight shifts in acid catalyzed PMSO glass, as compared to the solution, may be attributed to the interaction of the dye molecules with the acidic glass environment. It is noteworthy that only Polyphenyl 1 shows a blue-shift in its absorption, and it is the only di-anion molecule, while Coumarin 153 is (originally) neutral and Rhodamine 6G and Pyridin 1 are both cations. On the other hand, all the dyes exhibit distinct blue-shifts of the fluorescence maxima. These shifts should most probably be attributed to the caging effect of the glass, in agreement with observations made by others (11). The most prominent blue-shift, which is exhibited by Pyridine 1, may also be an outcome of replacement of the perchlorate anions by chloride.

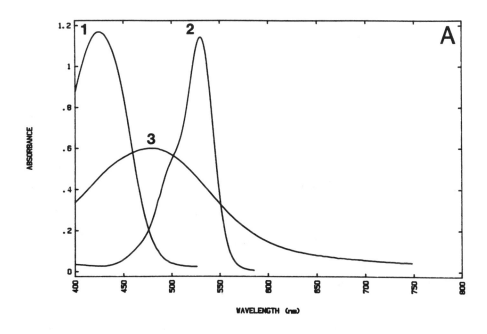

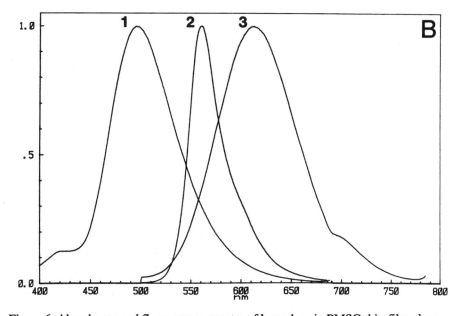

Figure 6. Absorbance and fluorescence spectra of laser dyes in PMSO thin-film glass.

Table V. Laser Dyes Absorbance and Fluorescence in Sol-Gel Glasses and in Solution

Dye	Medium	Absorbance Maximum [c]	Fluorescence Maximum [c,d]	
Rhodamine 6G	Ethanol	530	581 (e)	(308)
	PMSO (a)	532	557	(308)
	PMSO (b)	536	550	(308)
Coumarin 153	Ethanol	423	540 (e)	(308)
	PMSO (a)	428	504	(308)
	PMSO (b)	410	525	(308)
Pyridine 1	Ethanol	480	710 (e)	(510)
	PMSO (a)	480	615	(480)
	PMSO (b)	486	597	(480)
Polyphenyl 1	Ethanol	308	382 (f)	(308)
	PMSO (a)	298	374	(308)
	PMSO (b)	270	420	(270)

(a)   Polymethylsiloxane matrices prepared via HCl catalysis.
(b)   Polymethylsiloxane matrices prepared via Me_2NH catalysis.
(c)   Solution data taken from Lambdachrome dyes catalog **(45)**
(d)   Excitation wavelength denoted in brackets.
(e)   In methanol.
(f)   In ethylene glycol.

In base-catalysed glass (see former subsection) the absorbance and fluorescence also exhibit a mixed trend and as expected, the pattern is opposite to that of the acid-catalysed glass: a blue shift of the absorbance and a red shift of the fluorescence is exhibited by the Polyphenyl (anion) dye, in contrast to the slight red shifts in the absorbance and blue shift in the fluorescence of Rhodamine and Pyridine (cation) dyes. The intensity of fluorescence in Me_2NH-catalyzed glass was found to be usually 1-2 orders of magnitude lower than acid-catalyzed glass of comparable dye loading. The fluorescence intensity could be increased by prolonged extraction of the catalyst by water, implying that a substantial concentration of the amino-catalyst remains in the glass and quenches the fluorescence unless leached out. The effect of the residual catalyst on dyes of different ionic nature should be taken into consideration when new chromophore-glass systems are being planned and the appropriate catalyst is selected.

Conclusions

The new fast synthetic route to Sol-Gel glasses reported herein facilitates the preparation of supported glass thin-films and multilayered assemblies in the thickness range of 1-50 μm. These thin-films can be prepared with various indices of refraction, either in the form of pure glass or as dye-encaging layers, which enable the formation of planar thin-waveguide multi-layered assemblies.

The new route is established on the combination of three principles:

1. Minimizing the volume change upon curing: eliminate solvent, smallest alkoxide groups, low water-to-siloxane ratio.
2. Driving the reaction to linear polymerization: low water-to-monomer-ratio, homogeneity of intermediates by high-temperature hydrolysis

3. Blocking one substituent of the alkoxysilane monomer, to enable relaxation of the polymer gel.

Different aspects of these principles have been separately investigated or suggested by others, but have not been combined previously into a unified synthetic scheme.

Embodying of guest molecules in the sol-gel glasses prepared by this route is generally applicable with almost any (miscible) guest material. Relatively high concentrations of the latter can be embodied in the glass without significantly affecting either the clarity of the glass host or the dispersion of the guest material. Some tuning of the reaction conditions are required to maintain a very high concentration (>10% w/w) or large molecular weight (>600) embodied molecules.

The intense fluorescence of laser-dyes embodied in these thin-film glasses, the simplicity to cast these sol-gel glasses in a multilayered sequence of tailored indices of refraction and the waveguide properties observed in such assemblies imply that such multilayered films may be constructed into a surface-laser. Currently, we are examining these structures as the core of a two dimensional laser and other non-linear optical devices.

Acknowledgement

The authors would like to acknowledge the support of this work by a grant from the State of Texas, through Advanced Technology Program, # 003658-394. The authors would like to acknowledge the great help in launching this research project endowed by Prof. D. Avnir of the Hebrew University, Jerusalem, Israel, through detailed discussions of state-of-the-art sol-gel chemistry and physics. We would also like to acknowledge the help of Dr. Neil P. Desai and Prof. Jefferey A. Hubbell in carrying out fluorescence-microscopy measurements.

Literature Cited

1. "Glasses and Glass Ceramics from Gels", Proc. Int. Workshop, *J. Non-Cryst. Solids*, a. Gottardi, V., Ed., **1982**, *48*; b. Scholze, H., Ed., **1984**, *63*; c. Zarzycky, J., Ed., **1986**, *82*; d. Sakka, S., Ed., **1988**, *100*; e. Aegerter, M.A., Ed., **1990**, *121*.
2. "Better Ceramics Through Chemistry", Brinker, C.J., Clark, D.E., Ulrich, D.R., Eds., *Mater. Res. Soc. Symp. Proc.*, a. **1984**, *32;* b. **1986**, *73;* c. **1988**, *121*.
3. "Glasses for Optoelectronics", Righini, G.C., Ed., Proc. SPIE, **1989**, *1128*.
4. "Sol-Gel Technology for Thin Films, Fibers, Preforms, Electronics and Specialty Shapes", Klein, L.C., Ed., **1988**, Noyes Publ., Park Ridge, NJ.
5. "Sol-Gel Science: the Physics and Chemistry of Sol-Gel Processing", Brinker, C.J., Scherer, G.W., **1990**, Academic Press, San-Diego, CA.
6. Schmidt, H., in Ref. 1(d), p. 51.
7. Brinker, C.J., *Ibid*, p. 31.
8. Scherer, G.W., *Ibid*, p. 77.
9. Reisfeld, R., *J. Phys. Coll. C7* (suppl. 12), **1987**, *48*, 423.
10. Reisfeld, R., Brusilovsky, D., Eyal, M., Miron, E., Burstein, Z., Ivri, J., *Chem. Phys. Let.*, **1989**, *160*, 43.
11. Avnir, D., Kaufmann, V.R., Reisfeld, R., *J. Non-Cryst. Solids*, **1985**, *74*, 395.
12. Salin, F., Le Saux, G.,Georges, F., Brun, A., *Optic. Let.*, **1989**, *14*, 785.
13. Fay, W., Mizell, G., Thomas, M., *Proc. SPIE*, **1989**, *1104*, 259.
14. Knobbe, E.T., Dunn, B., Fuqua, P.D., Nishida, F., Zink, J.I., in *"Proc. 4th Int. Conf. Ultrastructure of Ceramics, Glasses and Composites"*, Wiley, New-York, **1989**, in press.

15. Yoldas, B.E., *J. Mater. Sci.*, **1977**, *12*, 1203.
16. Kobayashi, Y., Kurokawa, Y., Imai, Y., *J. Non-Cryst. Solids*, **1988**, *105*, 198.
17. Uchihashi, H., Tohge, N., Minami, T., *Nippon Seramikussu Kyokai Gakujutsu Ronbunshi (Sci. J. Ceram. Soc. Japan)*, **1989**, *97*, 396.
18. Sasaki, H., Kobayashi, Y., Muto, S., Kurokawa, Y., *J. Amer. Ceram. Soc.*, **1990**, *73*, 453.
19. Philipp, G., Schmidt, H., in Ref. 1(b), p. 283.
20. Knobbe, E.T., Dunn, B., Fuqua, P.D., Nishida, F.,Appl. Opt., **1990**, *29*, 2729.
21. Reisfeld, R., Chernyak, V., Eyal, M., Weitz, A., *Proc. SPIE*, **1988**, *1016*, 240.
22. Avnir, D., Kaufmann, V.R., *J. Non-Cryst. Solids*, **1987**, *92*, 180.
23. for example, Sakka, S., Kamiya, K., Makita, K., Yamamoto, Y., in Ref. 1(b), p. 223: c.f. Table I.
24. Sakka, S., *J. Non-Cryst. Solids*, **1988**, *99*, 110. [DMF]
25. Hench, L.L., in "Science of Ceramic Chemical Processing", Hench, L.L., Ulrich, D.R., Eds, **1986**, Wiley, p.52
26. Haruvy, Y., Webber, S.E., *Chem. Mater.*, **1991**, *3*, 501.
27. Hiromitsu, K., Sakka, S., *Chem. Mater.*, **1989**, *1*, 398.
28. Pettit, R.B., Ashley, C.S., Reed, S.T., Brinker, C.J., in Ref. 4, p. 80.
29. Mizuno, T., Nagata, H., Manabe, S., in Ref. 1(d), p. 236.
30. Dislich, H., Jacobsen, A., DB Pat. 1-494-872, **1965**.
31. Dislich, H., Jacobsen, A., *Angew. Chem.*, **1973**, *12*, 439.
32. Dislich, H., in Ref. 4, p. 50.
33. Sakka, S., Tanaka, Y., Kokubo, T., in Ref. 1(c), p. 24.
34. Haruvy, Y., *ESA J.*, **1990**, *14*, 109.
35. Schmidt, H., Rinn, G., Naß, R., Sporn, D., in Ref. 2(c), p. 743.
36. Sakka, S., in Ref. 2(a), p. 91 [80 deg]
37. Sakka, S., Kamya, K.,in Ref. 1(a), p. 31   [drop by drop]
38. Kaufman, V.R., Avnir, D., *Langmuir*, **1986**, *2*, 717.
39. Sakka, S., in Ref. 1(e), p. 417.
40. Brinker, C.J., Keefer, K.D., Schaefer, D.W., Assink, R.A., Kay, B.D., Ashley, C.S., in Ref. 1(b), p. 45.
41. Yoldas, B.E., in Ref. 1(b), p. 145.
42. Yoldas, B.E., in Ref. 1(c), p. 82.
43. Yoldas, B.E., J. Polym. Sci., Part 1A: Polym. Chem., **1986**,.*24*, 3475, and references therein.
44. Flory, P.J., "Principles of Polymer Chemistry", Cornell Univ. Press, **1953**.
45. Brackmann, U., Lambdachrome Laser Grade Dyes Data Sheets, **1986**, Lambda Physik GmbH.

RECEIVED January 16, 1992

INDEXES

Author Index

427

Affiliation Index

Subject Index

Production: Betsy Kulamer
Indexing: Deborah H. Steiner
Acquisition: Anne Wilson

Printed and bound by Maple Press, York, PA

Bestsellers from ACS Books

The ACS Style Guide: A Manual for Authors and Editors
Edited by Janet S. Dodd
264 pp; clothbound, ISBN 0–8412–0917–0; paperback, ISBN 0–8412–0943–X

Chemical Activities and Chemical Activities: Teacher Edition
By Christie L. Borgford and Lee R. Summerlin
330 pp; spiralbound, ISBN 0–8412–1417–4; teacher ed. ISBN 0–8412–1416–6

Chemical Demonstrations: A Sourcebook for Teachers,
Volumes 1 and 2, Second Edition
Volume 1 by Lee R. Summerlin and James L. Ealy, Jr.;
Vol. 1, 198 pp; spiralbound, ISBN 0–8412–1481–6;
Volume 2 by Lee R. Summerlin, Christie L. Borgford, and Julie B. Ealy
Vol. 2, 234 pp; spiralbound, ISBN 0–8412–1535–9

Writing the Laboratory Notebook
By Howard M. Kanare
145 pp; clothbound, ISBN 0–8412–0906–5; paperback, ISBN 0–8412–0933–2

Developing a Chemical Hygiene Plan
By Jay A. Young, Warren K. Kingsley, and George H. Wahl, Jr.
paperback, ISBN 0–8412–1876–5

Introduction to Microwave Sample Preparation: Theory and Practice
Edited by H. M. Kingston and Lois B. Jassie
263 pp; clothbound, ISBN 0–8412–1450–6

Principles of Environmental Sampling
Edited by Lawrence H. Keith
ACS Professional Reference Book; 458 pp;
clothbound; ISBN 0–8412–1173–6; paperback, ISBN 0–8412–1437–9

Biotechnology and Materials Science: Chemistry for the Future
Edited by Mary L. Good (Jacqueline K. Barton, Associate Editor)
135 pp; clothbound, ISBN 0–8412–1472–7; paperback, ISBN 0–8412–1473–5

Personal Computers for Scientists: A Byte at a Time
By Glenn I. Ouchi
276 pp; clothbound, ISBN 0–8412–1000–4; paperback, ISBN 0–8412–1001–2

Polymers in Aqueous Media: Performance Through Association
Edited by J. Edward Glass
Advances in Chemistry Series 223; 575 pp;
clothbound, ISBN 0–8412–1548–0

For further information and a free catalog of ACS books, contact:
American Chemical Society
Distribution Office, Department 225
1155 16th Street, NW, Washington, DC 20036
Telephone 800–227–5558

Other ACS Books

Biotechnology and Materials Science: Chemistry for the Future
Edited by Mary L. Good
160 pp; clothbound, ISBN 0–8412–1472–7, paperback, ISBN 0–8412–1473–5

Chemical Demonstrations: A Sourcebook for Teachers
Volume 1, Second Edition by Lee R. Summerlin and James L. Ealy, Jr.
192 pp; spiral bound; ISBN 0–8412–1481–6
Volume 2, Second Edition by Lee R. Summerlin, Christie L. Borgford, and Julie B. Ealy
229 pp; spiral bound; ISBN 0–8412–1535–9

The Language of Biotechnology: A Dictionary of Terms
By John M. Walker and Michael Cox
ACS Professional Reference Book; 256 pp;
clothbound, ISBN 0–8412–1489–1; paperback, ISBN 0–8412–1490–5

Cancer: The Outlaw Cell, Second Edition
Edited by Richard E. LaFond
274 pp; clothbound, ISBN 0–8412–1419–0; paperback, ISBN 0–8412–1420–4

Chemical Structure Software for Personal Computers
Edited by Daniel E. Meyer, Wendy A. Warr, and Richard A. Love
ACS Professional Reference Book; 107 pp;
clothbound, ISBN 0–8412–1538–3; paperback, ISBN 0–8412–1539–1

Practical Statistics for the Physical Sciences
By Larry L. Havlicek
ACS Professional Reference Book; 198 pp; clothbound; ISBN 0–8412–1453–0

The Basics of Technical Communicating
By B. Edward Cain
ACS Professional Reference Book; 198 pp;
clothbound, ISBN 0–8412–1451–4; paperback, ISBN 0–8412–1452–2

The ACS Style Guide: A Manual for Authors and Editors
Edited by Janet S. Dodd
264 pp; clothbound, ISBN 0–8412–0917–0; paperback, ISBN 0–8412–0943–X

Personal Computers for Scientists: A Byte at a Time
By Glenn I. Ouchi
276 pp; clothbound, ISBN 0–8412–1000–4; paperback, ISBN 0–8412–1001–2

Chemistry and Crime: From Sherlock Holmes to Today's Courtroom[A
Edited by Samuel M. Gerber
135 pp; clothbound, ISBN 0–8412–0784–4; paperback, ISBN 0–8412–0785–2

For further information and a free catalog of ACS books, contact:
American Chemical Society
Distribution Office, Department 225
1155 16th Street, NW, Washington, DC 20036
Telephone 800–227–5558